Solitons

JOIN US ON THE INTERNET VIA WWW, GOPHER, FTP OR EMAIL:

WWW: http://www.thomson.com
GOPHER: gopher.thomson.com
FTP: ftp.thomson.com   A service of I(T)P®
EMAIL: findit@kiosk.thomson.com

Optical and Quantum Electronics Series

Series editors

Professor G. Parry, University of Oxford, UK
Professor R. Baets, University of Gent, Belgium

This series focuses on the technology, physics and applications of optoelectronic systems and devices. Volumes are aimed at research and development staff and engineers involved in the application of optical technologies. Graduate textbooks are included, giving tutorial introductions to the many exciting areas of optoelectronics. Both conventional books and electronic products will be published, to provide information in the most appropriate and useful form for users.

1 **Optical Fiber Sensor Technology**
   Edited by K.T.V. Grattan and B.T. Meggitt

2 **Vision Assistant Software**
   A practical introduction to image processing and pattern classifiers
   C.R. Allen and N.C. Yung

3 **Silica-based Buried Channel Waveguides and Devices**
   François Ladouceur and John D. Love

4 **Essentials of Optoelectronics**
   With applications
   Alan Rogers

5 **Solitons**
   Nonlinear pulses and beams
   Nail N. Akhmediev and Adrian Ankiewicz

Solitons

Nonlinear pulses and beams

Nail N. Akhmediev and Adrian Ankiewicz

Optical Sciences Centre
The Australian National University
Canberra
Australia

CHAPMAN & HALL

London · Weinheim · New York · Tokyo · Melbourne · Madras

Published by Chapman & Hall, 2–6 Boundary Row, London SE1 8HN, UK

Chapman & Hall, 2–6 Boundary Row, London SE1 8HN, UK

Chapman & Hall GmbH, Pappelallee 3, 69469 Weinheim, Germany

Chapman & Hall USA, 115 Fifth Avenue, New York, NY 10003, USA

Chapman & Hall Japan, ITP-Japan, Kyowa Building, 3F, 2-2-1 Hirakawacho, Chiyoda-ku, Tokyo 102, Japan

Chapman & Hall Australia, 102 Dodds Street, South Melbourne, Victoria 3205, Australia

Chapman & Hall India, R. Seshadri, 32 Second Main Road, CIT East, Madras 600 035, India

First edition 1997

© 1997 Nail N. Akhmediev and Adrian Ankiewicz

Printed in Great Britain at T.J. International Ltd, Padstow, Cornwall

ISBN 0 412 75450 9

A catalogue record for this book is available from the British Library

Library of Congress Catalog Card Number: 96-72036

Printed on permanent acid-free text paper, manufactured in accordance with ANSI/NISO Z39.48-1992 and ANSI/NISO Z39.48-1984 (Permanence of Paper).

Contents

Foreword

The field of nonlinear optics has blossomed in recent years, as nonlinear materials have become available, and widespread applications have become apparent. This is particularly true for soliton and other types of nonlinear pulse transmission in optical fibers, since this form of light propagation can be used to realize the long-held dream of very high capacity dispersion-free communications. Moreover, in recent years, soliton-like beams in highly nonlinear materials like organic polymers and photorefractive materials have been studied, in the hope that they can be used in 'light-guiding-light' devices for fast switching purposes.

As with other areas of technology, successful advancement of photonics depends on an in-depth understanding of the underlying physical principles. In the early stages of the development of a subject, *ad hoc* approximations are sometimes introduced in lieu of any theory with a solid foundation. As time goes by, the need for a self-contained, self-consistent theoretical base develops. In this book, we aim to provide such a theory for nonlinear light propagation in waveguides, with particular emphasis on soliton propagation. A soliton can propagate along an optical fiber without change of shape, and thus can be regarded as natural 'bit' of information. A pulse of somewhat different shape can evolve into a soliton during propagation. This is especially true for recent systems with sliding-guiding filters. As well as being of deep fundamental interest, the practical application to soliton-based, long-distance fiber communications makes it imperative that the physical principles are clearly known.

The field of solitons has grown as a result of the combined efforts of mathematicians, specialists in numerical simulations and masters of experimental physics. The whole area, especially in recent years, has been strongly supported by applications such as optical telecommunications. This incredible success would not have been possible without each of these contributions. The results of these efforts have been presented in thousands of papers and quite a few books. These include the mathematical, the more practical and the purely applied.

In our treatment of the material, we develop a comprehensive mathematical treatment while relating it to physical principles. We present practical methods for describing the behaviour and stability of solitons. In dealing with integrable systems, we present our original approach for obtaining the solutions, in addition to describing known methods like the inverse-scattering technique and methods related to it. For systems which are Hamiltonian but not integrable, we rely on numerical simulations and the qualitative theory of dynamical systems with an infinite number of degrees of freedom. This allows us to describe, in closed form, the dynamics of solitons in birefringent fibers and in nonlinear couplers. Surprisingly, many non-integrable systems, both Hamiltonian and non-Hamiltonian, have exact solutions which have been found in recent years. Some of them are related to the symmetries of the system under consideration, while the nature of others remains unclear. Examples are the two-dimensional non-linear Schrödinger equation and the (1+1)-dimensional complex Ginzburg–Landau equation. In this book, we discuss the known cases, again with the emphasis being on the physics rather than the mathematical structure of the solutions.

This book is intended to be a self-contained text for physicists, applied mathematicians and engineers who wish to enter the field. In addition, it can be of benefit as a reference book to those researchers who are already familiar with the field and working in this area. Futhermore, it can be used as a graduate textbook in soliton theory and may be prescribed for class-room or self-study.

Chapter 1 of this book shows how the basic equations for solitons appear in optics in general and in several special cases in particular. These include fiber optics, lasers and light-guiding-light phenomena. As a result, we are concerned with the nonlinear Schrödinger equation and its variations and generalizations. The rest of the book deals with these equations, presenting methods of analysis and solutions which are important in practice. We consider, mainly, soliton-like pulses and beams in Hamiltonian systems. These include birefringent media, the influence of higher-order dispersion, and other phenomena. We consider stationary solutions and their stability, as these are needed in applications. Another issue is 'global' dynamics, which is beyond the reach of linear stability analysis. The 'global' dynamics becomes important when the initial pulse (or state) is unstable. This will be discussed with reference to phenomena like modulation instability and soliton propagation in birefringent fibers. The final chapter deals with non-Hamiltonian systems, namely, systems with gain and loss. In this case, the approach and methods are different from those in previous chapters. However, again, the main features can be described by considering stationary solutions, and their stability and interactions.

As with any other book of this type, the choice of material is related to the research interests of the authors. We leave aside resonance phenomena and self-induced transparency. Hence we do not deal with sine-Gordon type equations, which can also be important in some applications. However, at this stage of research, all possible interesting topics cannot be covered in one book. We hope that we have covered those areas which have the most exciting applications.

Acknowledgements

We are grateful to various individuals and organizations for supporting our research and assisting us with this book. Much of the material presented here has been developed by the authors at the Optical Sciences Centre, Australian National University, Canberra. The authors also acknowledge the support of the Australian Photonics Co-operative Research Centre.

Many research ideas have sprung forth in exciting discussions with Prof. Allan Snyder. John Love proposed the idea of writing this book. We have enjoyed many fruitful collaborations with our colleagues at ANU: Barry Luther-Davies, John Mitchell, Yuri Kivshar, Vsevolod Afanasjev, Peter Miller, Ole Bang, Alex Buryak and Elena Ostrovskaya. We also thank Wiesław Królikowski, Falk Lederer and François Ladouceur for their assistance.

CHAPTER 1

Basic equations

The effects considered in this book are based on the fact that materials exhibit nonlinear responses to an applied optical field. Roughly speaking, the polarization depends on the applied electric field according to some relation

$$\mathbf{P}(\mathbf{r}, t) = \hat{F}\mathbf{E}(\mathbf{r}, t) \tag{1.1}$$

where $\mathbf{P}$ is the vector of polarization, $\mathbf{E}$ is the electric field of the radiation and $\hat{F}$ is an operator (response function) which is not linear. In general, the operator is non-local in space ($\mathbf{r}$) and in time (t), in addition to being nonlinear. The observation of nonlinear effects became possible after the invention of lasers capable of generating optical fields strong enough to produce these effects in materials. We suppose, throughout this book, that the response is local in space. Hence, we will not take into account any effects related to spatial dispersion (Agranovich and Ginzburg, 1966), although we will consider effects in materials with interfaces and boundaries, so that the linear and nonlinear susceptibility tensors may depend on the spatial variable. The time dependence can be Fourier-transformed. Then relation (1.1) can be written in the frequency domain. In homogeneous materials it is tensorial, and to lowest order in the field E it can be written in the form (Butcher, 1965):

$$P_i(\mathbf{r}, t) = \chi_{ij}^{(1)} E_j + \chi_{ijk}^{(2)} E_j E_k + \chi_{ijkl}^{(3)} E_j E_k E_l + \ldots \tag{1.2}$$

where P_i and E_i are components of the vectors of polarization and electrical field, respectively, and the coefficients χ represent the lowest-order nonlinear susceptibilities.

Wave propagation in optics is described by Maxwell's equations, the vectorial form of which can be written (in c.g.s. units), for zero current density, ($\mathbf{J} = 0$) as follows:

$$\nabla \times \mathbf{E} = -\frac{1}{c} \frac{\partial \mathbf{B}}{\partial t}, \quad \nabla \times \mathbf{H} = \frac{1}{c} \frac{\partial \mathbf{D}}{\partial t},$$

$$\nabla \cdot \mathbf{D} = 0, \quad \nabla \cdot \mathbf{B} = 0. \tag{1.3}$$

Here $\mathbf{D}$ is the displacement and $\mathbf{B}$ is the magnetic induction. These must be complemented with the constitutive equations:

$$\mathbf{D} = \mathbf{E} + 4\pi\mathbf{P}, \quad \mathbf{B} = \mathbf{H} + 4\pi\mathbf{M}, \tag{1.4}$$

where $\mathbf{M}$ is the magnetic moment. We assume that the medium is non-magnetic ($\mathbf{M} = 0$). Eliminating the magnetic field $\mathbf{H}$, these equations can be reduced to a single vector wave equation:

$$\nabla^2 \mathbf{E} - \nabla(\nabla \cdot \mathbf{E}) - \frac{1}{c^2}\frac{\partial^2}{\partial t^2}\mathbf{D}^{(L)} - \frac{4\pi}{c^2}\frac{\partial^2}{\partial t^2}\mathbf{P}^{(NL)} = 0 \qquad (1.5)$$

where $\mathbf{D}^{(L)} = \mathbf{E} + 4\pi\mathbf{P}^{(L)} = \hat{\varepsilon}\mathbf{E}$ only includes the linear part of the polarization $\mathbf{P}^{(L)} = \hat{\chi}^{(1)}\mathbf{E}$. Here $\hat{\varepsilon}$ is the permittivity. This equation is quite general, and, together with the material equation, can be used for any particular case. In various problems, certain approximations must be made to reduce this system to simpler forms.

1.1 Stationary beams in homogeneous nonlinear media

The field of a monochromatic light beam propagating along a certain direction (which we choose to be the z axis) can be represented in the form

$$\mathbf{E} = \frac{1}{2}[\mathbf{A}(x, y, z)\exp(ikz - i\omega t) + c.c.], \qquad (1.6)$$

where $\mathbf{A}(x, y, z)$ is a slowly varying function of z on spatial scales comparable with the wavelength and *c.c.* means complex conjugate terms. A similar expression is valid for the polarization:

$$\mathbf{P} = \frac{1}{2}[\mathbf{p}(x, y, z)\exp(ikz - i\omega t) + c.c.]. \qquad (1.7)$$

Here $\mathbf{p}$ is a slowly varying function of z. The linear part of the displacement $\mathbf{D}$ can be written in the form

$$D_i^{(L)} = \varepsilon_{ij}^{(L)}E_j,$$

where

$$\varepsilon_{ij}^{(L)} = 1 + 4\pi\chi_{ij}^{(1)}$$

is the linear dielectric constant.

Soliton solutions of the full vector equation (1.5), including longitudinal components of the field, have been studied numerically by Eleonskii *et al.*, (1972a,b). However, in many practical cases, the longitudinal components are small. The second term in (1.5) mixes the longitudinal and transverse components of the field of the solution. Using $\nabla \cdot \mathbf{D} = 0$, the scalar product $(\nabla \cdot \mathbf{E})$ can be represented in the form

$$\nabla \cdot \mathbf{E} = -\frac{\mathbf{E} \cdot (\nabla\hat{\varepsilon})}{\hat{\varepsilon}}.$$

We assume that the changes in nonlinear dielectric constant (or refractive index) are small, so that

$$|\varepsilon_{ij}^{(NL)}| \ll |\varepsilon_{ij}^{(L)}|,$$

where

$$\varepsilon_{ij}^{(NL)} = 4\pi\chi_{ijkl}^{(3)} E_k E_l^*.$$

This is an extension of the weak-guidance approximation (Snyder and Love, 1983) to nonlinear self-guiding phenomena (Lugovoi and Prokhorov, 1974). In this case, the second term in (1.5) can be ignored, and the governing equation for isotropic media and monochromatic waves takes the form

$$\nabla^2 \mathbf{E} + \frac{\omega^2}{c^2}\varepsilon^{(L)}\mathbf{E} + \frac{4\pi\omega^2}{c^2}\mathbf{P}^{(NL)} = 0. \tag{1.8}$$

This approximation is equivalent to ignoring longitudinal fields, so $\mathbf{E}$ and $\mathbf{P}$ have only transverse components.

We also assume that the envelope $\mathbf{A}(x,y,z,t)$ changes slowly in the z-direction, so that $\left|\frac{\partial \mathbf{A}(\mathbf{r})}{\partial z}\right| \ll k|\mathbf{A}|$. Then (1.8) can be approximated by

$$2ik\frac{\partial}{\partial z}\mathbf{A} + \frac{\partial^2}{\partial x^2}\mathbf{A} + \frac{\partial^2}{\partial y^2}\mathbf{A} - \left(k^2 - \frac{\omega^2}{c^2}\varepsilon_{ij}^{(L)}\right)\mathbf{A} + \frac{4\pi\omega^2}{c^2}\mathbf{p}^{(NL)} = 0. \tag{1.9}$$

In isotropic media, the dielectric constant has three equal diagonal elements, $\varepsilon_{ii}^{(L)} = \varepsilon_0$. The nonlinear term for linearly polarized fields also has only one component, $p^{(NL)} = \frac{3}{4}\chi_{xxxx}|A|^2 A$. If we choose $k = \frac{\omega}{c}\sqrt{\varepsilon_0^{(L)}}$, then the scalar version of (1.9) (in two spatial variables), with $\mathbf{A}$ polarized along the x or y axis, is the nonlinear Schrödinger equation (NLSE),

$$i\frac{\partial}{\partial\xi}\psi + \frac{1}{2}\frac{\partial^2}{\partial\tau^2}\psi + |\psi|^2\psi = 0 \tag{1.10}$$

where $\tau = kx$, $\xi = kz$ and $\psi = \sqrt{\frac{3}{2}\pi\chi_{xxxx}^{(3)}/\varepsilon_0}A$. We can see that, in homogeneous nonlinear media without dispersion, the NLSE follows (after some approximations) directly from Maxwell's equations. The propagation of pulses in optical fibers is more complicated, as we have to take into account the material and waveguide dispersions.

1.2 Nonlinear terms

The nonlinear terms reduce to the single term $|\psi|^2\psi$ only in special cases such as linearly polarized waves in isotropic media (i.e. media which have the axis of infinite order, C_∞, as an element of a point symmetry). In general, this reduction does not apply. For example, for elliptically polarized waves, we have to take into account the nonlinear interaction between the two components of the field even in isotropic media. If the medium is birefringent, then waves with an arbitrary initial polarization necessarily become elliptically polarized on propagation, and again we have to take into account the interaction between the two components. For media with inversion symmetry, like fused silica, the lowest-order nonlinearity is cubic.

Here, we consider nonlinear responses which are local in time and space. In the weak-guidance approximation, only the transverse components of the field are of importance. Transverse (x, y) components of the nonlinear polarization vector $\mathbf{P}^{(NL)}$, in a Kerr medium, are determined by the third-order susceptibility tensor, $\hat{\chi}^{(3)}$, so that, in Cartesian coordinates, we have:

$$\mathbf{P}^{(NL)} = \mathsf{P}_x{}^{(NL)}\hat{x} + \mathsf{P}_y{}^{(NL)}\hat{y}.$$

We define the components at a given spatial point as

$$\mathsf{E}_i = \frac{1}{2}[E_i(r)e^{i\omega t} + c.c.]$$

$$\mathsf{P}_i = \frac{1}{2}[P_i(r)e^{i\omega t} + c.c.].$$

Then, for an arbitrary anisotropic medium, the nonlinear polarization can be written in the form:

$$P_i^{(NL)} = \frac{1}{4}\Big[3\chi_{xxxx}^{(3)}(\omega;\omega,\omega,-\omega)|E_i|^2 E_i$$

$$+2\,[\chi_{xxyy}^{(3)}(\omega;\omega,\omega,-\omega)+\chi_{xyyx}^{(3)}(\omega;\omega,-\omega,\omega)+\chi_{xyxy}^{(3)}(\omega;\omega,\omega,-\omega)]|E_j|^2 E_i$$

$$+[\chi_{xxyy}^{(3)}(\omega;-\omega,\omega,\omega)+\chi_{xyyx}^{(3)}(\omega;\omega,\omega,-\omega)+\chi_{xyxy}^{(3)}(\omega;\omega,-\omega,\omega)]E_j^2 E_i^*$$

$$+2\,[\chi_{xxxy}^{(3)}(\omega;\omega,-\omega,\omega)+\chi_{xxyx}^{(3)}(\omega;-\omega,\omega,\omega)+\chi_{xyxx}^{(3)}(\omega;\omega,\omega,-\omega)]|E_i|^2 E_j$$

$$+[\chi_{xxxy}^{(3)}(\omega;\omega,\omega,-\omega)+\chi_{xxyx}^{(3)}(\omega;\omega,-\omega,\omega)+\chi_{xyxx}^{(3)}(\omega;-\omega,\omega,\omega)]E_i^2 E_j^*$$

$$+3\chi_{xyyy}^{(3)}(\omega;\omega,\omega,-\omega)|E_j|^2 E_j\Big], \tag{1.11}$$

where the $\chi_{ijkl}^{(3)}$ terms are the non-zero components of $\hat{\chi}^{(3)}$, and indices i, j take the values $\{x, y\}$, such that $i \neq j$. The factors of 2 and 3 in (1.11) are due to permutational symmetry. The nonlinear susceptibilities grouped in square brackets cannot be measured separately, as they are intrinsically linked. Permutational symmetry suggests that they are equal to each other.

For certain classes of symmetry (in particular, in cubic crystals and isotropic materials) the last three terms are zero. The same is true in uniaxial crystals which have their axis of symmetry along the z direction. Restricting ourselves to these classes, then, after normalizing, (1.11) can be written in the form

$$P_i^{(NL)} \propto |E_i|^2 E_i + A\,|E_j|^2 E_i + B\,E_j^2 E_i^* \tag{1.12}$$

where

$$A = \frac{2\,[\chi_{xxyy}^{(3)}(\omega;\omega,\omega,-\omega)+\chi_{xyyx}^{(3)}(\omega;\omega,-\omega,\omega)+\chi_{xyxy}^{(3)}(\omega;\omega,\omega,-\omega)]}{3\chi_{xxxx}^{(3)}(\omega;\omega,\omega,-\omega)},$$

and

$$B = \frac{[\chi_{xxyy}^{(3)}(\omega;-\omega,\omega,\omega)+\chi_{xyyx}^{(3)}(\omega;\omega,\omega,-\omega)+\chi_{xyxy}^{(3)}(\omega;\omega,-\omega,\omega)]}{3\chi_{xxxx}^{(3)}(\omega;\omega,\omega,-\omega)}.$$

The three terms on the right-hand side of (1.12) are usually called self-phase modulation (SPM), cross-phase modulation (CPM) and four-wave mixing (FWM), respectively. The FWM term is responsible for energy exchange between the two linearly polarized components. In order to keep the total energy constant, the coefficient B must be the same in the equations for P_x and P_y.

In general, the expressions in square brackets in the numerators of A and B are not equal to each other because of different frequency dependencies. However, in the Kleinman approximation, when ω and 2ω are both far from any resonances of the medium, they are related. Then

$$A \approx 2B.$$

This occurs for non-resonant electronic mechanisms of nonlinearity. In this case, the nonlinearity in (1.12) is defined by a single parameter (i.e. by two parameters before the normalization). In media with isotropic molecules, there is an additional constraint:

$$\chi^{(3)}_{xxxx} = \chi^{(3)}_{xxyy} + \chi^{(3)}_{xyxy} + \chi^{(3)}_{xyyx}. \tag{1.13}$$

This equation is a consequence of orientational averaging over the molecules of the medium. The nonlinear coefficients in (1.12) are then:

$$A = 2/3, \quad B = 1/3.$$

This applies for fused silica.

In more general cases, when there is a delay in the response function, the sum of the nonlinear coefficients of the FWM and CPM terms is equal to the coefficient of the SPM term. For the normalized coefficients (Maker *et al.*, 1964)

$$A + B = 1.$$

If the mechanism of nonlinearity is very slow, e.g. if refractive index changes occur due to temperature changes in the material, then we cannot use the Kleinman approximation. We then have to use relaxation equations (as the response is delayed in time) which give zero FWM terms and

$$P_i^{(NL)} \propto |E_i|^2 E_i + A |E_j|^2 E_i. \tag{1.14}$$

The same thing applies if the change in the refractive index of the materials is related to the molecular re-orientation, as occurs in liquid crystals and all fluids with anisotropic molecules. There is no energy exchange between the field components and they influence each other only through the change of refractive index. The FWM terms can also be ignored in more involved examples when the exchange of energy is fast in comparison with other processes. For example, in birefringent optical fibers, these terms can be averaged out to zero (Menyuk, 1987) if the effects related to the birefringence are stronger than the nonlinear effects. In real materials, the CPM coefficient can take on a wide range of values. In most cases it is positive. For example, in $AlGa_xAs_{1-x}$ crystals at frequencies near half of the band gap, A is close to unity (Villeneuve *et al.*, 1995; Hutchings *et al.*, 1995). However, in media with molecular reorientational mechanisms of nonlinearity, it can even take negative values (Buckingham and Orr, 1967).

The case $A = 1$, $B = 0$ is a special one. The wave propagation is then described by integrable Manakov equations . There are no standard crystal classes where this applies. However, 'Manakov conditions' can be arranged experimentally. One of the possibilities is to use randomly birefringent fibers. Then, the averaged equations can be reduced to Manakov equations (Wai *et al.*, 1991; Evangelides *et al.*, 1992) and the soliton propagation has special properties. Another possibility is to use elliptically birefringent fibers, i.e. twisted birefringent fibers where both birefringence and optical activity are present. Then, a transformation from a basis of linearly polarized waves to one of elliptically polarized waves makes two terms equal to each other (Menyuk, 1989). However, the physical interpretation of nonlinear terms using an elliptically polarized basis is different from that where a linearly polarized basis is used (see below), and FWM terms are still present. Other ways of arranging Manakov conditions are also known (Kang *et al.*, 1996). If one of the field components, E_i, is zero (in the scalar approximation), then (1.11) simplifies, and can be written in scalar form:

$$P_i^{(NL)} \propto E_i |E_i|^2.$$

This is the form of the nonlinear term in the standard NLSE.

In some problems, e.g. those dealing with media having optical activity, it is more convenient to write down the equations in terms of circularly polarized components,

$$E_+ = \frac{E_x + iE_y}{\sqrt{2}},$$

$$E_- = \frac{E_x - iE_y}{\sqrt{2}},$$

rather than in terms of the linearly polarized ones. For this basis, the nonlinear polarization takes the form

$$P_i^{(NL)} \propto C\,E_i|E_i|^2 + D\,E_i|E_j|^2 + F\,E_j^2 E_i^* \qquad (1.15)$$

where the subscripts i, j now take one of the signs $(+)$ or $(-)$ and

$$P_\pm = \frac{P_x \pm iP_y}{\sqrt{2}}.$$

If the polarization is normalized such that $C = 1$, in the same way as (1.12), then the coefficients D and F take the forms

$$D = \frac{2(1+B)}{1+A-B}$$

$$F = \frac{1-A-B}{1+A-B}$$

Although the nonlinear terms in (1.15) have the same form as those in (1.12), their physical meaning is clearly different. In isotropic media, $F = 0$, so the last term in (1.15) is zero. This means that the nonlinear terms do not alter the sharing of energy between the field components. This is one of the advantages of using the circularly polarized components as a basis in the theory. However, in anisotropic media, in general all three terms are non-zero, and the representations are equally convenient from a mathematical point of view.

1.3 Pulse propagation in nonlinear media

Now we write (1.5) in the form:

$$\nabla^2 \mathbf{E} - \frac{1}{c^2} \frac{\partial^2}{\partial t^2} \mathbf{D} - \frac{4\pi}{c^2} \frac{\partial^2}{\partial t^2} \mathbf{P}^{(NL)} = 0 \qquad (1.16)$$

where $\mathbf{D} = \hat{\varepsilon}^{(L)} \mathbf{E}$ includes only the linear part of $\mathbf{P}$, and we have again ignored the $\nabla(\nabla \cdot \mathbf{E})$ term. In this section, we consider only isotropic media and linearly polarized waves. Then, each component of the vector field $\mathbf{E}$ in (1.16) has the same form, and can be written as a wave equation for a scalar field. Suppose that the envelope function in (1.6) depends not only on spatial variables, but on time as well:

$$E = A(x, y, z, t) \exp(ik_0 z - i\omega_0 t), \qquad (1.17)$$

so that (1.17) represents a pulse rather than a stationary beam. The values ω_0 and $k_0 = k(\omega_0)$ in (1.17) are the central frequency and the central wavenumber respectively. Equation (1.17) is an approximation of a unidirectional wavepacket, because, in representing the fields in this form, we ignore the pulses propagating along the z axis in the opposite direction.

If the pulses are short in comparison with the response time of the medium, then we have to take into account the dispersion of the medium, as well as the nonlinear terms. For simplicity, we suppose that only the dispersion of the linear response function is important. For local media, the delayed response can be written as

$$D(x, y, z, t) = \int_{-\infty}^{\infty} \varepsilon(t - t') \, E_j(x, y, z, t') \, dt',$$

where $\varepsilon(t - t')$ can be represented in the form:

$$\varepsilon(t - t') = \int_{-\infty}^{\infty} \bar{\varepsilon}(\omega) \exp[-i\omega(t - t')] d\omega$$

such that $\bar{\varepsilon}(\omega) = 1 + 4\pi \bar{\chi}^{(1)}(\omega)$. We note that $\varepsilon(t - t')$ is proportional to the delta function $\delta(t - t')$ if the response is instantaneous. The amplitude $A(x, y, z, t)$, in turn, can be represented as an integral of Fourier components:

$$A(x, y, z, t) = \int\limits_{-\infty}^{\infty} \bar{A}(x, y, z|\omega) \exp(-i\omega t)d\omega.$$

Then we obtain, for the displacement,

$$D(x, y, z, t) =$$

$$\exp(ik_0 z) \int\limits_{-\infty}^{\infty} dt' \int\limits_{-\infty}^{\infty} d\omega \, \bar{\varepsilon}(\omega) \exp[-i\omega(t-t')] \int\limits_{-\infty}^{\infty} d\omega' \bar{A}(\omega') \exp[-i(\omega'+\omega_0)t'].$$

Setting $\omega_1 = \omega - \omega' - \omega_0$, we note that $\int_{-\infty}^{\infty} dt' \exp(it'\omega_1)$ can be written as a generalized function, namely the delta function $\delta(\omega_1)$. Using this, we find

$$D(x, y, z, t) = \exp(ik_0 z - i\omega_0 t) \int\limits_{-\infty}^{\infty} d\omega \, \bar{\varepsilon}(\omega_0 + \omega)\bar{A}(x, y, z|\omega) \exp(-i\omega t).$$

Hence the second term in (1.16) can be written as

$$-\frac{1}{c^2}\frac{\partial^2 D}{\partial t^2} = \exp(ik_0 z - i\omega_0 t) \int\limits_{-\infty}^{\infty} d\omega \, k^2(\omega_0 + \omega)\bar{A}(\mathbf{r}, z|\omega) \exp(-i\omega t),$$

where

$$k^2(\omega_0 + \omega) = \frac{(\omega_0 + \omega)^2}{c^2}\bar{\varepsilon}(\omega_0 + \omega)$$

is the dispersion relation for linear waves. Expanding $k^2(\omega_0 + \omega)$ around ω_0 and keeping terms up to second order, we obtain

$$-\frac{1}{c^2}\frac{\partial^2 D}{\partial t^2} = \exp(ik_0 z - i\omega_0 t)\left[k_0^2 A + 2ik_0 k_0' \frac{\partial A}{\partial t} - (k_0'^2 + k_0 k_0'') \frac{\partial^2 A}{\partial t^2}\right],$$

where

$$k_0 = k(\omega_0), \qquad k_0' \equiv \left.\frac{\partial k}{\partial \omega}\right|_{\omega_0} = v_g^{-1} \quad \text{and} \quad k_0'' \equiv \left.\frac{\partial^2 k}{\partial \omega^2}\right|_{\omega_0},$$

in which v_g is the group velocity. Equation (1.16) can now be written in the form

$$2ik_0\frac{\partial A}{\partial z} + \frac{\partial^2 A}{\partial x^2} + \frac{\partial^2 A}{\partial y^2} + \frac{\partial^2 A}{\partial z^2} + 2ik_0 k_0' \frac{\partial A}{\partial t}$$

$$- (k_0'^2 + k_0 k_0'')\frac{\partial^2 A}{\partial t^2} + \frac{4\pi\omega_0^2}{c^2}\chi_{xxxx}|A|^2 A = 0.$$

We are interested in waves propagating one way, namely in the positive z

direction. We are ignoring all radiation propagating in the opposite direction which may appear as a result of scattering. In this case, it is convenient to write the equations in a frame moving with the group velocity of the pulse, viz.

$$\xi = k_0 z, \qquad \tau = t - k'_0 z,$$

Then we have:

$$\frac{\partial A(\xi, \tau)}{\partial \xi} = \frac{1}{k_0} \left(\frac{\partial A}{\partial z} + k'_0 \frac{\partial A}{\partial t} \right),$$

and for the second derivatives

$$\frac{\partial^2 A}{\partial z^2} - k'^2_0 \frac{\partial^2 A}{\partial t^2} = \left(\frac{\partial}{\partial z} - k'_0 \frac{\partial}{\partial t} \right) \left(\frac{\partial}{\partial z} + k'_0 \frac{\partial}{\partial t} \right) A \ll A.$$

Hence, dividing both sides of the equation by $2k_0^2$, we have

$$i\frac{\partial A}{\partial \xi} + \frac{1}{2} \frac{\partial^2 A}{\partial x^2} + \frac{1}{2} \frac{\partial^2 A}{\partial y^2} - \frac{k''_0}{2k_0} \frac{\partial^2 A}{\partial \tau^2} + \alpha |A|^2 A = 0, \qquad (1.18)$$

where $\alpha = 2\pi \omega_0^2 (\chi_{xxxx})/(k_0^2 c^2)$, $x = k_0 x$ and $y = k_0 y$.

This is the so-called '(3+1)-dimensional' nonlinear scalar equation which describes wave propagation in nonlinear isotropic homogeneous media. It takes into account transverse self-focusing (or self-defocusing) effects, as well as the dispersion of the pulse in the longitudinal direction. Depending on the sign of the group velocity dispersion, k'', a pulse can disperse (positive k''), or contract (negative k''), on propagation. For negative k'', this equation describes, for example, propagation of 'optical bullets' (Silberberg, 1990) or the collapse of beams and pulses. It can be generalized to include the effects of birefringence, waveguiding structures or more complicated nonlinear terms, as in section 1.2.

For plane waves, we can ignore the x and y dependencies in (1.18). With two additional normalizations, $\tau \to \sqrt{-2k_0/k''_0}\tau$ (with k''_0 negative) and $A \to \sqrt{\alpha} A$, we have the standard form of the (1+1)-dimensional NLSE:

$$i\frac{\partial A}{\partial \xi} + \frac{1}{2} \frac{\partial^2 A}{\partial \tau^2} + |A|^2 A = 0. \qquad (1.19)$$

In the normal dispersion regime ($\lambda < \lambda_0$, where λ_0 is usually within the range 1.3–1.55 μm in a silica single-mode fiber), we have k''_0 positive, leading to the standard form of the NLSE for this range of wavelengths:

$$i\frac{\partial A}{\partial \xi} - \frac{1}{2} \frac{\partial^2 A}{\partial \tau^2} + |A|^2 A = 0. \qquad (1.20)$$

Note that the value k_0 in (1.17) does not have to be exactly $k(\omega_0)$. We may also choose a frame with a velocity which is slightly different from the group velocity of the wavepacket. If this is the case, then (1.18) includes additional terms:

$$i\frac{\partial A}{\partial \xi} + \beta A + i\delta\frac{\partial A}{\partial \tau} + \frac{1}{2}\frac{\partial^2 A}{\partial x^2} + \frac{1}{2}\frac{\partial^2 A}{\partial y^2} - \frac{k_0''}{2}\frac{\partial^2 A}{\partial \tau^2} + \alpha|A|^2 A = 0,$$

where $\beta = [k_0^2 - k^2(\omega_0)]/(2k_0^2) \approx [k_0 - k(\omega_0)]/k_0$ is the phase velocity mismatch and $\delta = k_0''/2 - 1/v_g$ is the inverse group velocity mismatch. These terms are important when describing pulses with two orthogonally polarized components in birefringent media where the phase and group velocities of the two transverse components are slightly different. The nonlinear terms then also have a more complicated structure.

1.4 Dispersive effects in waveguides

In the previous section we took into account the material dispersion of the medium. Waveguides have their own dispersion due to their geometry. When a waveguide is dispersive, we have to take into account the pulse distortions which occur upon propagation in the guide. The derivation of the equation for nonlinear pulse propagation in such a medium is complicated. Here, we are interested in the physical reasons for the appearance of dispersive terms in the wave equation. Correspondingly, we first use a linear approach based on an approximate dispersion relation (Akhmanov and Khokhlov, 1964; Marcuse, 1979; Vinogradova et al., 1979).

Let us consider an 'initial value' problem of pulse propagation in a dispersive medium. At $z = 0$, the pulse has the shape $u(t,0) = f(t)$. Its spectrum, $F(\omega)$, is

$$F(\omega) = \frac{1}{2\pi}\int\limits_{-\infty}^{\infty} f(t)\exp(i\omega t)\,dt \qquad (1.21)$$

For a linear medium, the signal at every point z can be written as a linear superposition of waves of all frequencies:

$$u(z,t) = \int\limits_{-\infty}^{\infty} F(\omega)\exp[ik(\omega)z - i\omega t]d\omega. \qquad (1.22)$$

Here $k(\omega)$ describes the dispersion relation for waves in a given medium. This description is convenient because both material and geometrical dispersion related to waveguide can be included in this relation.

Usually, the spectral width of the signal is small compared with the central frequency, ω_0. Hence, the signal only has the component

$$F(\omega) = A_0(t)\exp(-i\omega_0 t),$$

where $A_0(t)$ is a slowly varying function of time ($|\frac{dA_0}{dt}| \ll \omega_0|A_0|$). A similar expression can be written for the signal at any other point z:

$$u(z,t) = A(z,t)\exp(ik_0 z - i\omega t),$$

where $k_0 = k(\omega_0)$. The envelope, $A(z,t)$, is a slow function of both the z and t variables. Now, $A(z,t)$ evolves according to:

$$A(z,t) = \frac{1}{2\pi} \int_{-\infty}^{\infty} A_0(t')\exp[-i(t-t')\Delta\omega]\,dt' \int_{-\infty}^{\infty} \exp\{i[k(\omega) - k_0]z\}d(\Delta\omega),$$

where $\Delta\omega = \omega - \omega_0$.

The dispersion relation near the central frequency ω_0 can be described using a Taylor series:

$$k(\omega) = k_0 + k'\Delta\omega + \frac{1}{2}k''(\Delta\omega)^2 + \frac{1}{6}k'''(\Delta\omega)^3 + \frac{1}{24}k''''(\Delta\omega)^4 + \dots \quad (1.23)$$

where

$$k_0 = k(\omega_0), \quad k' = \left(\frac{dk}{d\omega}\right)_{\omega_0}, \quad k'' = \left(\frac{d^2 k}{d\omega^2}\right)_{\omega_0},$$

$$k''' = \left(\frac{d^3 k}{d\omega^3}\right)_{\omega_0} \quad \text{and} \quad k'''' = \left(\frac{d^4 k}{d\omega^4}\right)_{\omega_0}.$$

A dispersion relation in this form can take into account both material and geometrical factors related to the waveguide. The inverse group velocity can be written as

$$v_g^{-1} = k' + k''\Delta\omega + \dots$$

so that k' is the reciprocal group velocity at the central frequency, while the higher-order terms determine the dispersion of the group velocity. The dispersion parameter

$$D_1 = \frac{2\pi c}{\lambda^2}\frac{d^2 k}{d\omega^2}$$

is usually used to describe the dispersion of the group velocity. It, in turn, depends on the wavelength. For a standard silica single-mode fiber, this dependence is given by the solid line in Fig. 1.1. It passes through zero at approximately 1.3 μm and becomes negative at longer wavelengths. In addition, silica fibers have attenuation minima around 1.3 μm and 1.5 μm. These facts are important for soliton propagation in optical fibers. The dispersion parameter is not constant. However, it may change only slightly over a certain range of wavelengths. This range can be made wider, as in the case of the dashed line in Fig. 1.1, in specially designed optical fibers (Cohen et al., 1982; Mamyshev et al., 1993). The whole shape of this curve can be important if the pulse spectrum is wide enough to cover a significant part of the curve.

For a dispersion-flattened fiber, the third-order dispersion is zero at a wavelength of around 1.45 μm. In this case we can just retain the second-order, third-order and the fourth-order dispersion terms in the equation

(Wai *et al.* 1986, 1987; Cavalcanti, 1991; Karlsson and Höök, 1994; Karpman, 1994). For pulses longer than one picosecond, the fourth-order dispersion is small in comparison with the second-order dispersion. For pulses in the femtosecond range, it can happen that the two of them are of the same order.

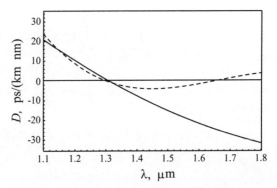

Figure 1.1 *(a) Schematic dependence of the group velocity dispersion, D, on the wavelength, λ, for a standard single-mode optical fiber (solid line) and for dispersion-flattened optical fiber (with specifically designed transverse profile of the refractive index) (dashed line).*

Taking (1.23) into account, we can write an equation which governs the evolution of the pulse envelope $A(z,t)$:

$$i\left(\frac{\partial A}{\partial z} + k'\frac{\partial A}{\partial t}\right) = \frac{k''}{2}\frac{\partial^2 A}{\partial t^2} + i\frac{k'''}{6}\frac{\partial^3 A}{\partial t^3} - \frac{k''''}{24}\frac{\partial^4 A}{\partial t^4}. \qquad (1.24)$$

It can be seen that, to zeroth order, when we ignore all the terms on the right-hand side of (1.24), the pulse moves, without distortion, at the group velocity

$$v_g = (k')^{-1}.$$

In the reference frame moving with the group velocity, with the transformations

$$\xi = z, \qquad \tau = t - k'z,$$

(1.24) takes the form

$$i\frac{\partial A}{\partial \xi} = \frac{k''}{2}\frac{\partial^2 A}{\partial \tau^2} + i\frac{k'''}{6}\frac{\partial^3 A}{\partial \tau^3} - \frac{k''''}{24}\frac{\partial^4 A}{\partial \tau^4}. \qquad (1.25)$$

Each term on the right-hand-side of (1.24) causes a particular type of distortion in the pulse. Numerical examples of Gaussian pulse propagation with higher-order dispersion terms have been given by Marcuse (1980). The main difference here is that the dispersion comes from combined sources – material plus waveguide dispersion. However, the nonlinear terms in this case must be incorporated in a different way. The lowest-order dispersion term ($\propto \frac{\partial^2 A}{\partial \tau^2}$) allows the initially symmetric pulse to remain symmetric. It increases the pulse width and introduces phase chirp. The nonlinearity of the fiber can compensate for this.

Third- and fourth-order dispersion are also important in laser systems generating short pulses in the femtosecond range. In these systems, higher-order dispersion terms can be adjusted by inserting additional dispersive elements into the cavity. It has been shown (Christov *et al.*, 1994) that fourth-order dispersion becomes important when the pulse duration is reduced to below 10 fs.

1.5 Pulse propagation in single-mode fibers

Single-mode fibers have become the main conduits for long-distance information transmission in recent years. Nonlinear effects in optical fibers are usually very weak, so that the changes in the refractive index due to nonlinear effects are generally much smaller than changes in the refractive index in the transverse profile of the waveguide. In this case, the transverse profile of the fundamental mode is defined mainly by linear effects, while the nonlinear effects only serve to compensate for the dispersion. In this case, the transverse and longitudinal variables can be separated (Hasegawa and Kodama, 1981):

$$E(r, z, t) = R(r)f(z, t)\exp(ikz - i\omega t) \qquad (1.26)$$

where $f(z, t)$ is a slowly varying function of z and t on spatial and time scales comparable with the wavelength and reciprocal frequency, respectively, and $R(r)$ is the transverse profile of the linear mode of the fiber. This latter function satisfies the equation

$$\nabla_\perp^2 R - [\beta^2 - \varepsilon(r)]R = 0.$$

The ansatz (1.26) becomes exact when nonlinearity is absent.

The small nonlinear term changes from the centre to the edges of the fiber. However, its influence on the longitudinal dynamics can be averaged across the fiber. By integrating over the cross-section of the fiber (Hasegawa and Tappert, 1973a; Jain and Tzoar, 1978a,b), we can obtain, with the same approximations as before,

$$i\frac{\partial A}{\partial \xi} - \frac{k''}{2}\frac{\partial^2 A}{\partial \tau^2} + 2\pi\alpha_1\chi^{(3)}|A|^2 A = 0, \qquad (1.27)$$

where $\alpha_1 = <R^4>/<R^2>$ and $<...>$ means averaging over the cross-section. The factor α_1 is of order unity. Solutions can be scaled with an arbitrary constant z_0, such that

$$\alpha_1\chi^{(3)}z_0 = 1, \qquad \xi = z/z_0, \qquad -k''z_0 = t_0^2.$$

This arbitrary constant can be chosen to measure the length, time and power in 'soliton units'. In particular, if we choose t_0 to be the full width (in picoseconds) of the pulse at half maximum intensity (Mollenauer *et al.*, 1986), then this defines the units for the other variables. The soliton period is

$$z_0 = 0.322 \frac{\pi^2 c t_0^2}{|D|\lambda^2} \qquad \text{(in kilometres)},$$

while its power is

$$P = \frac{\lambda}{4n_2 z_0} \qquad \text{(in watts)},$$

where $n_2 = \sqrt{\varepsilon^{(NL)}}$. The higher-order dispersive terms can be added as before. The propagation of bright solitons in fibers was first observed by Mollenauer *et al.* (1980), while dark solitons were first observed in fibers by Emplit *et al.* (1987).

For strong nonlinear terms, the full three-dimensional problem must be solved. Nonlinear effects influence the transverse field profile, as well as fiber dispersion (Sammut and Pask, 1991).

1.6 Birefringent fibers

For birefringent fibers, we can follow the same derivation as above for each linearly polarized component of the field. Choosing the average wave vector k_0 and the frame moving with the average group velocity, we can write the system of equations for the components in the form

$$iU_\xi + i\delta U_\tau + \beta U + \tfrac{1}{2}U_{\tau\tau} + (|U|^2 + A|V|^2)U + BV^2 U^* = 0,$$
$$iV_\xi - i\delta V_\tau - \beta V + \tfrac{1}{2}V_{\tau\tau} + (A|U|^2 + |V|^2)V + BU^2 V^* = 0,$$

(1.28)

where U and V are the slowly varying envelopes of the two linearly polarized components of the field along the x and y axes, δ is half the inverse group velocity difference, β is half the difference between the propagation constants, the coefficients A and B are defined in section 1.2, ξ is the normalized longitudinal coordinate and τ is the normalized retarded time.

If the fiber is twisted, it becomes optically active. The eigenfunctions of the linear problem are then circularly polarized waves rather than linearly polarized ones. These two eigenfunctions have different phase and group velocities. They can serve as the two eigenmodes of the nonlinear problem as well. Then, equations similar to (1.28) can be written for this fiber; however, U and V then represent the circularly polarized components of the field. The coefficients of the nonlinear terms must also be renormalized, as in section 1.2.

Generally speaking, in twisted fibers, both optical activity and birefringence may be present (Ulrich and Simon, 1979). The eigenmodes of the fiber are then elliptically polarized waves. Equations describing wave propagation in such fibers have the same form as before, but the coefficients of the nonlinear terms again have different values. Additional terms mixing the two modes can also be present. This case has been considered by Menyuk (1989).

Basically the same equations can be used to describe the instability of spatial solitons in planar waveguides (De Sterke and Sipe, 1991). The differences in interpretation are that U and V are then the TE (i.e. those where the electric field is along the y axis) and TM (i.e. those where the electric field is along the x axis) modes of the waveguide, and that δ then describes the difference in angles of propagation. Our analysis of equations (1.28) appears in Chapter 7.

1.7 Nonlinear guided waves

Nonlinear wave guidance is a general term for a plethora of wave phenomena in layered nonlinear media (Stegeman, 1985; Boardman and Egan, 1986). Some experimental investigations are included in the review by Michalache *et al.* (1989). For planar waveguides, we can suppose that the linear dielectric constants in (1.9) depend only on one spatial variable, so that $\varepsilon^{(L)} = \varepsilon(x)$. We can then use scalar equations only for TE modes. Its electric field can be written as

$$\mathbf{E} = \psi(x)\exp(ikz - i\omega t).$$

The equation governing the propagation of these waves can be written as

$$2ik\psi_z + \psi_{xx} - [k^2 - \epsilon(x)] + n_2(x)|\psi|^2\psi = 0, \tag{1.29}$$

where $n_2(x) = \frac{4\pi\omega^2}{c^2}\chi^{(3)}_{yyyy}(x)$, as the nonlinear refractive index also depends on the transverse coordinate, x. In this statement of the problem, we are interested in guided waves propagating along a layered medium (Fig. 1.2). Stationary solutions are the solutions of the following ordinary differential equation:

$$\psi_{xx} - [k^2 - \epsilon(x)] + n_2(x)|\psi|^2\psi = 0. \tag{1.30}$$

At low intensities, where we can ignore the nonlinear term in (1.30), the modal structure is the same as that in a linear waveguide. Dispersion relations which are power-dependent are called 'energy-dispersion' curves. These curves (Torres and Torner, 1993), characterizing the properties of nonlinear guided waves (stationary solutions) and their stability (Akhmediev, 1982; Moloney *et al.*, 1986; Akhmediev and Ostrovskaya, 1988; Ankiewicz and Tran, 1991; Mitchell and Snyder, 1993), are the main focus of this topic. The problem can be generalized to take into account TM modes of the waveguide or guides with two transverse dimensions (Moloney, 1987). The stationary solutions can then only be found numeri-

cally (Akhmediev *et al.*, 1989). We present some recent results on stationary
solutions and their stability in Chapter 12.

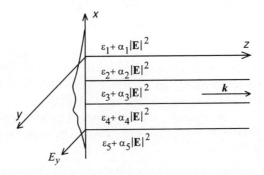

Figure 1.2 *Schematic plot of nonlinear guided wave propagation in a layered medium*

The evolution of solitons (the changes in beam shape along the z axis)
has been considered in a number of publications (e.g. Akhmediev *et al.*,
1985a; Aceves *et al.*, 1989). An 'effective particle' approach for solving
these problems can be found in the book by Moloney and Newell (1991).

It is also clear that self-focusing in the spatial case ($n_2 > 0$) corresponds
to anomalous dispersion and thus is described by the NLSE (1.19). Conversely, self-defocusing ($n_2 < 0$) corresponds to normal dispersion (1.20)
(with $\xi \rightarrow -\xi$).

1.8 The state of polarization of monochromatic optical waves

The electric field of a monochromatic plane wave can be described by two
complex transverse components, E_x and E_y:

$$\mathbf{E} = (\hat{\mathbf{e}}_x E_x + \hat{\mathbf{e}}_y E_y) \exp(ikz - i\omega t + i\phi_0), \qquad (1.31)$$

where the complex amplitudes can be written in the form

$$E_x = E_1 \exp(i\phi_1), \qquad E_y = E_2 \exp(i\phi_2). \qquad (1.32)$$

The state of polarization of such a field depends on the amplitudes of
these two components, E_1 and E_2, and the phase difference between them,
$\Delta\phi = \phi_2 - \phi_1$. The common phase, ϕ_0, of the total field is unobservable,
unless interference effects with other fields at the same frequency are considered. Thus the state of polarization of the field can be described by the
polarization ellipse

$$\left(\frac{E'_x}{E_1}\right)^2 + \left(\frac{E'_y}{E_2}\right)^2 - \frac{2E'_x E'_y}{E_1 E_2} \cos \Delta\phi = \sin^2 \Delta\phi, \qquad (1.33)$$

where $E'_x = \text{Re }(E_x)$ and $E'_y = \text{Re }(E_y)$. This ellipse is shown in Fig. 1.3a. Depending on the sign of $\Delta\phi$, the state of polarization is right-hand elliptic ($\Delta\phi > 0$) or left-hand elliptic ($\Delta\phi < 0$). We assume that $\Delta\phi$ may vary in the interval ($-\pi < \Delta\phi < \pi$).

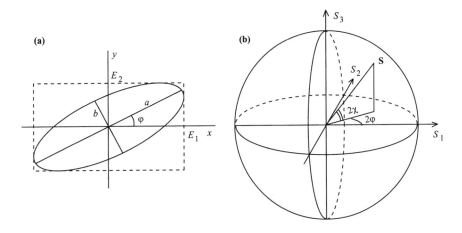

Figure 1.3 *(a) Polarization ellipse and (b) representation of the state of polarization on the Poincaré sphere.*

The orientation of the ellipse is defined by the angle φ between the major axis of the ellipse and the x axis. Thus we have

$$\tan 2\varphi = \frac{2E_1E_2 \cos\Delta\phi}{E_1^2 - E_2^2}. \tag{1.34}$$

The degree of ellipticity can be described by the ratio of the length of the semi-minor axis, b, to that of the semi-major axis, a:

$$\tan\chi = \frac{b}{a}\,\text{sign}(\Delta\phi). \tag{1.35}$$

Here the angle χ can be found from the equation

$$\sin 2\chi = \frac{2E_1E_2 \sin\Delta\phi}{E_2^2 + E_1^2}. \tag{1.36}$$

A convenient way to describe the state of polarization of a monochromatic field is to use the Stokes parameters, which can be defined as follows:

$$\begin{aligned}
S_0 &= |E_x|^2 + |E_y|^2 = E_1^2 + E_2^2,\\
S_1 &= |E_x|^2 - |E_y|^2 = E_1^2 - E_2^2,\\
S_2 &= E_xE_y^* + E_x^*E_y = 2E_1E_2 \cos\Delta\phi,\\
S_3 &= E_xE_y^* - E_x^*E_y = 2E_1E_2 \sin\Delta\phi.
\end{aligned} \tag{1.37}$$

It follows, from the definition of these parameters, that

$$S_0^2 = S_1^2 + S_2^2 + S_3^2.$$

(S_1, S_2, S_3) can be viewed as the components of a Stokes vector $\mathbf{S}$, having modulus S_0. They define, on the Poincaré sphere, a certain point which identifies the polarization ellipse uniquely.

Using definitions (1.37), we find

$$\sin 2\chi = \frac{S_3}{S_0} \tag{1.38}$$

and

$$\sin 2\varphi = \frac{S_2}{S_1}. \tag{1.39}$$

Hence, the inclination, 2χ, of the Stokes vector defines the degree of ellipticity, while its azimuth, 2φ, in the equatorial (S_1, S_2) plane, is double the angle of orientation of the major axis of the polarization ellipse.

Now, it is easy to extract the state of polarization from the location of the Stokes vector. The points on the Poincaré sphere corresponding to linearly-polarized light are located along the equatorial line, with an azimuthal angle equal to twice the angle of orientation in the plane of polarization. Circularly right and left polarized light corresponds to the north and south poles on the Poincaré sphere, respectively. Any other point corresponds to elliptically polarized light, with the orientation of the ellipse defined by the angle φ and the ellipticity related to the angle χ. The upper hemisphere corresponds to right and the lower hemisphere to left polarization. Stokes parameters can easily be measured in experiments, and the state of polarization can be deduced from them. We use this representation when considering soliton propagation in nonlinear birefringent fibers.

The approach of the Stokes vector and the Poincaré sphere can be applied to other problems with two complex variables related through a conserved quantity – the energy. Two complex variables are equivalent to four real variables, say two amplitudes and two phases. Due to the conservation of energy, only three of them are independent. Moreover, the absolute phase can be removed from the problem if we are not interested in it. As a result, only two variables are independent, and they can be mapped onto the surface of the sphere, allowing us, in this way, to reduce the phase space and to visualize complicated dynamics. An example is the theory of two-level atoms (Allen and Eberly, 1974), where the term 'Bloch vector' is used rather than 'Stokes vector'. Another problem which can be investigated in this way is wave propagation in dual-core fibers (Daino *et al.*, 1985). For continuous waves, this forms a dynamical system with two complex fields. We can express the fields in terms of Stokes parameters, but the physical meaning of these parameters will be different from that for the polarization vector. We use the Stokes parameters when considering birefringent fibers in Chapter 7 and nonlinear couplers in Chapter 8.

1.9 Nonlinear couplers

Optical nonlinear couplers are very useful devices which allow fast switching and signal coupling in optical communication links. Nonlinear couplers also have applications as intensity-dependent switches and as limiters. They can be used to multiplex two incoming bit streams onto one fibre, and also to demultiplex a single bit stream. Optical couplers can be made as planar devices (see Fig. 1.4a) using semiconductor material (Kang *et al.*, 1995), or dual-core, single-mode fibers (Friberg *et al.*, 1987) (see Fig. 1.4b). Solitons can propagate in each core. The coupling of energy from one guide to the other occurs because of the overlap of the evanescent fields between the cores.

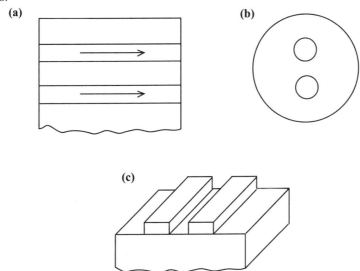

Figure 1.4 *Geometry of (a) planar (b) fiber and (c) rib-structure couplers.*

Taking into account the linear coupling between the fields, and assuming that the two interacting waveguides are equivalent, it can be shown that pulse propagation in this type of device is described by two coupled NLSEs (Trillo *et al.*, 1988):

$$iU_\xi + \frac{1}{2}U_{\tau\tau} + N(|U|^2)U + KV = 0$$

$$iV_\xi + \frac{1}{2}V_{\tau\tau} + N(|V|^2)V + KU = 0, \qquad (1.40)$$

where each of the equations describes pulse propagation, with complex amplitudes U and V, in a fiber in the anomalous dispersion regime, the terms KV and KU are added to each separate NLSE to describe linear coupling and $N(|U|^2)$ is the nonlinear response function. The coefficient K depends on the waveguide geometry, with the main factor being the distance between the guides. The equations are written in a normalized form for convenience of mathematical analysis. Clearly, this set of equations is a particular case of the general set of coupled NLSEs, when the cross-phase modulation term is zero.

These systems have various applications, such as intensity-dependent switches and devices for separating a compressed soliton from its broad 'pedestal'. Operation of these devices with continuous-wave (c.w.) inputs (i.e. without the dispersion terms in (1.40)) has been discussed by Maier, (1984); Jensen (1982); Ankiewicz (1988a,b) and Rowland (1991). The physics of nonlinear couplers with c.w. inputs has been presented by Snyder *et al.* (1991b). The theory establishes the basic principles of coupler operation and describes the main features of switching. As we show in Chapter 8, these features are, in some ways, similar to soliton-like pulse switching. This similarity, although it is incomplete, is a unique feature of solitons. However, quasi-c.w. (long) pulses in couplers behave quite differently from soliton-like pulses.

We note, that c.w. carries no information. To make the operation of non-linear devices as fast as possible, we have to use short pulses, rather than c.w. or quasi-c.w. inputs. In the pico- and femtosecond ranges, only solitons can carry information stably enough. Hence, we need to understand the response of these devices to soliton-like pulses. The stability properties of solitons in dual-core fiber devices have been studied by Wright *et al.* (1989). Since solitons may well soon be used in long-distance (especially submarine) fibre communications, it would be ideal if intensity-dependent switching could be performed in all-optical devices, instead of converting from optical to electrical signals for electronic switching and then regenerating the optical signals. Moreover, fast optical computers can be designed using soliton switching and logic components (Doran, 1987; Islam, 1992), perhaps involving couplers. The first experimental observations of soliton switching with short pulses have been reported by Friberg *et al.* (1987, 1988). We analyse and discuss pulse propagation in nonlinear couplers in Chapter 8.

1.10 Nonlinear fiber arrays

As a further generalization of the coupler with two cores, let us consider the multi-core fiber devices shown in Fig. 1.5. These can be of rib or ring geometry. Correspondingly, the set of equations describing them consists of N NLSEs with nearest-neighbour linear coupling. They can be used as more elaborate switching devices. The following are the equations for the ring geometry shown in Fig. 1.5c:

$$iU_\xi^{(1)} + \tfrac{1}{2}U_{\tau\tau}^{(1)} + |U^{(1)}|^2\, U^{(1)} + KU^{(N)} + KU^{(2)} = 0$$

$$iU_\xi^{(2)} + \tfrac{1}{2}U_{\tau\tau}^{(2)} + |U^{(2)}|^2\, U^{(2)} + KU^{(1)} + KU^{(3)} = 0 \qquad (1.41)$$

$$\cdot \quad \cdot \quad \cdot$$

$$iU_\xi^{(N)} + \tfrac{1}{2}U_{\tau\tau}^{(N)} + |U^{(N)}|^2\, U^{(N)} + KU^{(N-1)} + KU^{(1)} = 0$$

where $U^{(N)}$ is the optical field in the Nth core, and the coupling constant K is the same between all neighbouring fibers. The changes which must be made in these equations to describe the cases shown in Fig. 1.5a,b are obvious. We will analyse arrays in Chapter 9.

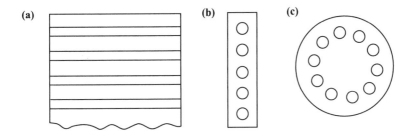

Figure 1.5 *Geometry of multi-core (a) planar, (b) rib and (c) ring fiber couplers*

1.11 Passively mode-locked lasers with fast saturable absorbers

Passive mode-locking of c.w. lasers is a reliable technique for short optical pulse generation. Theoretical modelling of passively mode-locked lasers is usually based on one of two different approaches. The first uses numerical simulations, taking into account the cyclical transmission of a light wave around a loop. It is the most straightforward way of modelling a system with arbitrary parameter values. The second, introduced by Haus (1975), is based on a continuous model, with the assumption that the pulse change is small during any one circuit of the laser. In this approximation, the dynamics of the field in a passively mode-locked laser is described by the complex Ginzburg–Landau equation (CGLE) (Moores, 1993):

$$i\psi_\xi + \frac{k''}{2}\psi_{\tau\tau} + |\psi|^2\psi = i\delta\psi + i\beta\psi_{tt} - i\epsilon|\psi|^2\psi + (i\mu - \nu)\,|\psi|^4\psi, \qquad (1.42)$$

where ξ is the distance along the axis of propagation (unfolding the resonator), τ is retarded time, k'' is the group velocity dispersion, $\delta = g - l$

is the net field linear gain coefficient (linear gain g minus loss l), and $\beta = (\frac{g}{\Omega_g^2} + \frac{1}{L_f \Omega_f^2})$ describes the effect of spectral limiting due to bandwidth-limited amplification and spectral filtering. Here, Ω_g is the bandwidth of the gain, and Ω_f is the bandwidth of the passive filtering effects in the laser over a filtering length L_f. The values ϵ and ν are nonlinear gain/loss coefficients describing a fast-saturable absorber (FSA). The term proportional to ν describes important physics. The cubic term (ϵ) is not sufficient to ensure dynamic stability for the laser. In this sense, the μ term gives a more realistic description of an artificial FSA.

One way to implement an FSA or intensity limiter in fiber lasers is by using nonlinear polarization rotation in a birefringent fiber. Linearly polarized light in a fiber then becomes elliptically polarized. The state of polarization changes, including a rotation of the ellipse. The final state of polarization depends on the intensity of the light. The transmission characteristics of this device depend strongly on the input intensity, so that significant fifth-order terms in (1.42) can easily be achieved. By properly adjusting the input and output polarizers relative to the principal axes of birefringence, we can obtain quite complicated transmission characteristics. The fiber laser, with passive mode-locking, used by Doerr *et al.* (1994a), is shown in Fig. 1.6. It contains the element described above, as well as an active erbium-doped fiber. We present an extensive analysis of the CGLE in Chapter 13.

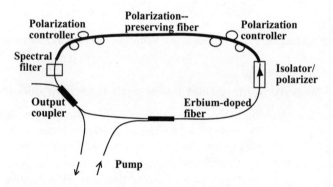

Figure 1.6 *Fiber laser with passive mode-locking*

1.12 Soliton-based optical transmission lines

Another optical system which is described by an extended CGLE is an optical fiber transmission line with bandwidth-limited gain and higher-order effects. The system may be described by the cubic or quintic Ginzburg-Landau equation (GLE) with terms such as the Raman term and frequency sliding added to the filtering:

$$i\psi_\xi + \frac{1}{2}\psi_{\tau\tau} + N(|\psi|^2)\psi = i\delta\psi + i\beta\psi_{\tau\tau} + \gamma_1\,\psi_\tau + t_R\,\psi\,(|\psi|^2)_\tau + i\,\beta_3\psi_{\tau\tau\tau}$$

$$- \beta_4\psi_{\tau\tau\tau\tau} - i\,s(|\psi|^2\psi)_\tau + i\mu\psi\,|\psi|^4 - i\epsilon\,|\psi|^2\psi. \tag{1.43}$$

The number of terms on the right-hand side which need to be considered depends on the guide material, and the intensity and width of the pulses being used in the fiber. Most of these terms are negligible for pulses wider than 1 ps, but when femtosecond pulses are being used in very high bit-rate communication systems, the small Raman delay and the higher-order effects on dispersion and gain must be taken into account. Thus the analysis becomes more complicated when narrower pulses are considered.

Here t_R is the Raman delay (a few femtoseconds) representing a non-instantaneous response. This term arises from a Taylor expansion, assuming that the delay is small. A typical spectral filtering device can be approximated by a spectrum of parabolic shape. If it is centred on the frequency ω_a, then the constant gain δ is replaced by $\delta[1 - (\omega - \omega_a)^2\,\beta/\delta]$, where β is the gain dispersion. Using the fact that $\omega - \omega_a$ corresponds to the operator $i\frac{d}{dt}$, we see that the term $i\delta\psi$ is replaced by $i(\delta\psi + \beta\psi_{tt})$ in the equation. If the gain were equal to its maximum value for all wavelengths, then, of course, β would be zero. Thus, in this application, the first two terms on the right-hand side of (1.43) can describe a parabolic gain spectrum of a rare-earth-doped fiber amplifier.

Equation (1.42) assumes that the gain maximum coincides with the soliton carrier frequency (i.e. that there is no 'detuning'). However, the soliton centre frequency, ω_0, may not be equal to ω_a. This introduces a 'detuning' parameter, γ_1, which is proportional to $\omega_0 - \omega_a$. Even in the presence of both Raman downshift and detuning, it is possible for a soliton to evolve to a new equilibrium frequency and amplitude.

The third- and fourth-order dispersion terms (involving β_3 and β_4, respectively), will be dealt with in Chapter 10.

The term in ϵ is two-photon absorption (or 'nonlinear gain' if ϵ is negative). A simple application of perturbation theory shows that nonlinear gain can balance the gain dispersion if $\beta = -2\epsilon$. In that case, the soliton propagates with constant amplitude. Self-steepening is given by the term with parameter s.

In the last term on the left-hand side, the nonlinearity function N depends only on intensity, $I = |\psi|^2$, and thus it effectively modifies the nonlinearity law from the standard Kerr-law term if $N \neq I$. This term can still be important for relatively broad pulses. If $N = I + \nu I^2$, then the index increase is a parabolic function of intensity. On the other hand, if

$$N = \frac{I}{1 + \gamma I} \tag{1.44}$$

then the index increase exhibits saturation. We will consider this as a separate effect in Chapter 4.

Higher-order gain is included by introducing the term in μ. Thus the term $(-\nu + i\mu)\,\psi\,|\psi|^4$ appearing in the GLE represents a parabolic nonlinearity with higher-order gain. We will not consider the full equation, (1.43), i.e. the GLE with additional terms, in this book. However, we note that, if terms on the right-hand side of (1.43) can be considered small, then perturbation theory provides some information on their effects; if this is not the case, then only numerical simulations are reliable. A particular case of (1.43), namely

$$i\psi_\xi + \frac{1}{2}\psi_{\tau\tau} + |\psi|^2\psi = -\,i\,s(|\psi|^2\psi)_\tau, \tag{1.45}$$

is integrable. This has been shown by using the inverse scattering technique (Kaup and Newell, 1978; Vysloukh and Cherednik, 1989). Using the Painlevé analysis, other cases of integrability can also be found. Another particular case is (Porsezian and Nakkeeran, 1996):

$$i\psi_\xi + \frac{1}{2}\psi_{\tau\tau} + 2|\psi|^2\psi = -\,i\,s\left[\psi_{\tau\tau\tau} + 6|\psi|^2\psi_\tau + 3\psi(|\psi|^2)_\tau\right], \tag{1.46}$$

where s is an arbitrary real coefficient.

1.13 Soliton X-junctions

High-intensity light increases the local refractive index of a material medium, and thus can set up its own waveguide. A spatial soliton created in this way can then guide a weaker beam. The idea has been suggested theoretically (Chiao et al., 1964; Snyder et al., 1991a) and proved experimentally (Luther-Davies and Yang, 1992a). Colliding solitons induce a complicated criss-cross (linear) waveguide structure (Snyder et al., 1995). Low-power light travelling along one waveguide will, in general, leak due to reflection and scattering. When viewed from this physical perspective, it is surprising that the waveguide parameters can ever be adjusted to result in exactly radiation–free transmission. But we know, from the results of inverse scattering, that the collision of one-dimensional self-guided beams in an isotropic Kerr medium is precisely radiation free. The physics that underlies this important result can be regarded as an extension of the special reflectionless character of the sech-squared profile waveguide or potential for incident linear waves. This latter point will be discussed in section 2.6.

This concept can be applied in various ways, such as making nonlinear Y-junctions (Luther-Davies and Yang, 1992b), X-junctions (Akhmediev and Ankiewicz, 1993b) or more complicated $N \times N$-port devices (Miller and Akhmediev, 1996b). These have certain advantages, such as being lossless. We also have the possibility of calculating the exact transmission characteristics.

In 'light-guiding-light' experiments, a weak beam can be separated from a strong pump if they have different frequencies or different polarizations. In the former case we can use spectral filtering to separate them, and in the latter case polarizers. However, each method has its own drawbacks. One of them, which is always present, is energy exchange between the two fields. Hence, the probe beam cannot propagate in the waveguide passively. In other words, the equation for the probe beam cannot be considered as an exactly linear one. Another problem is that the pump and the probe-beam 'see' different profiles of refractive index change (Torres-Cisneros *et al.*, 1993). Thus, the probe beam and the pump are not exactly matched. This can cause additional loss from the probe beam. As a consequence, one of the best possibilities for experimental implementation is to make permanent 'soliton collisions'. These are X-junctions created, for example, using ultraviolet light to induce refractive index changes in photosensitive materials, where the soliton junctions can then be used as linear devices. In these cases, available exact solutions can be programmed into the writing device, in order to control the local index by varying the intensity or speed of travel of the incident radiation beam. Opto-electronic chips, with many lossless junctions in a small area, could then be produced, where the index at each junction corresponds to the collision of two dark solitons. We shall analyse such switching devices in Chapter 6.

In this short introduction, we obviously cannot cover all possible physical models where the NLSE and its various modifications appear. We have restricted ourselves to those situations which will be considered in the following chapters.

CHAPTER 2

The nonlinear Schrödinger equation

We write the NLSE in the following form:

$$i\psi_\xi + \frac{1}{2}\psi_{\tau\tau} + |\psi|^2\psi = 0, \qquad (2.1)$$

This is the standard way of writing this equation in the mathematical literature. As we have already seen, the actual meaning of the independent variables ξ and τ is different for different problems. For spatial solitons, both variables are spatial, so no confusion can occur. For problems related to dispersive waves, and, in particular, to optical fibers, ξ is the distance along the fiber, and τ is the retarded time, i.e. a coordinate moving with the group velocity of the pulse. We will use this notation throughout this chapter, bearing in mind that the solutions relate to various physically different situations. As noted in Chapter 1, this form of the NLSE is for anomalous dispersion in the temporal (fiber) case and for self-focusing in the spatial case. In this chapter, we shall briefly describe methods for solving the NLSE and linear equations related to it. Clearly, we cannot cover the whole body of knowledge which exists today, and we refer the reader to existing books (e.g. Bullough, 1980; Novikov et al., 1984; Dodd et al., 1984; Newell, 1985; Ablowitz and Clarkson, 1991) for details. Here we only give those facts about the NLSE, its integrability and its consequences which will be needed to understand the material in the rest of this book.

2.1 One-soliton solution

The NLSE admits an infinite number of exact solutions. A complicated solution can be represented as a nonlinear superposition of simple ones. The structure of the superposition is defined by the spectrum of the inverse scattering technique (IST). One of the main points of interest is that (2.1) admits the soliton solution

$$\psi = \frac{e^{i\xi/2}}{\cosh\tau}. \qquad (2.2)$$

This simple solution can be extended to include more parameters by using the symmetries of the NLSE. Examples are given below.

2.2 Scaling transformation

If we know a solution of the NLSE, $\psi(\tau, \xi)$, then we can obtain a one-parameter family of solutions by the simple transformation

$$\psi'(\tau, \xi|q) = q\,\psi(q\tau, q^2\xi), \tag{2.3}$$

where q is a real parameter. This transformation can be applied to an arbitrary solution of the NLSE. We can use it to extend the one-soliton solution, (2.2), from a fixed solution to a one-parameter family of solutions:

$$\psi'(\tau, \xi|q) = \frac{q}{\cosh\ q\tau}\ e^{i\omega\xi} \tag{2.4}$$

where $\omega = q^2/2$. Clearly, the parameter q gives the amplitude of the soliton. It also determines the width of the soliton and its period in ξ.

2.3 Galilean transformation

The soliton (2.4) does not move along the τ axis, as its velocity in that frame is zero. We can introduce a finite velocity using the following transformation. If $\psi(\tau, \xi)$ is a solution of the NLSE, then

$$\psi'(\tau, \xi|V) = \psi(\tau - V\xi, \xi)\ \exp(iV\tau - iV^2\xi/2), \tag{2.5}$$

where V is interpreted as a velocity, is also a solution. The transformation (2.5) is a remnant of the cylindrical symmetry of the wave equation which was used to derive the NLSE. Clearly, the solutions of the wave equation are invariant relative to a rotation in the (τ, ξ) plane. However, this interpretation is not applicable for pulses in dispersive media. In that case, (2.5) describes the carrier frequency shift of the solution.

Applying (2.5) to (2.4), we obtain a two-parameter family of one-soliton solutions:

$$\psi(\tau, \xi|q, V) = \frac{q}{\cosh\ q(\tau - V\xi)}\ \exp[iV\tau + i(q^2 - V^2)\xi/2]. \tag{2.6}$$

This family still admits additions of trivial shifts in ξ, τ and phase φ.

2.4 Conserved quantities

For a dynamical system with a finite number of degrees of freedom to be integrable, the system has to have as many conserved quantities as degrees of freedom. The NLSE is a dynamical system with an infinite number of degrees of freedom, and corresponds to an infinite-dimensional Hamiltonian system. It is integrable, and, as a consequence, has an infinite number of conserved quantities (integrals). We now write down a few of them. The lowest-order integrals are the energy (or power, or the number of particles)

$$Q = \int_{-\infty}^{\infty} |\psi|^2 \, d\tau, \tag{2.7}$$

the momentum

$$M = i \int_{-\infty}^{\infty} (\psi_\tau \psi^* - \psi_\tau^* \psi) \, d\tau, \tag{2.8}$$

and the Hamiltonian

$$H = \frac{1}{2} \int_{-\infty}^{\infty} (|\psi_\tau|^2 - |\psi|^4) \, d\tau. \tag{2.9}$$

All of these quantities remain constant as ξ changes: $Q_\xi = 0$, $M_\xi = 0$ and $H_\xi = 0$. The first few of the higher-order conserved quantities can be found in the original work by Zakharov and Shabat (1971). We have chosen only these lowest-order ones because their analogues can be found for many non-integrable generalizations of the NLSE.

Due to Noether's (1918) theorem, each of the conserved quantities is related to a certain symmetry of the equation. A symmetry is a transformation which leaves the equation unchanged and transforms one solution into another solution. For example, energy conservation is related to a symmetry relative to the phase shift. If $\psi(\tau, \xi)$ is a solution, then

$$\psi'(\tau, \xi) = \psi(\tau, \xi) \, e^{i\theta}, \tag{2.10}$$

where θ is an arbitrary constant phase, is also a solution of the NLSE. The symmetry transformation related to the momentum is a shift of $\Delta\tau$ along the τ axis. Clearly, the NLSE remains unchanged under this transformation, while its solutions are transformed as follows:

$$\psi(\tau, \xi) \rightarrow \psi(\tau + \Delta\tau, \xi). \tag{2.11}$$

The conservation of the Hamiltonian is related to the symmetry of shifts along the ξ axis:

$$\psi(\tau, \xi) \rightarrow \psi(\tau, \xi + \Delta\xi). \tag{2.12}$$

The Galilean transformation (2.5) generates another conservation law:

$$\int_{-\infty}^{\infty} \tau \, |\psi(\tau, \xi)|^2 \, d\tau = -\frac{1}{2} \, M \, (\xi - \xi_0), \tag{2.13}$$

where ξ_0 is the integration constant. This relation shows that the 'centre of mass' of the pulse moves along the τ axis linearly with ξ. Then the momentum is proportional to the velocity of the centre of mass. Higher-order conservation laws of the type (2.13) can also be constructed. Those given here are the simplest and most important transformations related to the NLSE. Some of the more complicated ones are considered below.

The Hamiltonian plays an inportant role in the whole dynamics of the system. The NLSE can be written in canonical form in terms of the Hamiltonian:

$$i\frac{\partial \psi}{\partial \xi} = \frac{\delta H}{\delta \psi^*},$$ (2.14)

where the right-hand side is a Fréchet derivative. Some important properties (e.g. stability) of the solutions follow from this canonical form (Zakharov *et al.*, 1986). Each stationary solution gives an extremum of the Hamiltonian. It is stable if the extremum is a global or local minimum. Otherwise, the stationary solution is unstable. Note, however, that the sign of the Hamiltonian in (2.9) must be chosen correctly. This principle can be applied to more complicated Hamiltonian systems as well. Examples will be given in the following chapters.

2.5 Continuity equations

Conserved quantities are integrals of 'densities' which are functions of ψ and ψ^* and their partial derivatives. These densities satisfy continuity equations of the form

$$\frac{\partial \rho}{\partial \xi} + \frac{\partial j}{\partial \tau} = 0,$$ (2.15)

where ρ is a *density* and j is a *flux*.

We look at two particular cases. Firstly, if ρ is the energy density,

$$\rho_Q = |\psi|^2,$$ (2.16)

then the corresponding flux is

$$j_Q = \frac{i}{2}(\psi\psi_\tau^* - \psi_\tau\psi^*).$$ (2.17)

Secondly, when ρ is the Hamiltonian density,

$$\rho_H = \frac{1}{2}\left(|\psi_\tau|^2 - |\psi|^4\right),$$ (2.18)

the corresponding flux is

$$j_H = -\frac{1}{2}\left[\left(\frac{\partial \psi^*}{\partial \tau}\right)\left(\frac{\partial \psi}{\partial \xi}\right) + \left(\frac{\partial \psi}{\partial \tau}\right)\left(\frac{\partial \psi^*}{\partial \xi}\right)\right].$$ (2.19)

For the NLSE, there are infinitely many continuity equations which are related to the densities of conserved quantities. This is a consequence of the fact that the system described by the NLSE is Hamiltonian. Equation (2.15) can also be written in the integral form

$$\frac{\partial}{\partial \xi}\Upsilon = -[j(\tau_2) - j(\tau_1)]$$ (2.20)

where $\Upsilon = \int_{\tau_1}^{\tau_2} \rho \, d\tau$. This represents the fact that the rate of increase of energy in a given interval equals the net inflow across the boundaries. A finite number of density functions exist for other Hamiltonian generalizations of the NLSE.

2.6 Non-interaction with radiation

Solitons exist due to a balance between the dispersion (diffraction when τ is a spatial variable) and nonlinearity. This fact becomes obvious if we substitute the one-soliton solution (2.2) into the NLSE. Then the nonlinear term in the NLSE is exactly cancelled by the dispersion term. This simple consideration shows that solitons can exist due to this balance, but it tells us nothing about the interactions between the solitons, or of solitons with radiation. There are deep physical reasons for the existence of solitons described by the NLSE as stationary solutions. Here, we take an intuitive step to understand this phenomenon.

Let us consider the τ variable as the time, and consider the soliton in the frequency domain. It can be seen, from (2.4), that the parameter q defines the soliton amplitude. The soliton frequency shift is equal to half the amplitude squared and so must be positive. On the other hand, small-amplitude ($\mu \ll 1$) linear waves, described by $\psi = \mu \, \exp(ik\tau - i\omega\xi)$, which can exist in the system, are solutions of the NLSE without the nonlinear term ($\mu^3 \ll \mu$):

$$i\psi_\xi + \frac{1}{2}\psi_{\tau\tau} = 0 \ . \tag{2.21}$$

The dispersion relation for these waves is $\omega = -k^2/2$. This means that linear waves can exist only at negative frequencies when considered in the frequency domain. The conclusion is that solitons and linear waves are located in different regions of the frequency domain (as shown in Fig. 2.1), and so cannot interact. Hence, the energy of the localized state (soliton) cannot be dispersed away. This condition, viz. the lack of interaction with linear waves, is a necessary condition for the existence of solitons as stationary solutions. As we will see later in the book, some perturbations couple solitons with radiation, so that stationary solutions no longer exist.

These simple arguments are applicable in most cases to solitons in lossless media. One class includes the so-called 'gap solitons' (Chen and Mills, 1987; De Sterke and Sipe, 1994). They also exist in frequency-domain regions which are forbidden for linear waves.

It is well known (Kay and Moses, 1956; Morse and Feshbach, 1953), that sech-squared-shape potentials are 'reflectionless' for the linear Schrödinger equation. In linear theory, linear waves pass through these potential wells without any reflection, and only acquire additional phase changes. In nonlinear theory, solitons, to some extent, can be considered as sech-squared-

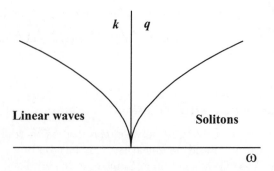

Figure 2.1 *Frequency-domain regions of existence for solitons and linear waves.
Note that these two regions do not overlap.*

shape potentials for linear waves. They do not change the energy of linear
waves, and they have the correct amplitude–width product to be reflectionless. In fact, not only do linear waves pass through the solitons, but solitons
can pass through each other without any change of shape. They simply acquire additional phase shifts. These observations allow us to conclude that
solitons serve as nonlinear 'modes' in the whole dynamics of the system
described by the NLSE, and, more generally, other integrable systems. The
concept of 'nonlinear modes' for the Korteweg–de Vries equation (Zabusky
and Kruskal, 1965) allowed for the development of the inverse scattering
technique for that equation (Gardner *et al.*, 1967).

2.7 Basics of integrability

The inverse scattering technique is the main tool for solving initial value
problems related to integrable equations, including the NLSE. For the
NLSE, the method has been developed by Zakharov and Shabat (1971).
There are several excellent books devoted to results in this exciting area
(e.g. Ablowitz and Clarkson,1991; Bullough, 1980; Newell, 1985). A Hamiltonian formulation of the theory of solitons is presented in Faddeev and
Takhtadjan (1987). A geometric interpretation of inverse scattering technique can be found in Anderson and Ibragimov (1979). Here, we only
present the principles of integrability and the inverse scattering transform
as far as necessary for understanding the material in the rest of this book.
We refer the reader to the above-mentioned books for more details.

The inverse scattering technique is based upon the fact that the NLSE
can be represented in the form of a compatibility condition between the

linear equations of the following set:

$$\mathbf{R}_\tau = -\mathbf{JR\Lambda} + \mathbf{UR},$$

$$\mathbf{R}_\xi = -\mathbf{JR\Lambda}^2 + \mathbf{UR\Lambda} - \tfrac{1}{2}(\mathbf{JU}^2 - \mathbf{JU}_\tau)\mathbf{R},$$

$$\tag{2.22}$$

where $\mathbf{R}, \mathbf{J}$ and $\mathbf{U}$ are the following matrices:

$$\mathbf{R} = \begin{bmatrix} r_{11} & r_{12} \\ r_{21} & r_{22} \end{bmatrix}, \quad \mathbf{U} = \begin{bmatrix} 0 & \psi \\ \phi & 0 \end{bmatrix}, \quad \mathbf{J} = \begin{bmatrix} i & 0 \\ 0 & -i \end{bmatrix}, \tag{2.23}$$

and

$$\mathbf{\Lambda} = \begin{bmatrix} \lambda_1 & 0 \\ 0 & \lambda_2 \end{bmatrix} \tag{2.24}$$

is an arbitrary diagonal complex matrix. The compatibility condition for these differential equations in $\mathbf{R}$, with non-constant coefficients depending on $\psi(\tau, \xi)$ and $\phi(\tau, \xi)$, is the equation

$$\mathbf{U}_\xi - \frac{1}{2}\mathbf{JU}_{\tau\tau} + \mathbf{JU}^3 = 0, \tag{2.25}$$

which is called 'split NLSE' because it defines a set of two nonlinear equations involving the two functions ψ and ϕ:

$$i\phi_\xi - \tfrac{1}{2}\phi_{\tau\tau} + \phi^2\psi = 0,$$

$$i\psi_\xi + \tfrac{1}{2}\psi_{\tau\tau} - \psi^2\phi = 0.$$

$$\tag{2.26}$$

The matrix partial differential equations (2.22) are simultaneously satisfied for all $\mathbf{\Lambda}$ if and only if $\psi(\tau, \xi)$ and $\phi(\tau, \xi)$ solve (2.26). Equations (2.26) reduce to a single NLSE in ψ for a special choice of functions involved in the transformation, e.g. $\phi = \pm\psi^*$.

The definition of the set of linear equations (2.22) is not unique. The ones presented here are different from those given in the original Zakharov and Shabat paper. We have chosen a matrix $\mathbf{R}(\tau, \xi)$ rather than a vector function for two reasons. Firstly, these matrix relations allow us to write the Darboux transformations in a simple matrix form. Secondly, with this choice, both the NLSE for anomalous dispersion and that for normal dispersion can be described using the same set. We set $r_{21} = s$, $r_{12} = s^*$ and $\lambda_2 = \lambda_1^*$. We define $\lambda = \lambda_1$. In the case of anomalous dispersion

$$\psi = -\phi^*, \quad r_{22} = -r_{11}^* \ (\equiv -r^*) \tag{2.27}$$

and equations (2.26) reduce to the NLSE for the function ψ. For the case of normal dispersion

$$\psi = \phi^*, \quad r_{22} = r_{11}^* \ (\equiv r^*) \tag{2.28}$$

and equations (2.26) reduce to the equation

$$i\psi_\xi + \frac{1}{2}\psi_{\tau\tau} - |\psi|^2\psi = 0 \tag{2.29}$$

and its complex conjugate.

We define $r = r_{11}$ and $s = r_{21}$. Then, for the anomalous dispersion case, using relations (2.27), we can preserve the full information content of matrix (2.22) with the following vector equation expressions:

$$R_\tau = \mathbf{L}\,R \tag{2.30}$$

$$R_\xi = \mathbf{B}\,R \tag{2.31}$$

where

$$\mathbf{L} = \begin{bmatrix} -i\lambda & \psi \\ -\psi^* & i\lambda \end{bmatrix}, \quad \mathbf{B} = \begin{bmatrix} -i\lambda^2 + i\frac{1}{2}|\psi|^2 & \lambda\psi + i\frac{1}{2}\psi_\tau \\ -\lambda\psi^* + \frac{i}{2}\psi_\tau^* & i\lambda^2 - \frac{i}{2}|\psi|^2 \end{bmatrix}$$

and

$$R = \begin{pmatrix} r \\ s \end{pmatrix}. \tag{2.32}$$

Thus R is the first column of the matrix $\mathbf{R}$. The operators $\mathbf{L}$ and $\mathbf{B}$ form a Lax pair for the NLSE, although they are different from the operators originally used by Zakharov and Shabat (1971).

For each solution, $\psi(\tau, \xi)$, of (2.1), there is a basis of two functions, r and s, parametrized by λ, which solve the set of linear equations (2.30) and (2.31). The compatibility condition in this case takes the form

$$\mathbf{L}_\xi - \mathbf{B}_\tau = \mathbf{BL} - \mathbf{LB}, \tag{2.33}$$

and its consequence is the NLSE. In fact, (2.33) should hold for each λ. It represents a cubic polynomial in λ. The coefficients of λ, λ^2 and λ^3 vanish identically. The vanishing of the constant term is equivalent to the NLSE in ψ. Equation (2.33) is sometimes called the *zero curvature condition*, as it has deep roots in differential geometry (Faddeev and Takhtadjan, 1987).

2.8 Solution of the initial value problem

The above equations establish the relation between the nonlinear equation and the set of linear differential equations. This representation allows us to solve the initial value problem. Thus, having an arbitrary pulse-like (zero at infinity) initial condition

$$\psi(\tau, \xi = 0) = f(\tau), \tag{2.34}$$

we can, first, solve the eigenvalue problem (2.30) with $\psi = f(\tau)$. The linear eigenfunctions, R, evolve (in ξ) in accordance with (2.31), but the eigenvalues do not change during this process. Knowledge of the asymptotics of the eigenfunctions for arbitrary ξ allows us to reconstruct the function ψ at arbitrary ξ.

Let us start with the eigenvalue problem (2.30) at $\xi = 0$. In explicit form,

$$r_\tau - i\lambda r = i\psi^* s$$

$$s_\tau + i\lambda s = i\psi r.$$

At infinity ($\tau \to \pm\infty$), where $\psi \to 0$, the solutions of this eigenvalue problem are obvious. We construct the solutions of (2.30), $u(\tau; \lambda)$, $v(\tau; \lambda)$ and $\bar{v}(\tau; \lambda)$, which, for real λ, satisfy the boundary conditions

$$u(\tau; \lambda) = \begin{pmatrix} 1 \\ 0 \end{pmatrix} \exp(i\lambda\tau) \quad \text{at} \quad \tau \to \infty,$$

$$v(\tau; \lambda) = \begin{pmatrix} 0 \\ 1 \end{pmatrix} \exp(-i\lambda\tau) \quad \text{at} \quad \tau \to -\infty,$$

and

$$\bar{v}(\tau; \lambda) = \begin{pmatrix} 1 \\ 0 \end{pmatrix} \exp(i\lambda\tau) \quad \text{at} \quad \tau \to -\infty.$$

Here, the function $\bar{v}$ is the adjoint of v. By definition, if v is the solution of (2.30) for λ_1 then

$$\bar{v} = \begin{pmatrix} s^* \\ -r^* \end{pmatrix}$$

is the solution of (2.30) for $\lambda = \lambda_1^*$. These two functions, $v(\tau; \lambda)$ and $\bar{v}(\tau; \lambda)$, comprise a complete set of solutions. Hence, we can write $u(\tau; \lambda)$ in terms of $v(\tau; \lambda)$ and $\bar{v}(\tau; \lambda)$:

$$u(\tau; \lambda) = a(\lambda)\bar{v}(\tau; \lambda) + b(\lambda)v(\tau; \lambda).$$

The coefficients $a(\lambda)$ and $b(\lambda)$ are transmission and reflection coefficients for the given initial condition, (2.34), for a given real λ. They satisfy

$$|a(\lambda)|^2 + |b(\lambda)|^2 = 1.$$

The functions u and v may be analytically continued to the upper half-plane of λ. Hence, the function $a(\lambda)$ also admits this continuation. The zeros of $a(\lambda)$, viz. λ_k ($k = 1, 2, ..., N$), in the upper half-plane of λ determine the set of the discrete eigenvalues of (2.30). The imaginary parts of λ define the soliton amplitudes while the real parts of λ define their velocities. At these points,

$$u(\tau; \lambda_j) = c_j\, v(\tau; \lambda_j), \quad j = 1, 2, ..., N.$$

The eigenfunctions of this eigenvalue problem change according to (2.31), but the eigenvalues λ_j are constant. If, at $\xi = 0$, the function R is the solution of (2.31) (i.e. it is the initial condition) then the solution of (2.31) at arbitrary ξ satisfies (2.30) with the same λ. In particular, $a(\lambda)$ does not depend on ξ. The evolution of the coefficients $b(\lambda)$ and c_j is described by

$$b(\lambda, \xi) = b(\lambda, 0) \exp(i\lambda^2\xi)$$

and

$$c_j(\xi) = c_j(0)\exp(i\lambda^2\xi).$$

Knowledge of these coefficients at any ξ allows us to reconstruct the 'potential', $\psi(\tau,\xi)$, at any ξ.

The solution of the NLSE is given by (Zakharov and Shabat, 1971)

$$\psi(\tau,\xi) = 2\sum_{k=1}^{N}\frac{c_k^* e^{-i\lambda_k^*\tau}}{[a'(\lambda_k)]^*}s(\tau,\lambda_k)^* + \frac{1}{i\pi}\int_{-\infty}^{\infty}\Phi_2^*(\lambda)\,d\lambda,$$

$$\int_{\tau}^{\infty}|\psi(\tau',\xi)|^2 d\tau' = -2i\sum_{k=1}^{N}\frac{c_k e^{i\lambda_k\tau}}{a'(\lambda_k)}r(\tau,\lambda_k) + \frac{1}{\pi}\int\Phi_1(\lambda)d\lambda.$$

where $r(\tau,\lambda_k)$, $s(\tau,\lambda_k)^*$, Φ_1 and Φ_2^* are the solutions of the simultaneous equations,

$$\Phi_1 - c(\tau,\lambda)\frac{1+T}{2}\;\Phi_2^* = -c(\tau,\lambda)\sum_{k=1}^{N}\frac{e^{-i\lambda_j^*\tau}}{\lambda-\lambda_k^*}\;\frac{c_k^*}{[a'(\lambda_k)]^*}\;s(\tau,\lambda_k)^*,$$

$$c^*(\tau,\lambda)\frac{1-T}{2}\;\Phi_1 + \Phi_2^* = c^*(\tau,\lambda) + c^*(\tau,\lambda)\sum_{k=1}^{N}\frac{e^{i\lambda_k\tau}}{\lambda-\lambda_k}\;\frac{c_k}{a'(\lambda_k)}\;r(\tau,\lambda_k),$$

$$r(\tau,\lambda_j)\,e^{-i\lambda_j\tau} + \sum_{k=1}^{N}\frac{e^{-i\lambda_k^*\tau}}{\lambda_j-\lambda_k^*}\frac{c_k^*}{a'(\lambda_k)}\;s(\tau,\lambda_k)^* = \frac{1}{2\pi i}\int_{-\infty}^{\infty}\frac{\Phi_2^*(\lambda)}{\lambda-\lambda_j^*}\,d\lambda,$$

and

$$s(\tau,\lambda_j)^* e^{i\lambda_j^*\tau} - \sum_{k=1}^{N}\frac{e^{i\lambda_j\tau}}{\lambda_j^*-\lambda_k}\frac{c_k}{a'(\lambda_k)}\;r(\tau,\lambda_k) = 1 + \frac{1}{2\pi i}\int_{-\infty}^{\infty}\frac{\Phi_1(\lambda)}{\lambda-\lambda_j^*}\,d\lambda,$$

in which $a'(\lambda_k)$ is the derivative of a with respect to λ_k,

$$c(\tau,\lambda) = \frac{b(\lambda)}{a(\lambda)}\;\exp(2i\lambda\tau),$$

T is the Hilbert transformation operator,

$$T\Phi = \frac{1}{i\pi}P\int_{-\infty}^{\infty}\frac{\Phi(\lambda')}{\lambda'-\lambda}d\lambda',$$

and P denotes principal value.

The final step in these calculations is highly non-trivial, and solutions have been found in analytical form in a limited number of special cases. The radiative part of the solution diffracts, and, in any constant velocity frame, its amplitude goes to zero as $\xi \to \infty$. Hence, asymptotically, the solution consists of a finite number of solitons. For an arbitrary initial condition, the scheme can be programmed on a computer, so that the eigenvalues can be found numerically (Blow and Doran, 1985; Vysloukh and Cherednik, 1986, 1988). When $b(\lambda) = 0$, there is no radiation component and

$$a(\lambda) = \prod_{i=1}^{N} \frac{\lambda - \lambda_i}{\lambda - \lambda_i^*}.$$

The solution, which then consists of N solitons, can be written analytically, provided that the eigenvalues λ_j and the constants c_j are known.

Periodic solutions for the NLSE have been considered in a number of publications (Its and Kotlyarov 1976; Ma and Ablowitz, 1981; Tracy *et al.*, 1984; Akhmediev *et al.*, 1985a; Previato, 1985; Forest and Lee, 1986; Akhmediev and Korneev, 1986). Particular solutions can be written in terms of elliptic Jacobi functions, or, more generally, in terms of periodic theta functions. However, the inverse scattering algorithm for periodic boundary conditions has not been generally developed. The reason for this is that most periodic potentials have Bloch spectra containing an infinite number of gaps. To solve the initial value problem, one needs to use concepts of Riemann surfaces of infinite genus and their corresponding theta functions.

2.9 Darboux transformations

For integrable nonlinear equations, there are approaches, called 'dressing methods', which allow us to obtain complicated solutions from simple ones. In particular, we can use these to construct multi-soliton solutions by starting with the trivial solution of the NLSE, $\psi = 0$.

Specifically, it has been shown by Matveev and Salle (1991) that, if we know a solution ψ_1, and, corresponding to it, a solution $\mathbf{R}_1$ of the linear equations (2.22), with a certain value of λ, then a different solution, ψ_2, and, corresponding to it, solutions of the linear equations, can be found from the following matrix relations:

$$\begin{aligned}
\mathbf{R}_2 &= \mathbf{R}_1\mathbf{\Lambda} - \sigma\mathbf{R}_1, \\
\mathbf{U}_2 &= \mathbf{U}_1 + [\mathbf{J}, \sigma],
\end{aligned} \tag{2.35}$$

where $[\mathbf{J}, \sigma] \equiv \mathbf{J}\sigma - \sigma\mathbf{J}$, $\sigma = \mathbf{R}_{1a}\mathbf{\Lambda}\mathbf{R}_{1a}^{-1}$ and $\mathbf{R}_{1a}$ is a solution of (2.22) with a different value of λ. The transformations in (2.35) are a particular form of the gauge transformations discussed in Faddeev and Takhtadjan (1987). They transform the matrices $\mathbf{R}$ and $\mathbf{U}$ into different forms but retain the zero curvature condition, so that the NLSE is invariant under these transformations. They relate solutions of a certain class to each other.

The point is that the transformation (2.35) can give solutions which are more complicated than the initial solution, ψ_1. For example, by starting from the zero solution, $\psi = 0$, solving (2.22) and using (2.35), we obtain a one-soliton solution of the NLSE with λ as the soliton parameter. Further applications of (2.35) give multi-soliton solutions. In this case, we need to solve (2.22) only once, for the trivial solution $\psi = 0$. Solutions of the linear equations (2.22) at higher iterations are given by (2.35).

For anomalous dispersion, starting from some seeding solution, ψ_0, and solving the linear set (2.22), we can find the functions r and s corresponding to it. The pair of functions, r and s, depends not only on λ, but also on two arbitrary integration constants (say, C and D). We label the set of these constants, together with λ, as $\chi = \{\lambda, C, D\}$. This set defines the parameters of a new solution of the NLSE. The new solution can be obtained from (2.35), which can be written in the form

$$\psi_1 = \psi_0 + \frac{2(\lambda_1^* - \lambda_1)r_0^* s_0}{|r_0|^2 + |s_0|^2}. \tag{2.36}$$

The equations for the new r and s are:

$$r_1(\chi_1, \chi_n) = \Delta[(\lambda_1^* - \lambda_1)s_0^*(\chi_1)r_0(\chi_1)s_0(\chi_n)$$

$$+(\lambda_n - \lambda_1)|r_0(\chi_1)|^2 r_0(\chi_n)$$

$$+(\lambda_n - \lambda_1^*)|s_0(\chi_1)|^2 r_0(\chi_n)],$$

$$s_1(\chi_1, \chi_n) = \Delta[(\lambda_1^* - \lambda_1)s_0(\chi_1)r_0^*(\chi_1)r_0(\chi_n) \tag{2.37}$$

$$+(\lambda_n - \lambda_1)|s_0(\chi_1)|^2 s_0(\chi_n)$$

$$+(\lambda_n - \lambda_1^*)|r_0(\chi_1)|^2 s_0(\chi_n)],$$

where $\Delta = [|r_0(\chi_1)|^2 + |s_0(\chi_1)|^2]^{-1}$ and $n \geq 2$.

This process can be continued. We can use the above equations, and change the indices to those of higher levels. The schematic diagram for constructing these higher-order solutions is presented in Fig. 2.2. The notation in this figure is the same as in the above formulae. The arrows in the diagram show the sequence of actions for constructing the new solutions. A pair of arrows pointing to some vector function, R, from the bottom, denotes the operator (2.37) and defines two vector functions from the previous stage used in this operator. A pair of arrows pointing to some solution, ψ, from the bottom, denotes (2.36) and defines the functions used in it for obtaining this solution.

Thus, we should first define the seeding solution, ψ_0. Then we need to solve the differential equations (2.22) to obtain the eigenfunctions $R(\chi)$ corresponding to this solution. Choosing the set of N parameters, χ_n, we have the set of N eigenfunctions $R_0(\chi_n)$. All of them are used in formulae (2.37) for constructing the set of $N - 1$ eigenfunctions of the second level,

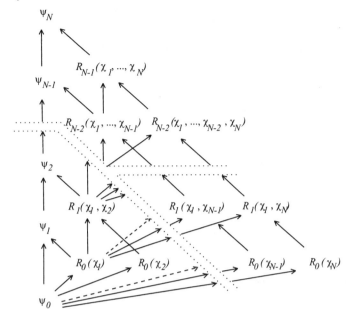

Figure 2.2 *Darboux transformations diagram for constructing higher-order solutions.*

but only one, viz. $R_0(\chi_1)$, is used in formula (2.36) for constructing the first-order solution, ψ_1. This step, like all the subsequent ones, consists solely of algebraic transformations. The rest of the steps are clear from the diagram.

The selection of the seeding solution, ψ_0, and of the constants χ_n, is determined by the actual physical problem to be solved. Two examples of the application of these transformations, starting from seeding solutions $\psi = 0$ and $\psi = e^{it}$, are given in Chapter 3.

2.10 Solution of the linearized NLSE

Some problems (Chapter 12) require knowledge of perturbations to the solutions of the equation (2.1). These can be obtained by linearizing around the solution of (2.1). Let us write the solution of (2.1) in the form

$$\psi(\tau,\xi) = \psi_0(\tau,\xi) + \mu g(\tau,\xi), \tag{2.38}$$

where ψ_0 is a solution of (2.1), μ is a small parameter and g is the perturbation function. Then, linearizing (2.1) around ψ_0, we obtain the equation

$$ig_\xi + \frac{1}{2}g_{\tau\tau} + 2|\psi_0|^2 g + \psi_0^2 g^* = 0. \tag{2.39}$$

Among particular exact solutions to this equation are $g = i\psi_0$ and $g = \frac{\partial \psi_0}{\partial \tau}$. However, the general solution is complicated.

It has been shown by Forest and Lee (1986) that (2.39) may be solved in terms of the so-called 'quadratic eigenfunctions' (2.32). Given two independent solutions of (2.30) and (2.31)

$$R_j = \begin{pmatrix} r_j \\ s_j \end{pmatrix}, \quad j = 1, 2,$$

we define

$$g = r_1 s_1 \quad \text{and} \quad h = r_2 s_2. \tag{2.40}$$

Then, the functions g and h respectively satisfy the linearized NLSE (2.39) and its complex conjugate

$$-ih_\xi + \frac{1}{2}h_{\tau\tau} + 2|\psi_0|^2 h + \psi_0^{*2}h^* = 0. \tag{2.41}$$

An explicit form of g will be given in Chapter 12.

2.11 Solution of the linear problem related to the NLSE

Sometimes we need to know the solution of the linear Schrödinger equation

$$i\varphi_\xi + \frac{1}{2}\varphi_{\tau\tau} + V(\tau, \xi)\varphi = 0 , \tag{2.42}$$

where the potential $V(\tau, \xi)$ represents an intensity found from a solution of the NLSE:

$$V(\tau, \xi) = |\psi(\tau, \xi)|^2.$$

This can be of interest in designing lossless X-junctions (Akhmediev and Ankiewicz, 1993b) and $N \times N$ switching devices in general (Miller and Akhmediev, 1996b). Clearly, there is always a solution φ which coincides with ψ. However, the general solution of (2.42) is highly non-trivial and the most interesting cases arise when φ differs from ψ. Miller and Akhmediev (1996b) have shown that the solution of (2.42) can be found from the linear vector functions, R. Given ψ, we use the solution r (from R in (2.32)) of (2.30) and (2.31). We define

$$\varphi(\tau, \xi, \lambda) = r \exp(-i\lambda\tau - i\lambda^2 \xi). \tag{2.43}$$

For this to satisfy (2.42), we require

$$ir_\xi + \frac{1}{2}\lambda^2 r + \frac{1}{2}r_{\tau\tau} - i\lambda r_\tau + r|\psi|^2 = 0. \tag{2.44}$$

For a given ψ, using r_ξ, r_τ and s_τ from (2.30) and (2.31) shows that (2.44) is satisfied, thus proving that, for any complex λ, the function φ satisfies (2.42). This result allows us to solve the linear Schrödinger equation for a large class of two-dimensional potentials, V, related to the solutions of the NLSE. Whenever the solution $\psi(\xi, \tau)$ of (2.1) can be specified exactly, the linear functions R can also be found exactly, using (2.30) and (2.31). For example, for $\psi = 0$, we have $r = \exp(-i\lambda\tau - i\lambda^2\xi)$, and so $\varphi = \exp(-2i\lambda\tau - 2i\lambda^2\xi)$ then satisfies (2.42).

CHAPTER 3

Exact solutions

Solving an initial value problem can be a complicated task, and, indeed is not always possible. This is especially true for periodic boundary conditions. However, if radiation is ignored, then we can deal with exact solutions which can approximate the solution of an initial value problem with a certain accuracy. There are many alternative approaches to the inverse scattering technique which can be used to obtain exact solutions of the NLSE, e.g. Hirota's method (Hirota and Satsuma, 1976) and symmetry reductions (Olver, 1986). The variational approach (Anderson, 1983) can be used to obtain approximate solutions. These methods are not limited to integrable equations like the NLSE, as they can be applied to non-integrable systems as well. Symmetry reductions have become powerful tools for analysing practically any differential equation.

In the following sections we shall present a special method for deriving exact solutions. The method consists of the reduction of the original equation, which has an infinite number of degrees of freedom, to a dynamical system with a finite number of degrees of freedom. For solutions which involve only solitons, this clearly can be done. In principle, the reduction itself can be done in many different ways. What is remarkable about the method described below is that it allows us to find both solitary wave and periodic solutions using the same procedure. Moreover, all these solutions are members of the same three-parameter family of solutions. The reduction allows us to obtain only a certain limited class of solutions which we call 'solutions of first order'. For higher-order solutions, we use standard (e.g.'dressing') techniques. Despite the fact that the approach here provides a relatively simple method only for the first-order and second-order solutions, it gives us a deeper understanding of the relationships between the periodic and solitary wave solutions.

3.1 Special ansatz

The possibility of reducing the number of degrees of freedom is a consequence of a special relation between the real and imaginary parts of the solution which has been found using extensive numerical simulations. Following (Akhmediev and Korneev, 1986, Akhmediev et al., 1987), we suppose that there is a linear relationship between the real and imaginary parts of the unknown NLSE solution $\psi = u + iv$:

$$u(\tau, \xi) = a_0(\xi)\, v(\tau, \xi) + b_0(\xi). \tag{3.1}$$

We assume further that the coefficients a_0 and b_0 in (3.1) depend only on the variable ξ; it is this that will ultimately enable us to find the solutions. If we convert to new functions $\phi(\xi), \delta(\xi)$, and $Q(\tau, \xi)$ by means of the change of variables $a_0 = \cot \phi$, $b_0 = -\delta / \sin \phi$, and $u = Q \cos \phi - \delta \sin \phi$, then the solution $\psi(\tau, \xi)$ will have the representation

$$\psi(\tau, \xi) = [Q(\tau, \xi) + i\delta(\xi)]\, \exp[i\phi(\xi)], \tag{3.2}$$

where only the function Q depends on the variable τ. It is remarkable that solutions having the property (3.1) actually exist. Moreover, they comprise a three-parameter family of solutions. We call them 'first-order' solutions because the relation (3.1) is linear in τ. Higher order solutions (which are defined by polynomials of higher order) also exist. This means that there is a symmetry in the NLSE which allows us to represent solutions in the form (3.1).

3.2 Reduction to a finite-dimensional dynamical system

Using the ansatz (3.2) in (2.1), and separating the real and imaginary parts, we obtain a system of equations for the function Q:

$$Q_{\tau\tau} - 2\delta_\xi - 2\phi_\xi Q + 2\delta^2 Q + 2Q^3 = 0, \tag{3.3}$$

$$Q_\xi - \phi_\xi \delta + \delta Q^2 + \delta^3 = 0. \tag{3.4}$$

Integrating (3.3) once gives

$$Q_\tau^2 + Q^4 + 2(\delta^2 - \phi_\xi)Q^2 - 4\delta_\xi Q = h, \tag{3.5}$$

where $h = h(\xi)$ depends only on ξ. If the system of equations (3.4) and (3.5) is to be compatible, the Frobenius relation $Q_{\tau\xi} = Q_{\xi\tau}$ must hold. Substituting (3.4) and (3.5) in this relation and then equating the coefficients of equal powers of Q, we obtain a system of three differential equations:

$$4\delta\delta_\xi + \phi_{\xi\xi} = 0, \tag{3.6}$$

$$h_\xi + 4\delta\delta_\xi \phi_\xi - 4\delta^3 \delta_\xi = 0, \tag{3.7}$$

$$\delta h + \delta_{\xi\xi} - 2\delta^3 \phi_\xi + \delta\phi_\xi^2 + \delta^5 = 0. \tag{3.8}$$

The coefficients of Q for the fourth and fifth powers vanish identically.

Equations (3.4)–(3.8) can be integrated successively. The first integral of (3.4) has the form

$$2\delta^2 + \phi_\xi = W/2, \tag{3.9}$$

where W is a constant of integration. Substituting ϕ_ξ from (3.9) in (3.7), we find the second integral of the system:

$$h + W\delta^2 - 3\delta^4 = H, \tag{3.10}$$

in which H is a further constant of integration. Finally, using (3.9) and (3.10), we find an integral of (3.8):

$$\delta_\xi^2 + (H + W^2/4)\delta^2 - 2W\delta^4 + 4\delta^6 = D, \tag{3.11}$$

where D is a third constant of integration. By means of the substitution $z = \delta^2$, (3.9) and (3.11) can be rewritten in the form

$$z_\xi^2 = -16z^4 + 8Wz^3 - (4H + W^2)z^2 + 4Dz, \tag{3.12}$$

$$\phi_\xi = W/2 - 2z. \tag{3.13}$$

With the use of (3.10), (3.5) takes the form

$$Q_\tau^2 = -Q^4 + (W - 6z)Q^2 + 2\frac{z_\xi}{\sqrt{z}}Q + (H - Wz + 3z^2). \tag{3.14}$$

Thus, apart from translations with respect to both variables, and rotation in the complex plane through a constant angle, the class of solutions determined by (3.1) or (3.2) is a three-parameter family of NLSE solutions, and it can be found by successively solving (3.12), (3.13) and (3.14).

3.3 Solutions of the dynamical system

To simplify subsequent calculations, we now convert from the parameters D, H and W, which determine the NLSE solutions, to three other equivalent parameters, α_1, α_2 and α_3, which are roots of the polynomial of fourth degree on the right-hand side of (3.12). The fourth root of this polynomial is obviously zero. These triplets of parameters are related to each other by Viète's relations for a cubic polynomial:

$$W = 2(\alpha_1 + \alpha_2 + \alpha_3), \tag{3.15}$$

$$H = 2(\alpha_1\alpha_2 + \alpha_2\alpha_3 + \alpha_1\alpha_3) - \alpha_1^2 - \alpha_2^2 - \alpha_3^2, \tag{3.16}$$

$$D = 4\alpha_1\alpha_2\alpha_3, \tag{3.17}$$

In the region of allowed parameters, determined below, each point in the space of the parameters $(\alpha_1, \alpha_2, \alpha_3)$ corresponds to a certain solution of the NLSE. Equation (3.12) can be expressed in terms of the new parameters as follows:

$$z_\xi^2 = -16z(z - \alpha_1)(z - \alpha_2)(z - \alpha_3). \tag{3.18}$$

We are only interested in real positive solutions of (3.18), since by definition $z = \delta^2$ with δ being real. Therefore, at least one of the roots, say α_3, must be positive. The two other roots are either real or complex conjugates. If real, we order them and will assume $\alpha_1 \leq \alpha_2 \leq \alpha_3$. In the case of complex roots, we convert from α_1 and α_2 to two other parameters, ρ and η, such that $\alpha_1 = \alpha_2^* = \rho + i\eta$.

After specifying the roots of the right-hand side of (3.18) and obtaining its solutions, we must then find the roots of the polynomial of fourth degree on the right-hand side of (3.14). By making use of (3.18), we rewrite this polynomial equation in the form

$$Q^4 - (W - 6z)Q^2 - 8\sqrt{(\alpha_1 - z)(\alpha_2 - z)(\alpha_3 - z)}\, Q$$
$$- (H - Wz + 3z^2) = 0. \tag{3.19}$$

By means of Ferrara's formulae (Korn and Korn, 1961), we may decompose this polynomial into the following two quadratic equations:

$$Q^2 \pm 2\sqrt{\alpha_3 - z}\, Q + \alpha_3 - \alpha_1 - \alpha_2 + z$$
$$\pm 2\sqrt{(\alpha_1 - z)(\alpha_2 - z)} = 0. \tag{3.20}$$

Their discriminants are

$$D_\pm = 4(\alpha_1 + \alpha_2 - 2z \mp 2\sqrt{(\alpha_1 - z)(\alpha_2 - z)}). \tag{3.21}$$

Further, it is necessary to analyse the case where the roots α_1 and α_2 are real separately from that where they are complex.

First we consider the case of real α_i. Then formula (3.21) can be written in the form

$$D_\pm = -4(\sqrt{z - \alpha_1} \pm \sqrt{z - \alpha_2})^2 \quad \text{for } z \geq \alpha_2,$$
$$\tag{3.22}$$
$$D_\pm = 4(\sqrt{\alpha_1 - z} \pm \sqrt{\alpha_2 - z})^2 \quad \text{for } z \leq \alpha_2.$$

It can be seen from formulae (3.22) that the polynomial equation (3.19) has four real roots only for solutions of (3.18) in the interval $0 \leq z \leq \alpha_1$, when all the parameters α_i are positive, and that it has one multiple real root in the case of coincident negative roots, $\alpha_1 = \alpha_2 \leq 0$ and $0 \leq z \leq \alpha_3$. In the remaining cases, the roots of the polynomial equation (3.19) are complex, and (3.14) does not have real solutions.

Thus we let all the α_i be positive. In this case, the solution of (3.18) can be described as the motion of a particle in the potential well $U(z) = 8z(z - \alpha_1)(z - \alpha_2)(z - \alpha_3)$, as depicted in Fig. 3.1a. The solution, on the interval $0 \leq z \leq \alpha_1$, can be expressed in terms of Jacobian elliptic functions:

$$z(\xi) = \frac{\alpha_1 \alpha_3\, \text{sn}^2(\mu\xi/2, k)}{\alpha_3 - \alpha_1\, \text{cn}^2(\mu\xi/2, k)}, \tag{3.23}$$

where $\mu = 4\sqrt{\alpha_2(\alpha_3 - \alpha_1)}$ and $k^2 = \alpha_1(\alpha_3 - \alpha_2)/[\alpha_2(\alpha_3 - \alpha_1)]$ is the modulus of the elliptic functions. The roots of the polynomial equation (3.19) are thus real and have the forms

$$Q_1 = \sqrt{\alpha_1 - z} + \sqrt{\alpha_2 - z} + \sqrt{\alpha_3 - z},$$

$$Q_2 = -\sqrt{\alpha_1 - z} - \sqrt{\alpha_2 - z} + \sqrt{\alpha_3 - z},$$

$$Q_3 = -\sqrt{\alpha_1 - z} + \sqrt{\alpha_2 - z} - \sqrt{\alpha_3 - z},$$

$$Q_4 = \sqrt{\alpha_1 - z} - \sqrt{\alpha_2 - z} - \sqrt{\alpha_3 - z}, \tag{3.24}$$

where the signs in the front of the roots are chosen so that $Q_4 \leq Q_3 \leq Q_2 \leq Q_1$. Equation (3.14) can now be written in the form

$$Q_\tau^2 = -(Q - Q_1)(Q - Q_2)(Q - Q_3)(Q - Q_4). \tag{3.25}$$

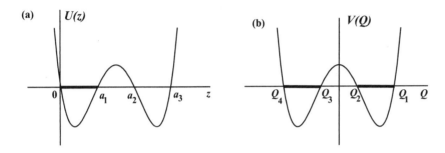

Figure 3.1 *Potential (a) $U(z)$ and (b) $V(Q)$. Bold lines show the regions where the functions $z(\xi)$ and $Q(\tau)$ can provide solutions $\psi(\tau, \xi)$.*

Equation (3.25) can be viewed as governing the oscillations of a particle in a potential well specified by the polynomial $V(Q) = \frac{1}{2}(Q-Q_1)(Q-Q_2)(Q-Q_3)(Q-Q_4)$, as shown in Fig. 3.1b. It has solutions in two intervals,

$$Q = \frac{Q_1(Q_2 - Q_4) + Q_4(Q_1 - Q_2)\,\mathrm{sn}^2(p\tau, k)}{(Q_2 - Q_4) + (Q_1 - Q_2)\,\mathrm{sn}^2(p\tau, k)}, \quad Q_2 \leq Q \leq Q_1, \tag{3.26}$$

and

$$Q = \frac{Q_4(Q_1 - Q_3) + Q_1(Q_3 - Q_4)\,\mathrm{sn}^2(p\tau, k)}{(Q_1 - Q_3) + (Q_3 - Q_4)\,\mathrm{sn}^2(p\tau, k)}, \quad Q_4 \leq Q \leq Q_3, \tag{3.27}$$

where $p = \sqrt{\alpha_3 - \alpha_1}$ and $k^2 = (\alpha_2 - \alpha_1)/(\alpha_3 - \alpha_1)$. To each solution, (3.26) and (3.27), there corresponds a solution of the NLSE. However, these solutions can be transformed into each other by shifts with respect to the variables τ and ξ. Formulae (3.23), (3.26) and (3.27) are not the only form of expression of the solutions as they can be written more simply in a number of special cases. If, for example, $Q_3 = Q_4$, then (3.26) can be expressed in the form

$$Q = \frac{2Q_1 Q_2 - Q_4(Q_1 + Q_2) + Q_4(Q_2 - Q_1)\,\cos(2p\tau)}{Q_1 + Q_2 - 2Q_4 + (Q_2 - Q_1)\,\cos(2p\tau)}.$$

We find the function ϕ by integrating (3.13). Taking into account (3.16), we obtain

$$\phi(\xi) = W\xi/2 - 4 \int_0^{\xi/2} z(\xi')\, d\xi'$$

$$= (\alpha_1 + \alpha_2 - \alpha_3)\xi + \frac{4\alpha_3}{\mu}\Pi(n; \mu\xi/2|k), \tag{3.28}$$

where

$$\Pi(n; \mu\xi/2|k) = \int_0^{\mu\xi/2} \frac{d\xi'}{1 - n\,\mathrm{sn}^2(\xi', k)} \tag{3.29}$$

is the incomplete elliptic integral of the third kind (Abramowitz and Stegun, 1964), and $n = \alpha_1/(\alpha_1 - \alpha_3)$. Since $\phi(\xi)$ occurs in solution (3.2) as an argument of trigonometric functions, the solutions of the NLSE of first order are, in the general case, conditionally periodic with respect to ξ (with a two-frequency basis). For brevity, in (3.23), (3.26), (3.27) and (3.28), we have omitted the obvious constants of integration ξ_0, τ_0 and ϕ_0, although with them, and everywhere below, we shall allow for the possibility of translating the variables $\xi \to \xi - \xi_0$, $\tau \to \tau - \tau_0$ and $\phi \to \phi - \phi_0$.

3.4 The case of complex roots

The case of negative $\alpha_1 = \alpha_2$ can be regarded as a special case of complex conjugate roots with vanishing imaginary part. Since Q in this case does not depend on τ, the NLSE solution is also independent of τ. It is easy to show that it has the form $\psi = -\sqrt{\alpha_3}\,\exp(i\alpha_3\xi)$.

Let the roots be complex, i.e. $\alpha_1 = \alpha_2^* = \rho + i\eta$. The root α_3 in this case can be positive or zero. The solution of (3.18) can be written in the form

$$z(\xi) = \frac{\alpha_3(1 - \nu)(1 - \mathrm{cn}(\mu\xi, k))}{2(1 + \nu\,\mathrm{cn}(\mu\xi, k))}, \tag{3.30}$$

where

$$\nu = \frac{f - g}{f + g}, \quad \mu = 4\sqrt{fg}, \quad k^2 = \frac{1}{2}\left[1 - \frac{\eta^2 + \rho(\rho - \alpha_3)}{fg}\right],$$

$$f = \sqrt{(\alpha_3 - \rho)^2 + \eta^2}, \quad g = \sqrt{\rho^2 + \eta^2}.$$

The discriminants of (3.20) are then

$$D_\pm = 8\left[(\rho - z) \mp \sqrt{(\rho - z)^2 + \eta^2}\right].$$

For all values of ρ and η, two roots of (3.19) are real, and two are complex:

$$Q_{1,2} = b \pm c_+, \quad Q_{3,4} = -b \pm ic_-, \tag{3.31}$$

where $b = \sqrt{\alpha_3 - z}$ and $c_\pm = \sqrt{2[\sqrt{(\rho - z)^2 + \eta^2} \pm (\rho - z)]}$. A solution of

(3.25) exists in the interval $Q_2 \leq Q \leq Q_1$:

$$Q = \frac{wQ_2 + yQ_1 + (yQ_1 - wQ_2)\,\mathrm{cn}(\sqrt{wy}\tau, k_1)}{w + y - (w - y)\,\mathrm{cn}(\sqrt{wy}\tau, k_1)}, \tag{3.32}$$

where

$$w^2 = (2b + c_+)^2 + c_-^2, \quad y^2 = (2b - c_+)^2 + c_-^2,$$

and

$$k_1^2 = \frac{1}{2} + \frac{2}{wy}(\rho - \alpha_3).$$

Integrating (3.13), with the use of (3.30), we obtain the expression for the function ϕ:

$$\phi = (2\rho - \alpha_3)\xi - \frac{4\alpha_3(\nu + 1)}{\mu k^2(\nu - 1)}\left[\frac{a}{y_+}\Pi\left(\frac{1}{y_+}; \mu\xi/2, k\right)\right.$$
$$\left. - \frac{1 + a}{y_-}\Pi\left(\frac{1}{y_-}; \mu\xi/2, k\right)\right], \tag{3.33}$$

where

$$y_\pm = \frac{\nu \pm \sqrt{\nu^2(1 - m) + m}}{m(\nu - 1)}, \quad m = k^2,$$

$$a = \frac{1 + y_+}{y_+ - y_-},$$

with k as defined for (3.30).

Equations (3.2), (3.23), (3.26)–(3.30), (3.32) and (3.33) determine the solutions of the NLSE of first order for all α_i in the region of permitted values. From these formulae we can select some special cases of particular practical interest.

3.5 Reduction of the number of parameters

Division of the roots α_i by some positive number q is equivalent to a transition from the solution $\psi(\tau, \xi)$ to a different solution $\psi'(\tau, \xi)$ corresponding to the roots $a_i = \alpha_i/q$. These two solutions are related by the scaling transformation (2.3). As we can choose q, we can, for example, set the value of one of the roots and seek a two-parameter family of solutions $\psi'(\tau, \xi)$, directly specifying the a_i. The third parameter is introduced into the solution by means of (2.3). We can always add this extra parameter to a family of solutions using it. Below, when considering special cases, we shall proceed in precisely this manner, and restrict ourselves to finding the solutions $\psi'(\tau, \xi)$ and using (2.3) to add the third parameter to the family. To do this, we set $q = 2\alpha_3$, so that the two parameters, a_1 and a_2, of the family of solutions, $\psi'(\tau, \xi)$, are related to the parameters of the required family of solutions, $\psi(\tau, \xi)$, by $a_1 = \alpha_1/2\alpha_3$, $a_2 = \alpha_2/2\alpha_3$, while the third parameter is $a_3 = 1/2$.

3.6 Special cases

First we consider cases where all the roots are real and positive. All possible solutions can be conveniently classified using the the plane of parameters (a_1, a_2) (Fig. 3.2). We only need to consider the single quadrant with $a_1 > 0$ and $a_2 > 0$. Moreover, due to the symmetry $a_1 <=> a_2$ we can restrict ourselves to just half of that quadrant, and due to a scaling transformation, we can restrict the values of the parameters a_1 and a_2 to being less than $1/2$.

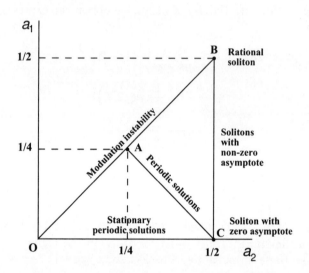

Figure 3.2 *Classification of solutions on the plane of parameters (a_1, a_2). Point A corresponds to (3.41), while the origin (O) corresponds to (3.38). Point B is (3.47) and point C is the standard sech-shape soliton (3.54). The line AC corresponds to (3.53). The line OB corresponds to (3.39). The line OC corresponds to (3.65) while the line BC corresponds to (3.63).*

3.7 Modulation instability

Suppose $0 \leq a_1 = a_2 \leq a_3 = 1/2$. In this case (line OB in Fig. 3.2), we obtain, from (3.23),

$$z(\xi) = \frac{a_1 \sinh^2(\beta\xi/2)}{\cosh^2(\beta\xi/2) - 2a_1}, \qquad (3.34)$$

where $\beta^2 = 8a_1(1 - 2a_1)$, and for the function ϕ, we obtain from (3.28),

$$\phi = \xi/2 + \arctan \gamma, \qquad (3.35)$$

where $\gamma = \sqrt{2a_1/(1 - 2a_1)} \tanh(\beta\xi/2)$. The expressions for the roots (3.24) take the forms

$$Q_{1,2} = \frac{\sqrt{1 - 2a_1}(\cosh(\beta\xi/2) \pm 2\sqrt{2a_1})}{\sqrt{2\cosh^2(\beta\xi/2) - 4a_1}},$$

$$Q_{3,4} = -\frac{\sqrt{1 - 2a_1}\,\cosh(\beta\xi/2)}{\sqrt{2\cosh^2(\beta\xi/2) - 4a_1}}.$$

One of the roots, $Q_3 = Q_4$, is a repeated one, and the solution corresponding to this case, $Q = Q_3$, does not depend on τ. The solution of (3.25), in the interval $Q_2 \leq Q \leq Q_1$, has the form

$$Q = \frac{\sqrt{1 - 2a_1}}{\sqrt{\cosh^2(\beta\xi/2) - 2a_1}}\,\frac{\cosh^2(\beta\xi/2) - 4a_1 + \sqrt{2a_1}\cos\ p\tau\ \cosh(\beta\xi/2)}{\sqrt{2}\,[\cosh(\beta\xi/2) - \sqrt{2a_1}\cos\ p\tau]},$$

(3.36)

where $p = \sqrt{2(1 - 2a_1)}$. The solution of the NLSE for this case can be simplified by means of the formula

$$\psi'(\tau,\xi) = \frac{Q - \gamma\sqrt{z(\xi)} + i(\gamma Q + \sqrt{z(\xi)}\ )}{\sqrt{1 + \gamma^2}}\ e^{i\xi/2}. \qquad (3.37)$$

For Q independent of τ, we obtain the stationary solution

$$\psi' = -\frac{1}{\sqrt{2}}\ e^{i\xi/2}, \qquad (3.38)$$

and for Q expressed by (3.36), we obtain the solution

$$\psi'(\tau,\xi) = \frac{(1 - 4a_1)\cosh(\beta\xi/2) + \sqrt{2a_1}\ \cos\ p\tau + i\beta\ \sinh(\beta\xi/2)}{\sqrt{2}\,[\cosh(\beta\xi/2) - \sqrt{2a_1}\ \cos\ p\tau]}\ e^{i\xi/2}.$$

(3.39)

Thus, the case $a_1 = a_2$ corresponds to two solutions. One of them, (3.38), describes a wave with constant amplitude, while the other, (3.39), describes exponential growth of perturbations (periodic in τ) superimposed on a constant amplitude, and a subsequent return, after attainment of the maximal modulus, to the original state, (3.38). This is the solution which describes modulation instability and its full evolution in ξ. The space–time profile of the solution is shown in Fig. 3.3. Obviously, for $\xi \to -\infty$, (3.39) coincides with (3.38) except for a constant phase shift:

$$\psi'(\tau,\mp\infty) = \frac{1 - 4a_1 \mp i\beta}{\sqrt{2}}\ e^{i\xi/2} = -\frac{1}{\sqrt{2}}\ e^{i\xi/2\mp i\varphi}. \qquad (3.40)$$

where

$$\varphi = \arccos(1 - 4a_1).$$

Hence, the manifold of initial and final points of evolution is a circle in the complex plane (Re ψ', Im ψ'), with radius $1/\sqrt{2}$.

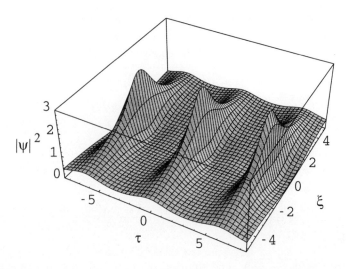

Figure 3.3 *Modulation instability (equation (3.41)) and its full evolution in ξ. Only three periods in τ are shown.*

The initial stages of this evolution can be studied using linear stability analysis with a c.w. input having small periodic modulations in τ. Alternatively, we can decompose the solution (3.39) around the points (3.40) for $\xi \to \mp\infty$:

$$\psi'(\tau,\xi) \approx \left[\frac{(1-4a_1)\mp i\beta}{\sqrt{2}} + 2\sqrt{2a_1}(2-4a_1\mp i\beta)(\cos\ p\tau)e^{\pm\beta\xi/2} \right] e^{i\xi/2}.$$

This decomposition is equivalent to the linear stability analysis of the c.w. initial stage. We shall return to this topic in the next section.

For $a_1 = 1/4$ (point A in Fig. 3.2), we have $\beta = 1$, and the solution (3.39) simplifies to

$$\psi'(\tau,\xi) = \frac{\cos\ \tau + i\sqrt{2}\ \sinh(\xi/2)}{2\ \cosh(\xi/2) - \sqrt{2}\ \cos\ \tau}\ e^{i\xi/2}. \tag{3.41}$$

In practical applications, this solution is the second most important. (The basic soliton solution is the most important.) Experimentally, the effects of modulation instability have been observed in fibers by Tai *et al.* (1986a,b). Let us consider the solution describing modulation instability in more detail.

3.8 Geometric interpretation of the solution

Historically, the solution (3.39) was first investigated using linear stability analysis of the c.w. solution (3.38). It is known (Bespalov and Talanov,

1966; Benjamin and Feir, 1967), that the stationary solution (3.38) is unstable with respect to long-wave periodic perturbations, and that these perturbations grow exponentially as the wave propagates (as ξ increases). To see this, let us remove the fast oscillatory factor $e^{i\xi/2}$ from the solutions (3.39) and (3.41) by setting

$$u = \psi \exp(-i\xi/2).$$

Then u is the solution of the reduced equation

$$iu_\xi - \frac{1}{2}u + \frac{1}{2}u_{\tau\tau} + |u|^2u = 0.$$

Now, let the input continuous wave be weakly modulated:

$$u = \frac{1}{\sqrt{2}}[1 + a(\xi) \ \cos \ \kappa\tau] \exp \ i\phi. \tag{3.42}$$

Substituting (3.42) into the reduced NLSE, linearizing it with respect to small $a(\xi)$ and taking into account that $a(\xi)$ is complex, we find that, for $\xi \to \pm\infty$,

$$a(\xi) = (\kappa/\sqrt{2} \pm i\sqrt{1 - \kappa^2/2}) \ \exp(\pm\delta_{gr}\xi) \tag{3.43}$$

where

$$\delta_{gr} = \frac{\beta}{2} = \kappa\sqrt{a_1} = \frac{\kappa}{\sqrt{2}}\sqrt{1 - \frac{\kappa^2}{2}} \qquad (\kappa = p) \tag{3.44}$$

is the growth rate of the perturbation with frequency κ. It is real in the frequency interval $0 < \kappa < \sqrt{2}$, as shown in Fig. 3.4. The two signs in (3.43) correspond to two different types of behaviour. Perturbations grow when δ_{gr} is positive and decay when it is negative. Growing and decaying trajectories start at different angles in complex space. Hence, every point on the circle $|u| = 1/\sqrt{2}$ is a saddle point. Note also that there is always a pair of side-bands, with positive and negative κ. This solution agrees with the decomposition of (3.39) given in the previous section.

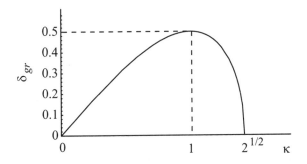

Figure 3.4 *Initial growth rate of modulation instability.*

The initial instability growth rate is a maximum when $\kappa = 1$ (i.e. $a_1 = 1/4$), and the full evolution of modulation instability in this case is given by (3.41). For visual illustration of the full evolution of modulation instability, we shall consider the motion of a point on the plane (Re ψ, Im ψ) as ξ varies (Fig. 3.5). The trajectories are drawn for $\tau = 0$, where the field is highest, and for $p\tau = \pi$, where the field is lowest. The dashed circle corresponds to the c.w. solution (which can have arbitrary phase ϕ). Hence, each point on this circle can serve as an initial condition. Let us suppose, to be specific, that $\phi = -\pi/2$ in (3.42). Then the initial point in the evolution is $(0, -1/\sqrt{2})$, which is a saddle point. There are two incoming trajectories and two outgoing ones, in accordance with (3.43). The full trajectory moves along the circle

$$(\text{Im } u)^2 + \left(\text{Re } u - \frac{1}{\sqrt{2}}\right)^2 = 2 - \kappa^2 = 4a_1 \qquad (3.45)$$

(the large circle on the right in Fig. 3.5), and has radius 1 when $\kappa = 1$. The maximum intensity reached is $(1/\sqrt{2} + 2\sqrt{a_1})^2$. This is highest (viz. 9/2) when $a_1 = 1/2$ (i.e. $\kappa = 0$), and will be considered as a 'rational soliton' in section 3.10. Note that its initial growth rate is zero. The reason for this is that the growth of instability in this limit is linear rather than exponential. The final stage of the evolution is motion back to the saddle point, i.e. it converges back to c.w. with $\phi = +\pi/2$. Trajectories starting at this point move back to the original point along the circle on the right.

Modulation instability is a recurrent phenomenon. It is symmetric in ξ and has its beginning and end at the same saddle point, apart from a phase shift, which, in this particular case, is equal to π. The phase shift is different for different values of the frequency κ. For arbitrary κ it is equal to

$$\Delta\phi = \phi_{final} - \phi_{initial} = 2\arccos(\kappa^2 - 1).$$

This phase shift is acquired by the c.w. as a result of the instability development. It is similar to the 'Berry' or 'geometric phase' but appears as a result of the nonlinear process. Examples of trajectories for two other values of κ are shown in Fig. 3.6. This figure shows clearly the difference in phase shifts for solutions with different κ. As $\kappa \to \sqrt{2}$, the phase shift goes to zero, as the radius of the trajectory approaches zero. As $\kappa \to 0$, the phase shift has a limit of 2π and the radius of the trajectory approaches $\sqrt{2}$.

The solution (3.39) describes a homoclinic orbit or separatrix. In the parameter space, it separates two qualitatively different types of periodic solution. These are shown, schematically, as the dotted curves in Fig. 3.5. Solutions of type B have maxima located at fixed τ, while for solutions of type A, the maxima alternate (so the left-hand and the right-hand parts of A-type trajectories correspond to different τ). Parameters a_i are real for B-type solutions and complex in the case of A-type solutions. Ana-

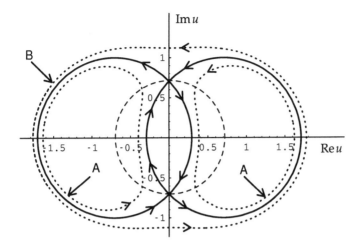

Figure 3.5 *Trajectories (bold circles) in the complex plane describing the modulation instability with the highest initial growth rate. The circle around the origin (dashed curve) is the manifold of initial conditions. Here $\kappa = 1$, i.e. $a_1 = 1/4$. The dotted lines schematically show two qualitatively different types of periodic solution close to the separatrix. The curves labelled A correspond to solution (3.53), while curve B corresponds to (3.57).*

lytic expressions for these periodic solutions are given later in this chapter (solutions (3.53) and (3.57)).

The continuous two-parameter family of periodic solutions, with the periods as solution parameters, exist in more complicated Hamiltonian systems, although they cannot always be written down analytically. The limiting cases of these solutions, when one of the periods goes to infinity, are soliton solutions or modulation instability. A particular example (viz. solutions of coupled NLSEs) has been considered by Haelterman and Sheppard (1994).

3.9 Evolution of spectral components

From a practical point of view, it is important to know how the spectral components evolve in the process of modulation instability. We are interested here in real physical spectra rather than in the spectra of the inverse scattering transform. In the case of solution (3.41), we can easily write down exact expressions for the spectrum. We restrict ourselves to the simplest case of maximal initial growth rate, although it can be done for the general case, (3.39), as well. We represent solution (3.41) in the form of the

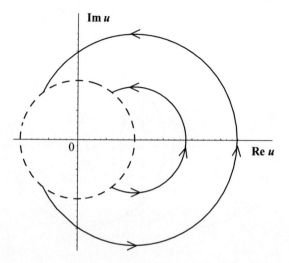

Figure 3.6 *Trajectories in the complex plane describing modulation instability with* $\kappa = \sqrt{2/5}$ *(outer circle) and* $\kappa = 2\sqrt{2/5}$ *(inner circle). The circle around the origin (dashed curve) is the manifold of initial conditions.*

Fourier expansion

$$\psi'(\tau, \xi) = f_0(\xi) + 2 \sum_{n=1}^{\infty} f_n(\xi) \, \cos(n\tau). \tag{3.46}$$

The coefficients in (3.46) can be readily calculated. They are:

$$f_0(\xi) = \frac{1}{2\pi} \int_0^{2\pi} \psi'(\tau, \xi) \, d\tau$$

$$= \left(\frac{\sqrt{2}[\cosh(\xi/2) + i \, \sinh(\xi/2)] - \sqrt{2\cosh^2(\xi/2) - 1}}{\sqrt{2\cosh^2(\xi/2) - 1}} \right) \frac{e^{i\xi/2}}{\sqrt{2}},$$

and

$$f_n(\xi) = \frac{1}{2\pi} \int_0^{2\pi} \psi'(\tau, \xi) \cos(n\tau) \, d\tau$$

$$= \frac{\cosh(\xi/2) + i \, \sinh(\xi/2)}{\sqrt{2\cosh^2(\xi/2) - 1}} \left[\sqrt{2} \, \cosh(\xi/2) - \sqrt{2\cosh^2(\xi/2) - 1} \right]^n e^{i\xi/2},$$

$$n \geq 1.$$

For the sum of the squared moduli of the coefficients $f_n(\xi)$, we have the relation

$$|f_0(\xi)|^2 + 2\sum_{n=1}^{\infty} |f_n(\xi)|^2 = \frac{1}{2\pi}\int_0^{2\pi} |\psi|^2 d\tau = \frac{1}{2},$$

which represents the conservation of energy.

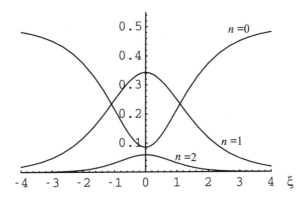

Figure 3.7 *Evolution of the pump* $|f_0(\xi)|^2$ *and spectral components* $2|f_n(\xi)|^2$ *during the process of modulation instability.*

The general evolution of the spectral components with ξ is shown in Fig. 3.7. It is obvious that, as $\xi \to \pm\infty$, all $f_n(\xi)$ vanish except for $|f_0(\xi)|^2$, which then approaches $1/2$. This means that all the energy is concentrated in the pump. For finite ξ, the energy of side-bands increases and the pump gradually becomes depleted. For arbitrary ξ, all $f_n(\xi)$ are non-zero, but the energy of the higher harmonics decreases with n in accordance with the law of geometric progression. At $\xi = 0$, the squares of the moduli of the coefficients are:

$$|f_0(0)|^2 = \frac{1}{2}(\sqrt{2} - 1)^2, \quad \text{and} \quad |f_n(0)|^2 = (\sqrt{2} - 1)^{2n}, \qquad n \geq 1.$$

The energy is then mainly concentrated in the two first side-bands. This fact means that sometimes it is possible to ignore higher-order spectral components and solve the problem using only three of them. This approach has been developed by Trillo and Wabnitz (1991). It allows us to reproduce the main qualitative features of the phenomenon in a simple way.

3.10 Rational solution

In the limit $a_1 \to 1/2$ (point B in Fig. 3.2), we obtain, from (3.39), a rational solution with power-law growth of the localized perturbation:

$$\psi'(\tau,\xi) = -\left[1 - 4\frac{1+i\xi}{1+2\tau^2+\xi^2}\right]\frac{e^{i\xi/2}}{\sqrt{2}}, \tag{3.47}$$

while in the case $a_1 = a_2 = 0$, (3.39) degenerates into the negative of (3.38). If $a_1 = a_2 = 0$, then only the repeated root $Q_3 = Q_4$ is real, and at this point only the stationary solution (3.38) exists. Solution (3.47) has been found by Akhmediev *et al.* (1985a). It can be considered as the limiting case of both the solution describing the modulation instability and the family of solitons with non-zero background (3.63). The intensity of the rational solution is shown in Fig. 3.8. Because of its soliton-like feature in ξ, this solution has been called an 'explode–decay solitary wave' (Nakamura and Hirota, 1985).

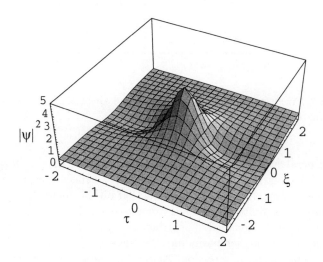

Figure 3.8 *Rational soliton*

3.11 Periodic solutions

We set $a_1 + a_2 = a_3 = 1/2$, corresponding to the line AC in Fig. 3.2. The resulting solutions are periodic in both τ and ξ. In this case, the function (3.23) can be transformed by means of a (lowering) Landen transformation

(Abramowitz and Stegun, 1964) to the form

$$z(\xi) = \frac{\kappa^2 \text{sn}^2(\xi/2, \kappa) \text{cn}^2(\xi/2, \kappa)}{2[1 - \kappa^2 \text{sn}^4(\xi/2, \kappa)]}, \tag{3.48}$$

where we have converted from the modulus $k = a_1/a_2$ of the elliptic function to a new modulus $\kappa = 2\sqrt{k}/(1+k)$. Furthermore, the expression for the function ϕ can also be written in the form of (3.35), where, however, γ is redefined as

$$\gamma = \frac{\text{sn}(\xi, \kappa) \, \text{dn}(\xi, \kappa)}{\text{cn}(\xi, \kappa)}. \tag{3.49}$$

The values of the roots (3.24) can be written in the form

$$Q_{1,2} = (1 \pm v_0)f_0, \quad Q_{3,4} = -(1 \mp w_0)f_0, \tag{3.50}$$

where

$$v_0 = \frac{\sqrt{1+\kappa}\,[1 - \kappa \, \text{sn}^2(\xi/2, \kappa)]}{\text{dn}(\xi/2, \kappa)}, \quad w_0 = \frac{\sqrt{1-\kappa}\,[1 + \kappa \, \text{sn}^2(\xi/2, \kappa)]}{\text{dn}(\xi/2, \kappa)},$$

$$\text{and} \qquad f_0 = \frac{\text{dn}(\xi/2, \kappa)}{\sqrt{2}\sqrt{1 - \kappa^2 \text{sn}^4(\xi/2, \kappa)}}$$

The solution of (3.25) which varies in the interval $Q_2 \leq Q \leq Q_1$ can conveniently be expressed, in the present case, in the form

$$Q = \frac{\kappa[\sqrt{1+\kappa}\,\text{sn}^2(\xi/2, \kappa)\,\text{dn}(\xi/2, \kappa) + \text{cn}^2(\xi/2, \kappa)\,A(\tau)]}{\sqrt{2}\sqrt{1 - \kappa^2\text{sn}^4(\xi/2, \kappa)}[\sqrt{1+\kappa} - \text{dn}(\xi/2, \kappa)\,A(\tau)]}, \tag{3.51}$$

where

$$A(\tau) = \frac{\text{cn}(\sqrt{(1+\kappa)/2}\,\tau, \, \sqrt{(1-\kappa)/(1+\kappa)})}{\text{dn}(\sqrt{(1+\kappa)/2}\,\tau, \, \sqrt{(1-\kappa)/(1+\kappa)})}. \tag{3.52}$$

The solution of the NLSE corresponding to (3.51) can also be calculated in accordance with formula (3.37); it takes the form

$$\psi'(\tau, \xi) = \frac{\kappa}{\sqrt{2}} \frac{A(\tau)\,\text{cn}(\xi/2, \kappa) + i\sqrt{1+\kappa}\,\text{sn}(\xi/2, \kappa)}{\sqrt{1+\kappa} - A(\tau)\,\text{dn}(\xi/2, \kappa)} e^{i\xi/2}. \tag{3.53}$$

This solution is periodic in both τ and ξ; it is illustrated, for the case $\kappa = 0.7$, in Fig. 3.9. In the limit $\kappa \to 1$, it reduces to solution (3.41) (point A in Fig. 3.2), while in the limit $\kappa \to 0$, it converges to the well-known basic bright soliton solution (point C in Fig. 3.2):

$$\psi'(\tau, \xi) = \frac{\sqrt{2}}{\cosh(\sqrt{2}\tau)} e^{i\xi}. \tag{3.54}$$

At the point $a_1 = 1/4$, $a_2 = 1/4$ of the parameter space, a_1 and a_2 can be analytically continued into complex space. Then $a_1 = a_2^* = 1/4 + i\eta$ and

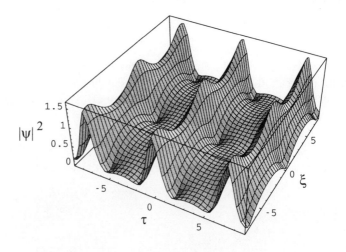

Figure 3.9 *Periodic solution (3.53) for* $\kappa = 0.7$.

$a_3 = 1/2$. The function $z(\xi)$ can then be represented in the form

$$z(\xi) = \frac{\text{sn}^2(\mu\xi/2, k)\text{dn}^2(\mu\xi/2, k)}{2[1 - k^2\text{sn}^4(\mu\xi/2, k)]}. \tag{3.55}$$

where $\mu = \sqrt{1 + 16\eta^2}$, $k = 1/\mu$. At the same time, ϕ is determined by (3.35), where

$$\gamma = \frac{k\, \text{sn}(\mu\xi/2, k)\, \text{cn}(\mu\xi/2, k)}{\text{dn}(\mu\xi/2, k)}. \tag{3.56}$$

The roots of the polynomial (3.19) can be expressed in the form (3.31), where

$$b = \frac{\text{cn}(\mu\xi/2, k)}{\sqrt{2[1 - k^2\text{sn}^4(\mu\xi/2, k)]}}, \quad d, c = \sqrt{\frac{1 \pm k}{2k}}\sqrt{\frac{1 \mp k\, \text{sn}^2(\mu\xi/2, k)}{1 \pm k\, \text{sn}^2(\mu\xi/2, k)}}.$$

It is convenient, in this case, to represent the solution of (3.25) in the form

$$Q = \frac{k^2\text{sn}^2\left(\frac{\xi}{2k}, k\right)\text{cn}\left(\frac{\xi}{2k}, k\right) + \sqrt{\frac{k}{1+k}}\, \text{cn}\left(\frac{\tau}{\sqrt{k}}, \sqrt{\frac{1-k}{2}}\right)\text{dn}^2\left(\frac{\xi}{2k}, k\right)}{k\sqrt{2}\sqrt{1 - k^2\text{sn}^4\left(\frac{\xi}{2k}, k\right)}\left[1 - \sqrt{\frac{k}{1+k}}\, \text{cn}\left(\frac{\tau}{\sqrt{k}}, \sqrt{\frac{1-k}{2}}\right)\text{cn}\left(\frac{\xi}{2k}, k\right)\right]}.$$

The complete solution of the NLSE then takes the form

$$\psi'(\tau, \xi) = \frac{\sqrt{\frac{k}{1+k}}\, \text{cn}\left(\frac{\tau}{\sqrt{k}}, \sqrt{\frac{1-k}{2}}\right)\text{dn}\left(\frac{\xi}{2k}, k\right) + i\,k\, \text{sn}\left(\frac{\xi}{2k}, k\right)}{k\sqrt{2}\left[1 - \sqrt{\frac{k}{1+k}}\, \text{cn}\left(\frac{\tau}{\sqrt{k}}, \sqrt{\frac{1-k}{2}}\right)\text{cn}\left(\frac{\xi}{2k}, k\right)\right]}e^{i\xi/2}. \tag{3.57}$$

This periodic solution is an analytic continuation of solution (3.53) for $\kappa > 1$ ($\kappa = 1/k$); it is shown in Fig. 3.10 for $k = 0.7$. As $k \to 1$, the solution (3.57) has formula (3.41) as its limit. Solutions (3.57) and (3.53) are located on different sides of the separatrix (3.41). Therefore, they are qualitatively different, as already discussed. This can be seen clearly by comparing Fig. 3.10 with Fig. 3.9.

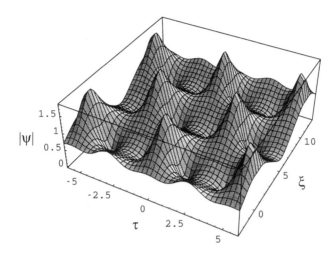

Figure 3.10 *Periodic solution (3.57) for $k = 0.7$.*

3.12 Solitons on a finite background

These solutions appear on the vertical line BC, viz. $a_2 = 1/2$ with $0 < a_1 < 1/2$ in the plane of parameters (Fig. 3.2). Here $z(\xi)$ can be expressed in terms of trigonometric functions:

$$z(\xi) = \frac{a_1 \sin^2(\mu\xi/2)}{1 - 2a_1 \cos^2(\mu\xi/2)}, \tag{3.58}$$

where $\mu = 2\sqrt{1 - 2a_1}$, and the function ϕ has the form

$$\phi = a_1\xi + \arctan\left[\frac{2}{\mu}\tan(\mu\xi/2)\right]. \tag{3.59}$$

The roots Q_i in (3.24) take the forms

$$Q_{2,3} = -f_1\sqrt{a_1}\,|\cos(\mu\xi/2)|,$$

$$Q_{1,4} = f_1(\sqrt{a_1}\,|\cos(\mu\xi/2)| \pm \sqrt{2}), \tag{3.60}$$

where $f_1 = \sqrt{1 - 2a_1}/\sqrt{1 - 2a_1 \cos^2(\mu\xi/2)}$, and the root $Q_2 = Q_3$ is repeated. It is convenient to write the solution to (3.25) in the form

$$Q = f_1 \frac{\sqrt{2}(1 - a_1 \cos^2(\mu\xi/2)) \mp \sqrt{a_1} \cos(\mu\xi/2) \cosh 2p\tau}{-\sqrt{2a_1} \cos(\mu\xi/2) \pm \cosh 2p\tau}, \qquad (3.61)$$

where $p = \sqrt{1/2 - a_1}$. The second solution, $Q = Q_2$, is stationary. The solution of the NLSE corresponding to this value of Q has the form

$$\psi' = -\sqrt{a_1}\, e^{ia_1\xi}, \qquad (3.62)$$

while the solution corresponding to (3.61) can be written as

$$\psi'(\tau,\xi) = \frac{2(1 - a_1)\cos(\mu\xi/2) \mp \sqrt{2a_1}\cosh 2p\tau + i\mu\,\sin(\mu\xi/2)}{-2\sqrt{a_1}\cos(\mu\xi/2) \pm \sqrt{2}\cosh 2p\tau}\, e^{ia_1\xi}.$$
$$(3.63)$$

The $a_1 \to 1/2$ limit of this expression (upper signs) gives the rational solution (3.47). Solution (3.63), which was found by Kawata and Inoue (1978) and Ma (1979), describes a soliton on a constant background. It can be considered as a nonlinear superposition of a soliton and a constant background (Akhmediev and Wabnitz, 1992). The soliton and the background have different frequencies in ξ, and this produces a beating between the soliton and the background. Hence the full solution, shown in Fig. 3.11, is periodic in ξ.

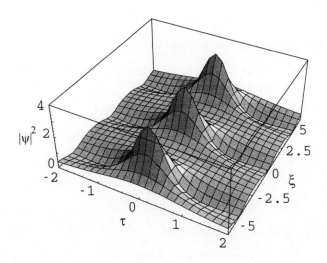

Figure 3.11 *Soliton on a finite background [equation (3.63) with $a_1 = 0.25$].*

This periodicity can be illustrated using the geometrical construction (Akhmediev and Wabnitz, 1992) presented in Fig. 3.12. Again, as in section 3.8, we remove the fast oscillatory factor $e^{ia_1\xi}$ from (3.63). Thus, we use the function $u(\tau, \xi) = \psi'(\tau, \xi)e^{-ia_1\xi}$. On the complex plane, the point A corresponds to the amplitude of the background, which, in this particular case, is negative and equal to $-\sqrt{a_1}$. At $\mu\xi/2 = \pi/2 + 2\pi N$ (point B in Fig. 3.12), the soliton is orthogonal to the background field (line OA in Fig. 3.12) and the whole solution can be separated into an exact soliton and the background:

$$u(\tau, \xi) = \left[-\sqrt{a_1} \pm i\frac{2p}{\cosh 2p\tau} \right].$$

The soliton centre point (at $\tau = 0$) rotates around the point E at the beat frequency $\mu/2$. We illustrate this in Fig. 3.12 for $a_1 = 1/4$. The point E $(\sqrt{a_1}, 0)$ is not the same as the background A $(-\sqrt{a_1}, 0)$. As a result, in the complex plane the field moves around a circle of radius $\sqrt{2}$ which is shifted from the origin. The intensity is lowest at the point D and highest at the point C. The resulting oscillations are seen in Fig. 3.11.

Note that the background field (which is c.w.) is unstable relative to perturbations. Therefore, we can view the whole solution as unstable, and, in long-distance evolution, the soliton part can be hidden in 'noise' created by this instability (Akhmediev and Wabnitz, 1992).

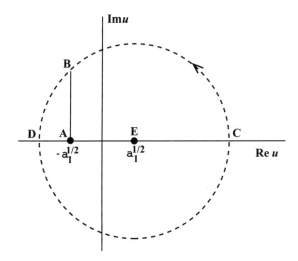

Figure 3.12 *Trajectory of the midpoint of the soliton (3.63), $u(0, \xi) = \psi'(0, \xi)e^{-ia_1\xi}$, on the complex plane of the field (dashed circle). Here $a_1 = 1/4$ and the circle radius is $\sqrt{2}$.*

When $a_1 \to 0$, solution (3.63) reduces to bright soliton solution (3.54).

Solution (3.39) can be called a soliton in the τ variable, in contrast to solution (3.63) (of which (2.2) is a particular case), which is a soliton in the ξ variable. Note the striking similarity between Figs 3.3 and 3.11 when one of them is rotated by 90°. Moreover, both of these solutions are limiting cases of the general periodic solution which has its two periods (along the ξ and τ axes) as arbitrary parameters. In the limit of infinite period, we obtain an ordinary soliton or modulation instability.

3.13 Stationary periodic waves

These solutions are located on the horizontal axis in Fig. 3.2: $a_1 = 0$, $0 \le a_2 \le 1/2$, i.e. line OC. In this case, a solution of the NLSE exists only for $z = 0$. The roots (3.24) do not depend on ξ:

$$Q_{1,2} = \frac{1 \pm k}{\sqrt{2}}, \quad Q_{3,4} = -\frac{1 \mp k}{\sqrt{2}}, \qquad (3.64)$$

where $k = \sqrt{2a_2}$. The solution of the NLSE can be expressed directly in the form

$$\psi'(\tau,\xi) = \pm\frac{1+k}{\sqrt{2}}\,\mathrm{dn}\left(\frac{1+k}{\sqrt{2}}\tau,\; \frac{2\sqrt{k}}{1+k}\right)\exp\left[i\frac{1+k^2}{2}\xi\right]. \qquad (3.65)$$

This solution describes a stationary envelope wave. Its profile does not require illustration. In the limit $a_2 \to 0$, solution (3.65) degenerates into the stationary solution (3.38) (i.e. the origin in Fig. 3.2), while as $a_2 \to 1/2$ the period of the dn function goes to infinity and the function (3.65) has the soliton solution (point C in Fig. 3.2) as its limit.

At the origin of the parameter space in Fig. 3.2, we may transform into complex parameters $a_1 = a_2^* = \rho + i\eta$. It is not possible to normalize the roots using $2a_3$, and we have to take $a_3 = 0$. Equation (3.18) then only has the trivial solution $z = 0$, and by means of (3.32) we can express the solution of the NLSE in the form

$$\psi(\tau,\xi) = \frac{kq_0}{\sqrt{2k^2 - 1}}\,\mathrm{cn}\left(\frac{q_0\tau}{\sqrt{2k^2 - 1}},\; k\right)\,\exp(iq_0^2\xi/2), \qquad (3.66)$$

where $q_0 = 4\rho$ and $k^2 = \frac{1}{2}[1 + \rho/\sqrt{\rho^2 + \eta^2}]$. The modulus of the elliptic function varies in the range $\frac{1}{2} \le m^2 \le 1$ for $\infty > \eta > 0$. This is the so-called '*cnoidal wave*'. It is an analytic continuation of the solution (3.62) for modulus larger than 1. Solution (3.66) describes a stationary periodic envelope wave. In contrast to solution (3.65), it has zeros in τ.

The four cases considered above exhaust all the possibilities for simplifying the general solution specified by (3.23), (3.26), (3.27) and (3.28), which determine, together with (3.2), the solution of the NLSE for real and complex parameters a_i.

3.14 Higher-order solutions

In principle, using the same approach as above, we can find higher-order solutions. We define solutions of the NLSE of nth order as solutions for which the real (u) and imaginary (v) parts satisfy a relation

$$P_n(u, v) = 0, \qquad (3.67)$$

where P_n is a polynomial of nth degree in u and v with coefficients that depend only on ξ. Specifying the order n of the polynomial, we can, in principle, construct a dynamical system corresponding to the given order and find its solution, as this is equivalent to finding a solution of the NLSE. However, such calculations are complicated, and it is easier to use dressing methods, such as Darboux transformations, to construct higher-order solutions. In the above classification, a two-soliton solution would correspond to the third-order solution. The class of second-order solutions has only a single available example, and we shall now derive it.

3.15 Second-order solution

We consider the simplest case of a curve of second order and assume that the points (u, v) lie on a circle with its centre at the origin and with a radius that depends on ξ:

$$u^2(\tau, \xi) + v^2(\tau, \xi) - R^2(\xi) = 0 . \qquad (3.68)$$

The solutions of the NLSE that satisfy condition (3.68) can be represented in the form

$$\psi(\tau, \xi) = R(\xi) \, \exp[i\Phi(\tau, \xi)] . \qquad (3.69)$$

Substituting (3.69) in (2.1) and separating into real and imaginary parts, we obtain the system of equations

$$2R_\xi + R\Phi_{\tau\tau} = 0,$$

$$2R^3 - R\Phi_\tau^2 - 2R\Phi_\xi = 0. \qquad (3.70)$$

After integration of the first equation and division of the second by R, the system takes the form

$$\Phi_\tau^2 = \frac{2S}{R} - \frac{4R_\xi}{R} \, \Phi,$$

$$2\Phi_\xi = 2R^2 - \Phi_\tau^2, \qquad (3.71)$$

where $S = S(\xi)$ is a constant of integration. Using the condition of compatibility, $\Phi_{\tau\xi} = \Phi_{\xi\tau}$, and equating the coefficients of different powers of Φ, we obtain the dynamical system

$$S_\xi R - 3SR_\xi - 2R^3 R_\xi = 0,$$

$$3R_\xi^2 - RR_{\xi\xi} = 0. \qquad (3.72)$$

The solution of this system,

$$R = C\xi^{-1/2},$$

$$S = -C^3\xi^{-1/2}\ln\,\xi,$$

(3.73)

where C is a constant of integration, together with Φ which is found by solving the equations in (3.71), viz.

$$\Phi = \tau^2/2\xi + C^2\ln\,\xi,$$

(3.74)

determines the NLSE solution

$$\psi(\tau,\xi) = \frac{C}{\sqrt{\xi}}\,\exp\left[i\left(\frac{\tau^2}{2\xi} + C^2\ln\,\xi\right)\right].$$

(3.75)

This is the only known solution of second order which exists for the NLSE. The solution is singular at $\xi = 0$ and undefined for negative ξ.

3.16 Multi-soliton solutions

In this section we will construct multi-soliton solutions using the Darboux transformation formulae given earlier. In principle, this technique allows us to write down the multi-soliton solution in a general form (Neugebauer and Meinel, 1984; Gagnon and Stiévanart, 1994; Matveev and Salle, 1991). To start with, we choose, as a seeding solution of the NLSE, the zero solution $\psi = 0$. Functions r and s, corresponding to this solution, can be found by solving (2.22); they have the form:

$$r = \exp[i\lambda_1(\tau - \tau_{01}) + i\lambda_1^2(\xi - \xi_{01})],$$

$$s = \exp[-i\lambda_1(\tau - \tau_{01}) - i\lambda_1^2(\xi - \xi_{01})],$$

(3.76)

where $\lambda_1 = a_1 + ib_1$, and τ_{01} and ξ_{01} are arbitrary parameters of the new soliton solution, to be obtained from (3.76). The real part of λ_1 defines the velocity of a soliton (a_1) and its imaginary part (b_1) defines an amplitude. Now, using (3.76) we obtain a one-soliton solution

$$\psi(\tau,\xi) = -\frac{2ib_1}{\cosh(2b_1\tau + 4a_1b_1\xi)}\,\exp\left[-2ia_1\tau - 2i(a_1^2 - b_1^2)\xi - 2i\varphi_1\right].$$

(3.77)

This is the standard form of a one-soliton solution.

In the second step of the construction of the two-soliton solution, a new set of parameters λ_2, τ_{02} and ξ_{02}, must be used. We can use the same set (3.76) to find the function r corresponding to the one-soliton solution (3.77). We note that we do not have to solve any linear differential equations. This function r will involve the new parameters, λ_2, τ_{02} and ξ_{02}. Now, the two-soliton solution follows from (2.36). For simplicity, we give the solution for equal amplitudes $b_1 = b_2 = b$ and with opposite velocities $a_1 = -a_2 = a$

(Akhmediev and Ankiewicz, 1993):

$$\psi(\tau,\xi) = -8iab \, \frac{A+iB}{D} \, \exp[-2i(a^2 - b^2)\xi + i(\varphi_1 + \varphi_2)], \qquad (3.78)$$

where

$$A = \cosh(4ab\xi)[a \, \cosh(2b\tau) \, \cos(2a\tau + \Delta\varphi) - b \, \sinh(2b\tau) \, \sin(2a\tau + \Delta\varphi)]$$

$$B = \sinh(4ab\xi)[a \, \sinh(2b\tau) \, \sin(2a\tau + \Delta\varphi) + b \, \cosh(2b\tau) \, \cos(2a\tau + \Delta\varphi)]$$

$$D = a^2\cosh(4b\tau) + (a^2 + b^2)\cosh(8ab\xi) - b^2 \cos(4a\tau + 2\Delta\varphi),$$

and $2\Delta\varphi = 2(\varphi_1 - \varphi_2)$ is the phase difference between the two solitons. This solution is shown in Fig. 3.13 for the case $a = 0.25$, $b = 0.1$ and $\Delta\varphi = 0$. When $2\Delta\varphi$ is zero or 2π, the solitons are initially in phase. When $2\Delta\varphi = \pi$, the solitons are initially out of phase.

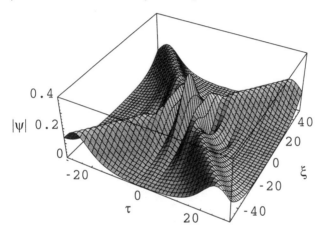

Figure 3.13 *Collision of two solitons with equal amplitudes and phases* $\Delta\varphi = 0$.

As noted in Chapter 1, if a refractive index distribution with the form of $|\psi|^2$, with ψ given by (3.78), can be written on a photosensitive material using lithographic or ultraviolet methods, then the device produced can operate as a useful X-junction.

3.17 Breathers

An interesting particular case of the two-soliton solution, (3.78), is that with zero soliton velocities, and with solitons located at the same position, $\tau_{01} = \tau_{02} = 0$:

$$\psi(\tau,\xi) = \frac{4i(b_2^2 - b_1^2)[b_1\cosh(2b_2\tau) \, \exp(2ib_1^2\xi) - b_2\cosh(2b_1\tau) \, \exp(2ib_2^2\xi)]}{(b_1 - b_2)^2 C_+ + (b_1 + b_2)^2 C_- - 4b_1 b_2 \cos\phi},$$

$$(3.79)$$

where $C_\pm = \cosh[2(b_2 \pm b_1)\tau]$ and $\phi = 2(b_2^2 - b_1^2)\xi$. This solution is periodic in ξ. Its periodicity is due to the nonlinear interference between the two solitons which compose the solution. The frequency of this beating along the ξ axis is exactly the difference between the spatial frequencies of two solitons, i.e. $2(b_2^2 - b_1^2)$. The corresponding period is $\pi/(b_2^2 - b_1^2)$. The solution admits arbitrary amplitudes b_1 and b_2, except $b_1 = b_2$.

When b_1 and b_2 are close to each other, the solution describes two pulses separated from each other and periodically attracted. These two pulses are frequently interpreted as solitons. Their maximum separation, $d \approx 2\ln[1/(b_2^2 - b_1^2)]$, depends on the amplitudes of the two solitons, and the transverse profile at this point can be approximated by $|\psi| \approx \text{sech}(\tau - d/2) + \text{sech}(\tau + d/2)$ (Gordon, 1983). A similar approximation can be written for the three (and more) soliton solutions with equally spaced b_i^2 ($b_3^2 - b_2^2 = b_2^2 - b_1^2$, etc.) (Akhmediev *et al.*, 1994a).

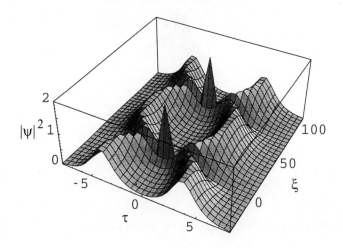

Figure 3.14 *Two-soliton solution (equation (3.79)) for $b_1 = 0.5$ and $b_2 = 0.54$.*

The case $b_1 = b_2$ is degenerate. The pulses diverge from each other and the beat period goes to infinity. Thus, only one impact region remains. The limiting case of solution (3.79), for $b_1 = b_2 (= b)$, can be written in the form:

$$\psi(\tau,\xi) = \frac{8ib[2b\tau \sinh(2b\tau) - \cosh(2b\tau) - 4ib^2\xi \cosh(2b\tau)]}{\cosh(4b\tau) + 1 + 8b^2\tau^2 + 32b^4\xi^2} \exp(2ib^2\xi).$$

$$(3.80)$$

Here, there is a single maximum which occurs at $(0,0)$; its amplitude is $|\psi(0,0)| = 4b$. This clearly differs from the case of two in-phase solitons of the same velocity (or which are parallel in the spatial domain), which coalesce periodically. Degenerate N-soliton solutions have been presented by Gagnon and Stiévenart (1994).

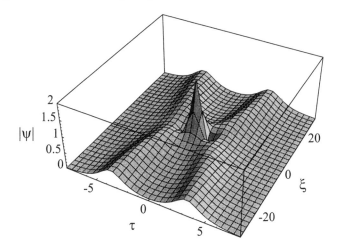

Figure 3.15 *Degenerate two-soliton solution (3.80) for equal amplitudes and zero velocities.*

If the amplitudes are $b_1 = 1/2$ and $b_2 = 3/2$, then we have

$$\psi(\tau,\xi) = 4\frac{\cosh 3\tau + 3\ \cosh \tau\ \exp(4i\xi)}{\cosh 4\tau + 4\ \cosh 2\tau + 3\ \cos\ 4\xi}\ \exp(i\xi/2)\ . \tag{3.81}$$

This periodic solution is a 'breather' (Fig. 3.16). At $\xi = 0$ it takes the simple form $2/\cosh \tau$. This solution was found first by Satsuma and Yajima (1974), and has been observed in experiments in optical fibers and in planar waveguides. Thus, a pulse which initially has the familiar sech shape, but has double the amplitude of a fundamental soliton, will evolve according to (3.81). As can be seen in Fig. 3.16, the pulse periodically becomes much narrower than it was originally.

More generally, an N-soliton solution (higher-order breather), with pulse amplitudes $b_i = (2i - 1)/2$ for $i = 1, ..., N$, has the shape $\psi = N$ sech τ when $\xi = 0$ (Satsuma and Yajima, 1974) if the soliton phases, φ_i, are chosen properly. In particular, this occurs for alternating phases: $\varphi_i = 0$ for i odd and $\varphi_i = \pi$ for i even. The above-mentioned narrowing is even greater for $N > 2$. This effect has been used to good advantage in soliton compression experiments, where very narrow output pulses are required.

On the other hand, if all phases are zero, than the amplitude of the N-soliton solution at $(\xi = 0, \tau = 0)$ is equal to N^2 (Akhmediev and Mitskevich, 1991). The total solution is periodic and the pulse again returns to this shape periodically. This effect can be used for an extremely high degree of pulse compression.

A nonlinear superposition of two- and more generally N-soliton solutions on a background (3.63) can also be constructed. This has been done by Bélanger and Bélanger (1996).

A nonlinear coupler can be used to combine low-order solitons to produce higher-order ones – e.g. two fundamental solitons can be combined to produce a two-soliton solution of the form (3.81) (Peng and Ankiewicz, 1992).

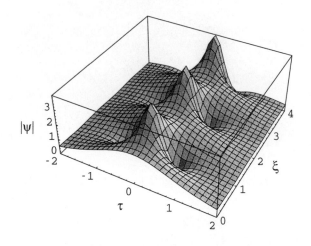

Figure 3.16 *Two-soliton solution (breather) (equation (3.81))*.

3.18 Modulation instability with two pairs of initial side-bands

Modulation instability, which was defined by the first-order solutions, can be excited using a single-frequency pump and a single pair of side-bands. What will happen if there is more than one pair of side-bands at the input, and they all are inside the gain curve depicted in Fig. 3.17?   In this instance, all of them will be amplified, due to four-wave mixing processes. The total effect cannot be described by the simple formula (3.39) because it is a nonlinear superposition of these elementary processes, in the same way that a multi-soliton solution is a nonlinear superposition of one-soliton solutions. This higher-order solution can be constructed using a Darboux transformation.

To do this, we commence by using the c.w. solution of the NLSE as the seeding solution:

$$\psi = \exp(i\xi + i\theta). \tag{3.82}$$

The arbitrary phase factor in (3.82) can be omitted. The functions r and s,

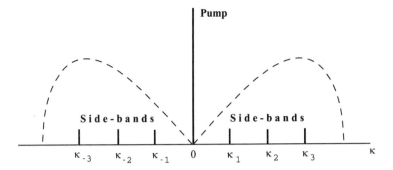

Figure 3.17 *Modulation instability with several pairs of side-bands. The dotted lines indicate the gain spectrum.*

corresponding to solution (3.82), are readily found by integration of (2.22):

$$r = \{C \, \exp\left[i(2\chi + \kappa_1\tau + \lambda_1\kappa_1\xi)/2\right]$$

$$-D \, \exp[-i(2\chi + \kappa_1\tau + \lambda_1\kappa_1\xi)/2]\} \, \exp(-i\xi/2),$$

$$s = \{C \, \exp[i(-2\chi + \kappa_1\tau + \lambda_1\kappa_1\xi)/2]$$

$$+D \, \exp[-i(-2\chi + \kappa_1\tau + \lambda_1\kappa_1\xi)/2]\} \, \exp(i\xi/2),$$

(3.83)

where $\kappa_1 = 2\sqrt{1 + \lambda_1^2}$ and $\cos 2\chi = \kappa_1/2$.

We assume that the parameter λ_1 is purely imaginary, so that $\lambda_1 = i\nu_1$, and we replace the constants C and D with real parameters τ_{01} and ξ_{01}:

$$C = \exp\{[\delta_1\xi_{01} - i\kappa_1(\tau_{01} + \pi/2)]/2\},$$

$$D = \exp\{[-\delta_1\xi_{01} - i\kappa_1(\tau_{01} + \pi/2)]/2\},$$

$$\delta_1 = \kappa_1\nu_1.$$

(3.84)

Then (3.83) can be written in the form

$$r = \{\exp[(i2\chi + i\kappa_1(\tau - \tau_{01} - \pi/2) - \delta_1(\xi - \xi_{01}))/2]$$

$$- \exp[(-i2\chi - i\kappa_1(\tau - \tau_{01} - \pi/2) + \delta_1(\xi - \xi_{01}))/2]\} \, \exp(-i\xi/2),$$

$$s = \{\exp[(-i2\chi + i\kappa_1(\tau - \tau_{01} - \pi/2) - \delta_1(\xi - \xi_{01}))/2]$$

$$+ \exp[(i2\chi - i\kappa_1(\tau - \tau_{01} - \pi/2) + \delta_1(\xi - \xi_{01}))/2]\} \, \exp(i\xi/2).$$

(3.85)

In this case, the solution obtained, with the aid of (2.36), is

$$\psi = \left[\frac{\kappa_1^2\cosh\delta_1(\xi - \xi_{01}) + 2i\kappa_1\nu_1\sinh\delta_1(\xi - \xi_{01})}{2[\cosh\delta_1(\xi - \xi_{01}) - \nu_1\cos\kappa_1(\tau - \tau_{01})]} - 1\right]e^{i\xi}, \qquad (3.86)$$

where $\kappa_1 = 2\sqrt{1 - \nu_1^2}$. Apart from scaling with the factor $q = -\sqrt{2}$ using (2.3), and the notation ($\nu_1 = \sqrt{2a_1}$), it is identical to solution (3.39). The quantity κ_1 then determines the frequency of the initial ($\xi \to -\infty$) modulation (side-band), whereas δ_1 is the growth rate of the instability. Note that the rescaling changes the range of instability by a factor of $\sqrt{2}$ ($0 < \kappa_1 < 2$) and the growth rate by a factor of 2 relative to Fig. 3.4.

The selection of the parameters, λ_N, in the next stage, in order to obtain the hierarchy of the solutions of the NLSE, is made in the same way as in (3.84):

$$\lambda_N = i\nu_N, \quad \kappa_N = 2\sqrt{1 - \nu_N^2}, \quad \delta_N = \kappa_N\nu_N,$$

$$C = \exp\{[\delta_N\xi_{0N} - i\kappa_N(\tau_{0N} + \pi/2)]/2\},$$

$$D = \exp\{[-\delta_N\xi_{0N} - i\kappa_N(\tau_{0N} + \pi/2)]/2\}.$$

The periodicity of the solution with respect to the variable τ is retained if we assume that $\kappa_N = N\kappa_1$. On the other hand, we need $0 < \kappa_N < 2$ to ensure that the growth rates of all harmonics remain real (Fig. 3.17). When $N=2$, the solution can still be written in an analytic form:

$$\psi = \left[1 - \frac{G + iH}{D}\right]\exp(i\xi + i\phi), \qquad (3.87)$$

where

$$G = \frac{\kappa^2}{4\delta_1}\cosh\delta_1(\xi - \xi_{01})\cos 2\kappa(\tau - \tau_{02}) + \frac{2\kappa^2}{\delta_2}\cosh\delta_2(\xi - \xi_{02})\cos\kappa(\tau - \tau_{01})$$

$$+ \frac{3\kappa^3}{2\delta_1\delta_2}\cosh\delta_1(\xi - \xi_{01})\cosh\delta_2(\xi - \xi_{02}),$$

$$H = \frac{1}{2}\sinh\delta_1(\xi - \xi_{01})\cos 2\kappa(\tau - \tau_{02}) + \sinh\delta_2(\xi - \xi_{02})\cos\kappa(\tau - \tau_{01})$$

$$- \frac{\kappa}{\delta_1\delta_2}[\delta_1\sinh\delta_1(\xi - \xi_{01})\cosh\delta_2(\xi - \xi_{02}) - \delta_2\cosh\delta_1(\xi - \xi_{01})\sinh\delta_2(\xi - \xi_{02})],$$

$$D = \frac{3}{4\kappa}\left[\cos\kappa(\tau + \tau_{01} - 2\tau_{02}) + \frac{1}{9}\cos\kappa(3\tau - 2\tau_{02} - \tau_{01})\right]$$

$$+ \frac{1}{2\delta_1}\cosh\delta_1(\xi - \xi_{01})\cos 2\kappa(\tau - \tau_{02}) + \frac{1}{\delta_2}\cosh\delta_2(\xi - \xi_{02})\cos\kappa(\tau - \tau_{01})$$

$$- \frac{2}{3\kappa}\left[\frac{\kappa^2(2\kappa^2 - 5)}{2\delta_1\delta_2}\cosh\delta_1(\xi - \xi_{01})\cosh\delta_2(\xi - \xi_{02})\right.$$

$$\left. + \sinh\delta_1(\xi - \xi_{01})\sinh\delta_2(\xi - \xi_{02})\right],$$

where we have set $\kappa_1 = \kappa$ and the constants ϕ, τ_{01}, τ_{02}, ξ_{01} and ξ_{02} are arbitrary. They can be related to the initial amplitudes of the side-bands. These quantities are, in effect, the centres of two elementary solutions, (3.86), whose nonlinear interaction leads to the solution (3.87). When the centres coincide, $\xi_{01} = \xi_{02} = 0$, and solution (3.87), with phase $\phi = 0$, is symmetric relative to reversal of the sign of ξ:

$$\psi(\xi, \tau) = \psi^*(-\xi, \tau).$$

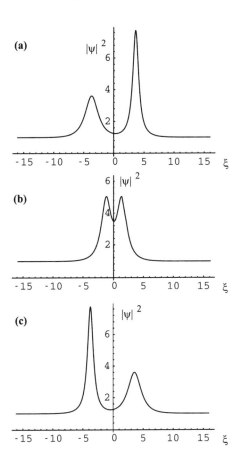

Figure 3.18 *Evolution of the field* $\psi(\xi)$, *at* $\tau = 0$, *in the process of modulation instability with two initial side-bands of modulation. Parameters are* $\kappa_1 = 2/\sqrt{5}$, $\kappa_2 = 2\kappa_1$, $\tau_{01} = \pi/\kappa$, *and* $\tau_{02} = \pi/2\kappa$. *Here* (ξ_{01}, ξ_{02}) *equals (a)* $(3, -3)$, *(b)* $(0, 0)$ *and (c)* $(-3, 3)$.

If the centres of the elementary solutions are separated by a large distance along the ξ axis, (3.87) breaks up into a sum of two elementary solutions, where each is of the form (3.86) and each has its own phase, just as in the case of two-soliton solutions of the NLSE when the distances between the soliton centres exceed the characteristic dimensions of each of them. This can clearly be seen in Fig. 3.18.

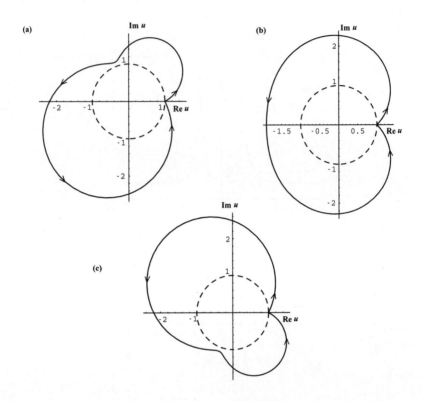

Figure 3.19 *Trajectories on the complex plane (Re $u(\xi,0)$, Im $u(\xi,0)$) for the solution describing modulation instability with two initial side-bands. Parameters are the same as in Fig. 3.18.*

In the top plot (a), the modulation instability of the first side-bands develops first (first peak). Then, when this evolution is almost complete and a return to the initial state has occurred, the modulation instability of the second side-bands develops. These two processes are separated in ξ by $\Delta\xi = \xi_{02} - \xi_{01}$. When $\Delta\xi = 0$, we have a nonlinear superposition of two elementary processes.

The same process is shown in Fig. 3.19 on the complex plane (Re u, Im u), where $u = \psi e^{i(\pi-\xi)}$. At $\xi \to -\infty$, the initial point corresponds to the c.w. solution with unit amplitude. At large values of $\Delta\xi$, the full trajectory is a combination of two parts, each of which is an arc of a circle corresponding to modulation instability with a single pair of side-bands

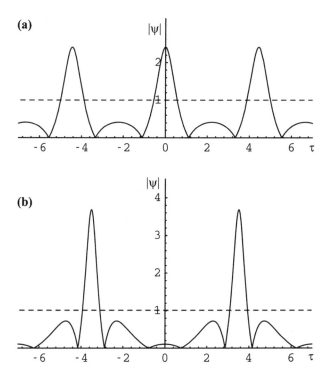

Figure 3.20 *Comparison of pulse shapes at $\xi = 0$ for (a) modulation instability with a single pair of initial side-bands (equation (3.86) with $\kappa_1 = \sqrt{2}$ and $\tau_{01} = 0$) and (b) two pairs of initial side-bands (equation (3.87) with $\kappa = 2/\sqrt{5}$ and $\tau_{01} = \tau_{02} = 0$). The dashed lines show the amplitude of the initial c.w.*

(Fig. 3.19a,c). Note that the total phase shift is the simple sum of the two phase shifts generated by the elementary solutions, and is thus independent of $\Delta\xi$. In this particular case ($\kappa_1 = 2/\sqrt{5}$), the total phase shift is exactly 2π.

The pulse profiles obtained as a result of modulation instability are shown in Fig. 3.20. For comparison, we give the pulse profiles for the case of one pair of initial side-bands (equation (3.86)) and two pairs of initial side-bands (equation (3.87)). In the former case, the frequency is chosen to be at the maximum of the growth-rate curve, while in the latter case the fundamental frequency and its second harmonic have the same growth rates. We can see that two frequencies of initial modulation give a periodic train of pulses which are narrower and taller than those in the case of a single frequency. This occurs, however, only for a special choice of the phase difference between the harmonics. Other cases can be found in Akhmediev *et al.* (1985a).

Solutions corresponding to three or more pairs of side-bands can also be obtained using the same technique. The analysis can be found in Akhmediev and Mitskevich (1991).

CHAPTER 4

Non-Kerr-law nonlinearities

When the nonlinear term in the NLSE differs from that of the Kerr law, the resulting equation,

$$i\psi_\xi + \frac{1}{2}\psi_{\tau\tau} + N(|\psi|^2)\psi = 0, \qquad (4.1)$$

is no longer integrable. Here we assume that the nonlinearity, N, depends only on the local instantaneous light intensity. Instead of an infinite number of conservation laws, the equation now has only a few. Although stationary pulses exist, and some solutions can be written in analytic form, their behaviour differs from that of solutions of the NLSE. As a result, the behaviour of the whole system with arbitrary initial conditions becomes different. In particular:

1. In contrast to solitons of the NLSE, soliton-like solutions can be stable or unstable, depending on the parameter of the family (Kaplan, 1985; Mitchell and Snyder, 1993).

2. A collision of two stable solitons is inelastic, in contrast to the collision of solitons of the NLSE. The amplitudes of the output pulses can differ from the amplitudes of input solitons. There is a loss of energy from the soliton component of the solution.

3. The number of pulses after the collision can be different from the number of solitons before the collision (Cowan et al., 1986; Gatz and Herrmann, 1992b; Snyder and Sheppard, 1993).

4. Part of the energy of the colliding solitons can be transformed into radiation (Abdulloev et al. 1976; Bona et al., 1980).

5. Superposition states of two solitons (e.g. solutions like the breather) do not exist.

Strictly speaking, because of these differences, the pulse solutions of non-integrable systems are not solitons in the mathematical sense. However, the term 'soliton' has been used in many cases for the solutions of nonintegrable systems, and this broader definition has become common. Hence, we also use this term in the rest of the book, but this remark should always be borne in mind.

Non-integrability is not necessarily related to the nonlinear term. Higher-order dispersion or birefringence, for example, also make the system non-integrable, while it remains Hamiltonian. However, terms associated with

higher-order dispersion or birefringence introduce other physical effects to
the behaviour of one-soliton solutions. If the part of the equation which
corresponds to the deviation can be considered small, then many features
of integrable systems can still be observed. We will not cover the whole
diversity of phenomena possible in non-integrable Hamiltonian systems.
Instead, we will give some exact solutions for soliton solutions and will give
an example which allows us to make quantitative calculations related to
soliton interaction and the resulting radiation effects.

4.1 Stationary solutions

For Hamiltonian systems, where the energy and Hamiltonian are conserved,
soliton-like solutions comprise a one-parameter family of solutions

$$\psi(\tau,\xi|q) = f(\tau|q)\ e^{iq\xi+i\varphi}$$

with variable amplitude, as was the case for the NLSE. The Galilean transformation (2.5) adds one more parameter to this family, viz. the velocity
v. Hence, one-soliton solutions can be described by two parameters.

We assume the function $f(\tau)$ to be real. This can always be arranged
with a proper choice of the phase, φ, unless the soliton is moving. (If the
soliton has a phase chirp, i.e. $\varphi = \varphi(\tau)$, then it cannot be stationary.) So

$$\frac{1}{2}f_{\tau\tau} - qf + N(f^2)f = 0. \tag{4.2}$$

Now, we can consider the function $f(\tau)$ as being the position of a particle
moving in a potential $R(f^2)$:

$$\frac{f_\tau^2}{2} + R(f^2) = 0$$

(or $f_{\tau\tau} = -\frac{\partial R}{\partial f}$), where the potential is

$$R(I) = \bar{F}(I) - qI, \tag{4.3}$$

and the intensity is

$$I = |\psi|^2 = f^2, \tag{4.4}$$

The function $\bar{F}(I)$ is defined in such a way that the nonlinearity function
N is related to $\bar{F}$ simply by $N = d\bar{F}/dI$, and thus

$$\bar{F}(I) = \int_0^I N(I')\,dI'. \tag{4.5}$$

This definition ensures that $\bar{F}(0) = R(0) = 0$. For physically realistic laws,
the potential decreases from zero as the intensity increases from zero. For
a soliton to exist, the potential must reach a minimum and then increase
up to zero again. We may state the soliton existence condition formally by

saying that we must have $R > 0$ for some positive intensity. The maximum intensity of the soliton, I_m, is then the first positive zero of $R(I)$.

4.2 Some examples

For the Kerr law the solution is obvious. We now consider some examples.

4.2.1 Kerr law

Here the increase in refractive index is linear in intensity – thus $N(I) = I$ and $P = I^2/2$. The potential is just $\bar{F} = I(\frac{I}{2} - q)$ and so the 'particle', which starts (at rest) at $\tau = -\infty$, travels to the potential minimum and then up to the point where R reaches zero again, viz. the point where $I = I_m = 2q$. This occurs at $\tau = 0$, and the particle then retraces its path, past the minimum and back to $I = 0$ (at $\tau = \infty$). Thus solitons exist for all values of q. This is not necessarily the case for other nonlinearity laws, and so the term 'soliton' is sometimes reserved for pulses in Kerr-law media, while terms like 'solitary waves' are used for pulses in other cases. Here we continue to use the extended definiton of 'soliton'.

4.2.2 Parabolic law

Here

$$N(I) = I + \nu I^2. \tag{4.6}$$

If ν is negative, the potential has one maximum, and it occurs at $I = I_1$, where

$$I_1 = -\frac{1}{2\nu}[1 + \sqrt{1 + 4\nu q}]. \tag{4.7}$$

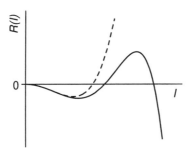

Figure 4.1 *Schematic diagram showing the potential for the parabolic nonlinearity:* $\nu > 0$ *(dashed line),* $\nu < 0$ *(solid line)*

The condition $R(I_1) \geq 0$ then demands $q\nu \geq -\frac{3}{16}$ for solitons to exist. If ν is positive, there is no maximum and solitons can thus exist for all

$\nu > 0$ (as in the Kerr law example). In either case, we may write down the maximum intensity (at the centre of the pulse):

$$I_m \;=\; \frac{3}{4\nu}\left[\sqrt{1+\frac{16}{3}\nu q}\;-\;1\,\right].\qquad(4.8)$$

Of course, in the Kerr law limit ($\nu \to 0$), q can approach infinity.

4.2.3 Power law

Taking

$$N(I) = I^b \quad (b > 0) \qquad(4.9)$$

leads to

$$R(I) \;=\; I\left(\frac{I^b}{b+1}-q\right).\qquad(4.10)$$

In this case the solitons exist for all q, and the maximum intensity is

$$I_m \;=\; [q\,(b+1)]^{1/b}.\qquad(4.11)$$

This case has been considered by Snyder and Mitchell (1993).

4.2.4 Dual power laws

We may generalize to some extent by considering:

$$N(I) = I^b + \nu I^{2b}, \qquad b > 0. \qquad(4.12)$$

Thus $b = 1$ is the parabolic case, (4.6), while $\nu = 0$ gives the power-law example, (4.9) above. The required conditions can still be reduced to quadratic equations for this form, and we find that we need

$$q\nu \;\geq\; -\,\frac{2b+1}{4(b+1)^2}\qquad(4.13)$$

for solitons to exist. The maximum intensity is then

$$I_m \;=\; \left[\frac{2b+1}{2\nu}\left\{\sqrt{\frac{1}{(b+1)^2}+\frac{4\nu q}{2b+1}}-\frac{1}{b+1}\right\}\right]^{1/b}\qquad(4.14)$$

4.2.5 Threshold nonlinearity

A simple example of a nonlinearity which admits exact solutions is the threshold nonlinearity (Snyder *et al.*, 1991a):

$$N(I) = \begin{cases} 0, & I < I_0 \\[2mm] n_2 - n_1, & I \geq I_0 \end{cases}\qquad(4.15)$$

where I_0 is the threshold.

4.3 Saturable media

All real materials exhibit saturation at sufficiently high intensities. We shall consider some consequences of this for switching in Chapter 8. Saturation can be modelled by:

$$N(I) = \frac{I}{1 + \gamma I}. \tag{4.16}$$

The potential is:

$$R(I) = (\gamma^{-1} - q)I - \gamma^{-2} \ln(1 + \gamma I). \tag{4.17}$$

We clearly need $\hat{\gamma} \equiv q\gamma < 1$ for solitons to exist. The relation between this and the maximum intensity is given by:

$$\hat{\gamma} = 1 - \frac{\ln(1 + x)}{x} = \frac{x}{2} - \frac{x^2}{3} + \frac{x^3}{4} + \cdots, \tag{4.18}$$

where x is defined as γI_m. For relatively small $\hat{\gamma}$, we have $x = 2\hat{\gamma}(1 + \frac{4}{3}\hat{\gamma} + \frac{14}{9}\hat{\gamma}^2 + \cdots)$, and so $I_m \approx 2q(1 + \frac{4}{3}\hat{\gamma})$.

A saturating case which is qualitatively similar (see 8.17) is (Ankiewicz et al., 1993):

$$N(I) = \frac{1}{\gamma}[1 - \exp(-\gamma I)]. \tag{4.19}$$

This also requires $\hat{\gamma} = q\gamma < 1$ for solitons, while the nonlinearity coefficient in terms of the maximum intensity is

$$\hat{\gamma} = 1 - \frac{1 - e^{-x}}{x} = \frac{x}{2} - \frac{x^2}{6} + \frac{x^3}{24} + \cdots \tag{4.20}$$

In both cases, (4.16) and (4.19), we have $x \to \infty$ as $\hat{\gamma} \to 1$.

4.4 Stability of solitary waves

Equation (4.1), although non-integrable, still has three basic conserved quantities. The energy Q is the same as (2.7) and the momentum is still given by (2.8), while the Hamiltonian is

$$H = \int_{-\infty}^{\infty} \left(\frac{1}{2}|\psi_\tau|^2 - \bar{F}(I)\right) d\tau \tag{4.21}$$

where $\bar{F}$ is given by (4.5).

Equation (4.1) can be written in a canonical form (Faddeev and Takhtadjan, 1987):

$$i\psi_\xi = \frac{\delta H}{\delta \psi^*}, \qquad i\psi_\xi^* = -\frac{\delta H}{\delta \psi}. \tag{4.22}$$

Equations (4.21) and (4.22) define a Hamiltonian dynamical system on an infinite-dimensional phase space of two complex functions U, V which decrease to zero at infinity; it can be analysed using the theory of Hamiltonian systems. This means that the behaviour of the solutions is defined, to a large extent, by the singular points of the system (i.e. stationary solutions of (4.1)), and depends on the nature of these points (as determined by the stability of its stationary solutions).

For Hamiltonian systems, the stationary solutions play a pivotal role in the dynamics. To find them, we represent the field in the form:

$$\psi = \bar{\psi}(\tau, q)e^{iq\xi}, \tag{4.23}$$

where q is the parameter of this family of solutions, and $\bar{\psi}$ is a real function.

Now the equation for finding stationary solutions, in the variational formulation, can be written as:

$$\delta(H - qQ) = 0. \tag{4.24}$$

This variational formulation of the problem also defines the stability of the stationary states: that is, for any fixed Q, the stationary state is stable if the corresponding H has a local minimum, with q being a Lagrange multiplier. Using this principle, we can study stability by plotting H versus Q in a diagram. For Hamiltonian systems, there may be one or several branches of solutions. For any fixed Q, the state which has the lowest H and which corresponds to a one-soliton solution can be considered stable.

This approach can be extended to other Hamiltonian systems (more complicated than (4.1)). These include pulse propagation in birefringent media, pulses and beams in three-dimentional space, etc. Specific examples are given in other chapters. For some nonlinearity laws, the family of one-soliton solutions can have stable and unstable branches (Kaplan, 1985).

For stationary solutions, there is a one-to-one correspondence between the Hamiltonian H, energy Q and the parameter of the family q. This means that, instead of H and Q, the stability can be reformulated in terms of the two other parameters. These can be, for example, Q and q. It has been shown in a number of publications (Vakhitov and Kolokolov, 1974 (for a Kerr medium); Mitchell and Snyder, 1993; Grillakis *et al.*, 1987) that lowest-order solutions are stable if the slope of the curve $\frac{\partial Q}{\partial q}$ is positive, and unstable if this slope is negative or zero. Strictly speaking, this criterion is valid only for bright solitons in self-focusing media, although the medium can be stratified perpendicular to the direction of propagation (Jones and Moloney, 1986).

The above criterion cannot be applied to higher-order (multi-soliton) solutions. In homogeneous systems (infinite space) all higher-order solutions are unstable. In inhomogeneous systems this is not true. Examples of the stability of higher-order guided wave solutions will be given in Chapter 12. Stability can be specified in terms of two other variables as well, if they are related to H, Q and q on a one-to-one basis.

4.5 Solitons for parabolic nonlinearity law

The generalized NLSE with quintic terms is:

$$i\frac{\partial \psi}{\partial \xi} + \frac{1}{2}\frac{\partial^2 \psi}{\partial \tau^2} + |\psi|^2\psi + \nu|\psi|^4\psi = 0, \tag{4.25}$$

where ν is the fifth-order nonlinear susceptibility. It describes propagation for the parabolic nonlinearity law (4.6). The Kerr law nonlinearity is the $\nu = 0$ case. The one-soliton solution of (4.25) is:

$$\psi(\xi,\tau) = \frac{2\sqrt{q}\ \exp(iq\xi)}{\sqrt{1 + \sqrt{1 + (16/3)\nu q}\ \cosh(2\sqrt{2q}\tau)}}, \tag{4.26}$$

where q is the soliton parameter. The solution is valid for $-3/16 < \nu q < \infty$. At the lower limit, the soliton energy becomes infinite. The peak intensity of the pulse, $|\psi(\xi,0)|^2$, is then given by (4.8). In the limit $\nu = 0$, solution (4.26) reduces to the standard NLSE soliton.

The energy carried by the soliton (4.26) is

$$Q = \int_{-\infty}^{\infty} |\psi|^2 d\tau = \sqrt{\frac{3}{2\nu}}\ \arctan\left(4\sqrt{\frac{\nu q}{3}}\right)$$

in the case of $\nu > 0$ and

$$Q = \frac{\sqrt{-3}}{\sqrt{2\nu}}\mathrm{arctanh}\left(4\sqrt{-\nu q/3}\right)$$

in the case of $-3/16q < \nu < 0$. If $|\nu|$ is small, then

$$Q \approx 2\sqrt{2q}\left(1 - \frac{16}{9}\nu q + ...\right).$$

This expression shows that the small parameter in the system is $\gamma = \nu q$.

4.6 Internal friction between solitons

The details of the interaction between soliton-like solutions during collision in non-integrable systems are still open to debate. Numerical simulations (Snyder and Sheppard, 1993) show that even the slightest change from the Kerr nonlinearity results in the two solitons annihilating each other, merging or creating many new solitons, depending on the initial inclination of the two solitons and their shapes. An analytical study of the inelastic interaction for the Peregrine–Benjamin equation has been carried out by Kodama (1987). In this section, we consider the two-soliton interaction in a system which is close to being integrable (Buryak and Akhmediev, 1994a). Radiation from the interaction area appears and the solitons exchange a part of their energies during such processes. Solitons interact inelastically, not only in collision, but even if they are initially located at the same place. They emit radiation continuously until both of them have energy levels suitable for interaction. This process can be considered as internal friction between two solitons.

We are interested in soliton interaction in the system described by (4.25). At small $|\nu|$, the interaction is minimal, and would not be noticeable for a small length of interaction, e.g. in collisions. Hence, we consider the case when the length of interaction goes to infinity. The solution corresponding to this case is the breather (3.79). We now consider what will happen with the breather if the system is slightly perturbed from integrability. In the case of non-zero ν, solution (3.79) no longer holds. However, for small ν, the solution still can be represented as the beating of the two solitons. A near-periodic pattern still persists (Snyder *et al.*, 1995), but the quintic term causes coupling to radiation modes. Because of losses due to the radiation, the amplitudes of the two solitons change during the interaction process in such a way that the total energy gradually decreases. This process can be viewed as 'internal friction' between the two interacting solitons causing radiation. To lowest order, the solution has the form

$$\psi(\xi,\tau) = \psi_0(\xi,\tau) + f_{rad}(\xi,\tau), \tag{4.27}$$

where ψ_0 is the solution (3.79) and f_{rad} is the radiation part. Asymptotically, in the limit $\tau \to \pm\infty$, the radiation can be represented as a superposition of linear plane waves

$$f_{rad} = \sum_n a_n \exp[i(\omega_n\xi \pm \sqrt{-\omega_n}\tau)], \tag{4.28}$$

where ω_n ($\omega_n < 0$) and a_n are, respectively, the frequencies in ξ and amplitudes of the radiation modes.

To determine the energy flow from the two-soliton solution, we consider the invariants of (4.25). Using (4.6) and (4.21), the Hamiltonian can be written in the form

$$H = \int_{-\infty}^{\infty} \left(\frac{1}{2}|\psi_\tau|^2 - \frac{|\psi|^4}{2} - \nu\frac{|\psi|^6}{3}\right)d\tau. \tag{4.29}$$

For example, for the one-soliton solution, (4.26), with ν negative, we obtain

$$H = \frac{3\sqrt{6}}{32(-\nu)^{3/2}}\tanh\left(Q\sqrt{\frac{-2\nu}{3}}\right) + \frac{3Q}{16\nu}. \tag{4.30}$$

For $\nu \to 0$, this reduces to $H = -Q^3/24$, i.e. the Kerr law soliton form. We shall return to this idea when considering couplers in Chapter 8. Conservation laws in the form of the continuity equations are given by (2.15), with $\rho_Q = |\psi|^2$, $\rho_H = \frac{1}{2}|\psi_\tau|^2 - |\psi|^4/2 - \nu|\psi|^6/3$ and with j_Q and j_H given by (2.17) and (2.19), respectively. Integration of (2.15) over the interval

($\tau_1 = -\tau_0$, $\tau_2 = \tau_0$), which includes the two-soliton part of the general solution (Fig. 4.2), gives the equations for the conservation of energy and Hamiltonian in the integral form (2.20). In explicit form these are:

$$\tfrac{d}{d\xi} Q_{TS} = -[j_Q(\tau_0) - j_Q(-\tau_0)]$$

$$\tfrac{d}{d\xi} H_{TS} = -[j_H(\tau_0) - j_H(-\tau_0)], \tag{4.31}$$

where Q_{TS} and H_{TS} are integrals over the central part of the solution.

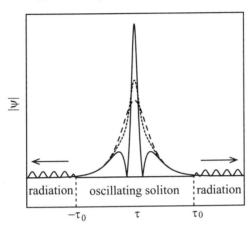

Figure 4.2 *Radiation emission from oscillating soliton*

The radiation part of the solution can be represented as a Fourier series of linear waves (4.28). Substituting (4.28) into (4.31), and taking into account only those waves which are moving away from the two-soliton part of the solution gives

$$\tfrac{d}{d\xi} Q_{TS} = -4 \sum_n \sqrt{-\omega_n} \, |a_n|^2,$$

$$\tfrac{\partial}{\partial\xi} H_{TS} = -4 \sum_n \omega_n \sqrt{-\omega_n} \, |a_n|^2. \tag{4.32}$$

To solve system (4.32), one has to express all the unknown quantities in terms of two basic variables. The most convenient variables are the parameters of the two partial solitons, q_1 and q_2. We assume that they change adiabatically in ξ. Hence, the values Q_{TS} and H_{TS} can be expressed approximately in terms of q_1 and q_2 in the same form as the two-soliton solution of the NLSE:

$$Q_{TS} = 2\sqrt{2}(\sqrt{q_1} + \sqrt{q_2}), \qquad H_{TS} = -\tfrac{2}{3}\sqrt{2}(q_1^{3/2} + q_2^{3/2}). \tag{4.33}$$

Now, we have to find $\omega_n(q_1, q_2)$ and $a_n(q_1, q_2)$. To do this, we use some simplifying assumptions. We assume the existence of two small parameters

$|\nu| \ll 1$ and $\varepsilon = q_2/q_1 \ll 1$. Thus, we assume that the amplitude $A_1 = \sqrt{2q_1}$ of one component of the soliton is much larger that the amplitude $A_2 = \sqrt{2q_2}$ of the other component.

According to (4.28), the frequencies ω_n of radiation are determined by the negative frequencies of a source of radiation, i.e. by the negative frequencies of the two-soliton solution (3.79). These frequencies are given by

$$\omega_n = q_2 - n(q_1 - q_2), \qquad (4.34)$$

where $n = 1, 2, 3, \ldots$. Expanding solution (3.79) in a Fourier series in ξ, we find that the component at the lowest frequency $\omega_1 = -(q_1 - 2q_2)$ has the largest amplitude. Thus the most of the radiation is emitted at this lowest frequency.

The amplitudes a_n of the plane waves can be obtained approximately using the small parameters ν and ϵ. The radiation appears because of the last term in (4.25), and this is proportional to ν. Therefore, in the linear approximation, the amplitudes of the radiation waves must also be proportional to ν. Substituting (4.27) into (4.25), and keeping only terms which are linear in ν, we obtain the equation

$$i\frac{\partial f_{rad}}{\partial \xi} + \frac{\partial^2 f_{rad}}{\partial \tau^2} + F(\xi,\tau)f_{rad} + G(\xi,\tau)f_{rad}^* = -\nu P(\xi,\tau), \qquad (4.35)$$

where $F(\xi,\tau) = 2|\psi_0|^2$, $G(\xi,\tau) = \psi_0^2$, $P(\xi,\tau) = |\psi_0|^4\psi_0$, and f_{rad}^* is the complex conjugate of f_{rad}. Due to the adiabaticity of q_1 and q_2, each of the functions F, G and P can be considered to be periodic in ξ, and can be represented in the form of an infinite series:

$$F(\xi,\tau) = \sum_{n=-\infty}^{\infty} F_n(\tau)\, e^{i(q_2-q_1)n\xi}$$

$$G(\xi,\tau) = e^{i2q_2\xi} \sum_{n=-\infty}^{\infty} G_n(\tau)\, e^{i(q_2-q_1)n\xi} \qquad (4.36)$$

$$P(\xi,\tau) = e^{iq_2\xi} \sum_{n=-\infty}^{\infty} P_n(\tau)\, e^{i(q_2-q_1)n\xi}.$$

We also expand $f_{rad}(\xi,\tau)$ in a Fourier series in ξ, where we only retain negative frequencies ω_n, which correspond to radiation:

$$f_{rad}(\xi,\tau) = e^{iq_2\xi} \sum_{n=-\infty}^{\infty} g_n(\tau)\, e^{i(q_2-q_1)n\xi} \qquad (4.37)$$

After substituting (4.36) and (4.37) into (4.35) and collecting the coefficients of the factors $e^{iq_2\xi}e^{i(q_2-q_1)\xi}$, we obtain an infinite set of equations:

$$\frac{d^2g_n}{d\tau^2} - \omega_n g_n + \sum_{j=1}^{\infty} F_{n-j}(\tau)\, g_j(\tau) + \sum_{j=1}^{\infty} G_{n+j}(\tau)\, g_j^*(\tau) = -\nu P_n(\tau), \qquad (4.38)$$

where $1 < n < \infty$. This infinite set of ordinary differential equations (4.38) defines the radiation at various frequencies ω_n. The terms on the right-hand side of (4.38) can be considered as source terms. As we explained above, we are interested in radiation at the frequency ω_1, and thus need to find only $g_1(\tau)$. It is possible to separate the equation for $g_1(\tau)$ from all others using a perturbation approach. Making the substitution $g_1(\tau) = \nu\epsilon^2 q_1^{3/2} b(t)$ (where $t = \tau\sqrt{q_1}$) in (4.38) and keeping the terms up to order $\nu\epsilon^2$, we find the following equation for $b(t)$:

$$\frac{d^2 b}{dt^2} + b + \frac{4b}{\cosh^2 t} = -\frac{4\sqrt{2}}{\cosh^3 t} + \frac{24\sqrt{2}}{\cosh^5 t} - \frac{4\sqrt{2}}{\cosh^7 t}. \tag{4.39}$$

This equation can be solved analytically. A particular solution for the boundary conditions $b(-\infty) = 0$ and $\frac{db}{dt}(-\infty) = 0$ is shown in Fig. 4.3.

Two linearly independent solutions of the homogeneous equation are the associated Legendre functions of the first kind $P_\mu^{\pm i}(z)$, where $z = \tanh(t)$, and $\mu = (\sqrt{17} - 1)/2$. Now the solution of the inhomogeneous equation (4.39) can be written in the form

$$b(z) = \left[\int_{-1}^{z} \frac{f(z') P_\mu^{-i}(z')}{W[P_\mu^{-i}(z'), P_\mu^i(z')]} \, dz' \right] P_\mu^i(z)$$

$$- \left[\int_{-1}^{z} \frac{f(z') P_\mu^i(z')}{W[P_\mu^{-i}(z'), P_\mu^i(z')]} \, dz' \right] P_\mu^{-i}(z) \tag{4.40}$$

where

$$f(z) = -4\sqrt{2}\left[(1 - z^2)^{-1/2} - 6(1 - z^2)^{1/2} + 6(1 - z^2)^{3/2}\right],$$

and

$$W[P_\mu^{-i}(z'), P_\mu^i(z')] = P_\mu^{-i}(z') \frac{\partial P_\mu^i(z')}{\partial z} - P_\mu^i(z') \frac{\partial P_\mu^{-i}(z')}{\partial z}$$

is the Wronskian. Solution (4.40) is shown in Fig. 4.3. We define b_0 as the amplitude of the oscillations in solution (4.40), in the limit $t \to \infty$. In fact

$$b_0 = \frac{4\sqrt{2}\,\pi^{5/2}}{\cosh(\pi/2)\,\sqrt{\sinh(\pi)}\,\Gamma(\delta)\,\Gamma(\sigma)\,|\Gamma(\delta + i/2)\,\Gamma(\sigma + i/2)|} = 1.986...,$$

where Γ denotes the gamma function, $\delta = (3 + \sqrt{17})/4$ and $\sigma = (3 - \sqrt{17})/4$. Thus, the amplitude of radiation at the main frequency ω_1 is given by

$$a_1 = \nu\sqrt{q_1}\,q_2 b_0 \tag{4.41}$$

up to first order in ν and up to second order in ε.

After neglecting radiation at frequencies other than ω_1, and substituting

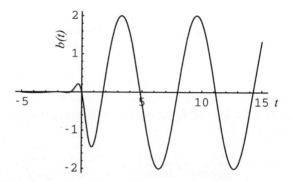

Figure 4.3 *Radiation mode*

expressions (4.33), (4.34) and (4.41) into (4.32), we obtain the forms

$$\frac{d}{d\xi}(\sqrt{q_1} + \sqrt{q_2}) = -\nu^2 b_0^2 \sqrt{q_1 - 2q_2} q_1 q_2^2,$$

$$\frac{d}{d\xi}(q_1^{3/2} + q_2^{3/2}) = 3\nu^2 b_0^2 (\sqrt{q_1 - 2q_2})^3 q_1 q_2^2.$$

(4.42)

These equations have first integral

$$2\sqrt{q_1(\xi)} + \sqrt{q_2(\xi)} = K,$$

(4.43)

where $K = 2\sqrt{q_1(0)} + \sqrt{q_2(0)}$. Expression (4.43) gives the relative dynamics of the soliton parameters of two interacting solitons. Substituting (4.43) into any equation of system (4.42), we get a solution in quadratures:

$$\xi = \frac{4}{\nu^2 b_0^2 K^6} \int_{\sqrt{q_2(\xi)}/K}^{\sqrt{q_2(0)}/K} \frac{dy}{y^4(1-y)^2\sqrt{1-2y-7y^2}}.$$

(4.44)

From (4.44) we can get the simple asymptotic results

$$q_2(\xi) = \frac{1}{[\beta\xi + q_2^{-3/2}(0)]^{2/3}},$$

(4.45)

where $\beta = 3\nu^2 b_0^2 K^3/4$, and

$$\sqrt{q_1(\xi)} = K/2 - \sqrt{q_2(\xi)}/2.$$

(4.46)

These expressions give the parameters for each interacting soliton as functions of propagation distance. Solutions (4.45) and (4.46) show that, during propagation, the amplitude of the smaller soliton decreases as ξ increases, while the amplitude of the larger soliton increases. Only the smaller soliton loses energy as a result of the internal friction. The larger soliton actually increases its energy. The energy lost from the smaller soliton is shared equally between the radiation and the energy increase in the larger soliton.

This result is similar to that obtained for breather interactions in a non-linear lattice. In multiple random collisions, the energy exchange tends to favour the growth of the larger excitation. In the model used by Dauxois and Peyrard (1993), it was a discreteness-induced phenomenon. In our case, the reason is the explicit deviation from integrability. We can reformulate the philosophical conclusion of Dauxois and Peyrard (1993): the world of solitons in non-integrable systems is as merciless for the weak as the real world – the larger solitons grow at the expense of the smaller.

The same relation between the loss and energy exchange takes place when the two solitons collide. However, the angle of the collision must be small enough, so that at least a few beats between the solitons take place when they are interacting. If this condition is fulfilled, the amplitude of the smaller soliton decreases, while that of the larger one increases. Radiation is also emitted from the impact area. At large angles of collision, where the interaction length is shorter than the beating period, the solitons are almost unchanged after the collision.

The difference between the cases $\nu > 0$ and $\nu < 0$ is in the interaction forces which act between the solitons. If $\nu > 0$ two solitons attract each other. If $\nu < 0$ the solitons repel each other and move apart. In this case both solitons have non-zero energy after the interaction between them has ceased. It is known that two solitons can merge after collision (Snyder and Sheppard, 1993). The process of the merging of two solitons into one soliton can be explained in our model as the survival of the soliton with the higher amplitude and the decay of the smaller one. This process occurs when $\nu > 0$. If ν is negative, the two solitons repel each other and move apart while the smaller soliton still has an energy comparable with the larger one.

In fact, the force between two equal amplitude Kerr law solitons is proportional to $\cos \theta$, where θ is their phase difference. Thus, in-phase solitons attract, while out-of-phase ($\theta = \pi$) solitons repel. If $\theta = \pi/2$ (quadrature case), then initially there is no force between them.

CHAPTER 5

Normal dispersion regime

Since the pioneering work of Zakharov and Shabat (1973) and Hasegawa and Tappert (1973b), dark optical solitons have been an active topic of research both theoretically (Blow and Doran, 1985; Gradeskul et al., 1990; Kivshar, 1993) and experimentally (Emplit et al.,1987; Krökel et al., 1988; Weiner et al., 1988; Tomlinson et al., 1989; and Zhao and Bourkoff, 1989). Dark solitons have been recently used by Nakazawa and Suzuki (1995) to carry information in optical transmission lines over a long distance. All of these works deal with solitary pulses of the NLSE in the normal dispersion regime. Periodic solutions have attracted less attention, even though they could be also important.

For the NLSE in the anomalous dispersion regime, a full classification of the solutions of first order, using a simple ansatz, has been presented in Chapter 3. This ansatz has also been used by Mihalache and Panoiu (1992a]) and Gagnon (1993) to analyse solutions in the normal dispersion regime. This method allows us to classify the exact solutions of the NLSE, including periodic solutions. It depends upon the order of the polynomial relating the real and imaginary parts of the solution. If the real and imaginary parts of the solutions are related through a polynomial of first order, then they can be called solutions of first order. Using this relation a three-parameter family of solutions can be obtained.

Below, following Akhmediev and Ankiewicz (1993c), we present a full classification of finite first-order exact solutions, including periodic solutions, of the NLSE with normal dispersion:

$$i\psi_\xi - \frac{1}{2}\psi_{\tau\tau} + |\psi|^2\psi = 0. \tag{5.1}$$

The derivation of this equation was discussed in Chapter 1. When applied to propagation in optical fibers, ξ is the (normalized) distance along the fiber and τ is the retarded time (meaning that the reference frame is moving at the pulse group velocity). On the other hand, for spatial propagation in planar structures, ξ is the (normalized) distance along the waveguide, while τ is the transverse dimension.

5.1 General form of the solution

Again, we use the ansatz (3.2) directly with the function $z(\xi)$:

$$\psi(\tau,\xi) = [Q(\tau,\xi) + i\sqrt{z(\xi)}]\ \exp[i\phi(\xi)]. \tag{5.2}$$

We omit some lengthy calculations which are analogous to those in Chapter 3. We find that $z(\xi)$ is the solution of the equation:

$$z_\xi^2 = -16z(z - \alpha_1)(z - \alpha_2)(z - \alpha_3), \qquad \alpha_1 < \alpha_2 < \alpha_3,\ \alpha_3 > 0, \tag{5.3}$$

where α_1, α_2 and α_3 are the three parameters which we use. We consider them to be real and positive. The solutions of (5.3) are always bounded. The function Q is now the solution of the equation:

$$Q_\tau^2 = (Q - Q_1)(Q - Q_2)(Q - Q_3)(Q - Q_4), \tag{5.4}$$

where the roots of the polynomial on the right-hand side are given by (3.24).

The difference from (3.25) is that the polynomial on the right-hand side of (5.4) has the opposite sign. Hence, the potential well (Fig. 5.1b) specified by the polynomial $V(Q) = -\frac{1}{2}(Q-Q_1)(Q-Q_2)(Q-Q_3)(Q-Q_4)$ is inverted relative to the potential in Fig. 3.1. This potential decreases without limit when $Q > Q_1$ and $Q < Q_4$, and its only minimum occurs between the roots Q_3 and Q_2. Hence, all finite solutions can be written as

$$Q = \frac{Q_2(Q_1 - Q_3) - Q_1(Q_2 - Q_3)\ \mathrm{sn}^2(p\tau, k)}{(Q_1 - Q_3) - (Q_2 - Q_3)\ \mathrm{sn}^2(p\tau, k)}, \tag{5.5}$$

where $Q_3 \leq Q \leq Q_2$, $p = \sqrt{\alpha_3 - \alpha_1}$, and $k^2 = (\alpha_3 - \alpha_2)/(\alpha_3 - \alpha_1)$, assuming that the roots Q_3 and Q_2 are real. Solution (5.5) represents all cases of interest (finite solutions). However, it can be written in a different form by shifting the argument or by using a double argument transform. The roots Q_i are real only if all three α_i are positive and $0 < z < \alpha_1$. In this case, at least Q_1 is positive. The only difference between this case and that for anomalous dispersion in Chapter 3 is the sign on the right-hand side of the expresssion for Q_τ^2. Thus, the solutions of this equation are located in intervals which are complementary to those of Chapter 3.

Now, as can be seen from Fig. 5.1, only the solution located between the roots Q_3 and Q_2 is finite. In contrast to the case of anomalous dispersion, (5.4) can have not only finite solutions, but also additional singular solutions located outside the roots Q_1 and Q_4. We shall consider only solutions which are finite everywhere. Singular ones probably have physical interest only in special cases such as layered media, where they can be matched with other types of solutions on boundaries in such a way as to avoid the singularities. When the parameters α_1 and α_2 are complex, the potential $V(Q)$ has only two real roots, and the solutions of (5.4) can only be singular. Taking this into account, we restrict ourselves to the cases where all three parameters α_i are real and the roots Q_i are also real.

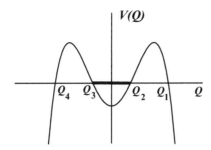

Figure 5.1 *Potential $V(Q)$. The bold line shows the regions where the functions $Q(\tau)$ can provide solutions $\psi(\tau, \xi)$ which are finite everywhere.*

This occurs if the solution of (5.3) for $z(\xi)$ is located between the zero root and the root α_1. We obtain:

$$z(\xi) = \frac{\alpha_1 \alpha_3 \, \text{sn}^2(\mu\xi, k)}{\alpha_3 - \alpha_1 \, \text{cn}^2(\mu\xi, k)}, \tag{5.6}$$

where $\mu = 2\sqrt{\alpha_2(\alpha_3 - \alpha_1)}$, and $k^2 = \alpha_1(\alpha_3 - \alpha_2)/[\alpha_2(\alpha_3 - \alpha_1)]$. The function $\phi(\xi)$ is then given by

$$\phi(\xi) = (\alpha_1 + \alpha_2 - \alpha_3)\xi + \frac{2\alpha_3}{\mu} \Pi(n; \mu\xi | k), \tag{5.7}$$

where Π is defined by (3.29).

Formulae (5.2), (5.5), (5.6) and (5.7) provide the general finite first-order solutions of (5.1). They form a three-parameter family of solutions. The squares of the absolute values of the solutions are periodic, with single periods along the τ and ξ axes. These two periods can be changed independently by adjusting the parameters, so we obtain a general periodic solution with the two periods as independent parameters. The third parameter allows us to adjust the amplitude. The period in τ can be extracted from (5.5):

$$T_\tau = \frac{2}{\sqrt{\alpha_3 - \alpha_1}} K\left[k = \sqrt{\frac{\alpha_3 - \alpha_2}{\alpha_3 - \alpha_1}}\right], \tag{5.8}$$

where K is the complete elliptic function of the first kind, while the period in ξ is found from (5.6):

$$T_\xi = \frac{1}{4\sqrt{\alpha_2(\alpha_3 - \alpha_1)}} K\left[\sqrt{\frac{\alpha_1(\alpha_3 - \alpha_2)}{\alpha_2(\alpha_3 - \alpha_1)}}\right], \tag{5.9}$$

Both periods become infinite on the line $\alpha_1 = \alpha_2$. This will lead to soliton solutions, as we will see later. However, they can be treated as limiting cases of periodic solutions. All three parameters can be scaled by the same constant, for example q. Thus, if we set $a_1 = \alpha_1/q^2$, $a_2 = \alpha_2/q^2$, and

$a_3 = \alpha_3/q^2$ then the whole solution will be scaled in such a way that $\psi' = q\,\psi(\tau',\xi')$, $\xi' = q^2\xi$, and $\tau' = q\tau$. It agrees with (2.3), the general NLSE scaling transformation for both anomalous and normal regimes. This transformation allows us to restrict ourselves to a two-dimensional space of parameters when analysing particular cases, because the scaling does not affect the physical nature of the solution. The general solution can be simplified and written in terms of elementary functions in some special cases.

If we consider the case where α_1 and α_2 are complex conjugates, as in section 3.4, then solutions (3.30) and (3.33) would still apply. However, the equation in Q only admits solutions with $Q \leq Q_1$ and $Q \geq Q_2$ in the normal dispersion regime, so that $Q(\tau)$ would become infinite at large τ. Thus, solutions which are finite everywhere, and are analogous to those of section 3.4, do not exist for normal dispersion.

5.2 Particular cases

The solution of the NLSE in the form given by (5.2), (5.5), (5.6) and (5.7) is general, and describes a variety of different physical phenomena. Thus, we need to classify these solutions and simplify them, giving forms which are convenient to use. To do this, we classify the solutions with respect to parameters a_1, a_2 and a_3. We shall reduce the three-parameter family of solutions to several one-parameter-family solutions. Each of these families describes particular physical phenomena. Let us consider the plane of parameters (a_1, a_2) (Fig. 5.2). Because of the symmetry of these parameters in (5.3), we can see that if they are interchanged, then the solution will retain its form. For reasons of convenience, we have arranged them in such a way that $a_1 < a_2 < a_3$. This means that we need only consider the solutions inside the triangle OAB in Fig. 5.2. All other solutions of the three-parameter family can be written down just by using permutations of these three parameters. We have listed all cases which can be simplified in this figure. At the points of intersection, the solutions of different one-parameter families clearly must coincide. These intersections are designated in Fig. 5.2 by circles. We shall now present the solutions.

5.3 Periodic solutions

Let $a_1 + a_2 = a_3$. Then, (5.6) can be written as:

$$z(\xi) = \frac{a_1 a_3\,\mathrm{sn}^2(\mu\xi, k)}{a_2 + a_1\,\mathrm{sn}^2(\mu\xi, k)}, \qquad (5.10)$$

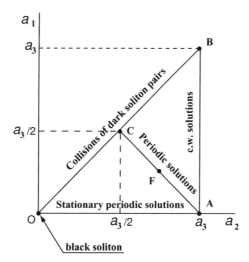

Figure 5.2 *Illustration of the location of solutions of various types in the plane of parameter space* (a_1, a_2), *with* a_3 *arbitrary. The c.w. solutions (line AB) are given by (5.34), while the stationary periodic solutions are given by (5.24). Equation (5.18) describes the periodic solutions (AC), and (5.45) is for the collision of two solitons (OB). Point C (5.52) is a particular case of the latter. The black soliton at the origin is given by (5.56).*

where $\mu = 2a_2$, and $k = a_1 / a_2$. The Landen (Gauss) transform provides the following transformation:

$$\text{sn}(\mu\xi, k) = \frac{2a_2 \ \text{sn}(a_3\xi, \kappa)\text{cn}(a_3\xi, \kappa)}{a_3 \ \text{dn}(a_3\xi, \kappa)}, \tag{5.11}$$

where $\kappa = 2\frac{\sqrt{a_1 a_2}}{a_3}$. Hence:

$$z(\xi) = a_3 \ \frac{\kappa^2 \ \text{sn}^2(a_3\xi, \kappa) \ \text{cn}^2(a_3\xi, \kappa)}{1 - \kappa^2 \ \text{sn}^4(a_3\xi, \kappa)}. \tag{5.12}$$

This can be converted (Gradshteyn and Ryzhik, 1962) to the double argument form:

$$z(\xi) = \frac{a_3\kappa^2}{2} \ \frac{\text{sn}(a_3\xi, \kappa) \ \text{cn}(a_3 \ \xi, \kappa) \ \text{sn}(2a_3\xi, \kappa)}{\text{dn}(a_3\xi, \kappa)}. \tag{5.13}$$

The equation for the phase derivative becomes

$$\phi_\xi = 2a_3 - 2a_3 \ \frac{\kappa^2 \ \text{sn}^2(a_3\xi, \kappa) \ \text{cn}^2(a_3 \ \xi, \kappa)}{1 - \kappa^2\text{sn}^4(a_3\xi, \kappa)}$$

$$= \frac{2a_3 \ \text{dn}^2(a_3\xi, \kappa)}{1 - \kappa^2\text{sn}^4(a_3\xi, \kappa)}$$

$$= a_3 \left[1 + \mathrm{dn}(2a_3\xi, \kappa)\right] \tag{5.14}$$

and the phase itself is now:

$$\phi = a_3\,\xi + \frac{1}{2}\arcsin[\mathrm{sn}(2a_3\xi, \kappa)] \tag{5.15}$$

If we let $\theta = \arcsin[\mathrm{sn}(2a_3\xi, \kappa)]$, then we can write

$$\sqrt{a_3 - z} = \frac{\sqrt{a_3}\,\mathrm{dn}(a_3\xi, \kappa)}{\sqrt{1 - \kappa^2\mathrm{sn}^4(a_3\xi, \kappa)}} = \sqrt{\frac{a_3}{2}\left[1 + \mathrm{dn}(2a_3\xi, \kappa)\right]} \tag{5.16}$$

and similar expressions for the other terms. We note that $z(\xi)$ of (5.13) can also be simplified:

$$z(\xi) = \frac{1}{2}a_3\kappa^2\left[1 + \mathrm{cn}(2a_3\xi, \kappa)\right]\mathrm{sn}^2(a_3\xi, \kappa). \tag{5.17}$$

Thus the periodic solution of the NLSE reduces to the following simple form:

$$\psi(\tau, \xi) = \kappa\sqrt{a_3}\,\frac{\mathrm{cn}(a_3\,\xi, \kappa) - i\sqrt{1+\kappa}\,D(\tau)\,\mathrm{sn}(a_3\,\xi, \kappa)}{\sqrt{1+\kappa}\,D(\tau) + \mathrm{dn}(a_3\,\xi, \kappa)}\,\exp(i\,a_3\,\xi) \tag{5.18}$$

where $D(\tau) = \mathrm{dn}\left(\sqrt{a_3(1+\kappa)}\,\tau, \sqrt{\frac{2\kappa}{1+\kappa}}\right)$. This is a two-parameter family of solutions, but, by using the scaling transformation $\psi' = \psi/\sqrt{a_3}$, $\xi' = a_3\,\xi$ and $\tau' = \sqrt{a_3}\,\tau$, it can be reduced to a one-parameter family of solutions located along the line AC in Fig. 5.2. Then equation (5.18) can be written down with κ being the only parameter. Thus:

$$|\psi'(\tau, \xi)|^2 \tag{5.19}$$

$$= \kappa^2\,\frac{\mathrm{cn}^2(\xi', \kappa) + (1+\kappa)\mathrm{sn}^2(\xi', \kappa)\,\mathrm{dn}^2\left(\sqrt{1+\kappa}\,\tau', \sqrt{\frac{2\kappa}{1+\kappa}}\right)}{\left[\sqrt{1+\kappa}\,\mathrm{dn}\left(\sqrt{1+\kappa}\,\tau', \sqrt{\frac{2\kappa}{1+\kappa}}\right) + \mathrm{dn}(\xi', \kappa)\right]^2}.$$

We note that the minimum possible value of the numerator is $\kappa^2(1-\kappa)$. This is strictly positive when $\kappa \ll 1$, thus showing that, if we exclude the point C itself, then the intensity is never zero along the line AC. The period in ξ' is then $2K(\kappa)$, while the period in τ' is $(2/\sqrt{1+\kappa})\,K\left(\sqrt{\frac{2\kappa}{1+\kappa}}\right)$. Changing the parameter will change the period of the function along the τ and ξ axes. As we move from A to F to C, we note that κ increases from 0 to 1, and thus, correspondingly, the periods in τ and ξ both increase. At point C, these periods become infinite and the periodic nature is lost. In Fig. 5.3 we present the periodic solution at point F of Fig. 5.2. At $\xi = 0$, the field amplitude is found from (5.18):

$$\psi(\tau, \xi = 0) = \frac{\kappa\sqrt{a_3}}{\sqrt{1+\kappa}\,\mathrm{dn}\left(\sqrt{a_3(1+\kappa)}\,\tau, \sqrt{\frac{2\kappa}{1+\kappa}}\right) + 1}. \tag{5.20}$$

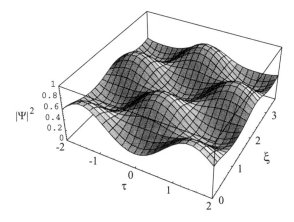

Figure 5.3 *An example of a periodic solution at point F of Fig. 5.2 ($a_1 = 1/2$, $a_2 = 3/2$, $a_3 = 2$).*

If κ is small (i.e near A on the line AF in Fig. 5.2) then we can expand the dn function and observe that we obtain a slight modulation superimposed on a fixed amplitude:

$$\psi(\tau, \xi = 0) \approx \frac{\kappa\sqrt{a_3}}{2}\left(1 - \frac{\kappa}{4}\cos(2\sqrt{a_3(1+\kappa)}\,\tau)\right). \tag{5.21}$$

It is clear that this is a small ripple on a constant value. The field evolves periodically with ξ and the energy swaps between the periods in τ.

5.4 Stationary periodic solution

We now consider the case $a_1 = 0$. In this case we necessarily have $z = 0$. Hence, the Q function does not depend on ξ. The phase function can be written in the form

$$\phi(\xi) = (a_2 + a_3)\xi. \tag{5.22}$$

The Q function can be transformed to

$$Q = \frac{2k\sqrt{a_3}}{1+k}\,\mathrm{sn}\left(\frac{2\sqrt{a_3}}{1+k}\,\tau, k\right), \tag{5.23}$$

where $k = (\sqrt{a_3} - \sqrt{a_2})/(\sqrt{a_3} + \sqrt{a_2})$. Thus the periodic stationary solution is:

$$\psi(\tau, \xi) = \frac{2k\sqrt{a_3}}{1+k}\,\mathrm{sn}\left(\frac{2\sqrt{a_3}}{1+k}\,\tau, k\right)\exp\left[2ia_3\frac{1+k^2}{(1+k)^2}\xi\right]. \tag{5.24}$$

Apart from the phase factor $\exp(2ia_3\frac{1+k^2}{(1+k)^2}\xi)$, this solution does not depend on ξ. Defining $\psi' = \psi/\sqrt{a_3}$ and $\tau' = \sqrt{a_3}\,\tau$ shows that $|\psi'|^2$ depends only on one parameter, namely, k. Thus, the absolute value of the solution is stationary and periodic. This solution corresponds to a cnoidal wave. In an optical fiber, this would represent an unchanging periodic signal propagating along the link. When $k = 0$ (point A in Fig. 5.2), it reduces to the trivial zero solution. Along the line AO in Fig. 5.2, we see that k increases and reaches 1 at the origin. Then we have the black soliton,

$$\psi' = \tanh(\tau') \exp(i\xi'), \tag{5.25}$$

where $\xi' = a_3 \xi$. When k is close to 1, the solution is essentially a train of black solitons, with the distance between them increasing as k approaches 1.

For k approaching 1, this limit of the solution looks like a sequence of black solitons with a constant background. In this case, the period of $|\psi'|^2$, viz. $2K(k)$, increases without limit, while the full width at half-height of each single dip is 1.7627. If we define σ as the ratio of this period to the full width, then for k near 1:

$$\sigma = 0.5673 \ \ln\!\left(\frac{16}{1 - k^2}\right). \tag{5.26}$$

Alternatively, the k required to produce a given ratio is found from

$$k^2 \approx 1.16 \ \exp(-1.7627\sigma). \tag{5.27}$$

As with the case of bright solitons in the anomalous dispersion case (Desem and Chu, 1987), we note that if $\sigma > 5$, then the individual pulses will have very little interaction. In a black soliton pulse train, each pulse necessarily has a phase difference of π from each of its neighbours. This contrasts with a train of bright solitons, where neighbours can have an arbitrary phase difference.

5.4.1 Continuous wave solutions (independent of τ)

We now turn to the case $a_2 = a_3$. The function $z(\xi)$ (5.6) now becomes:

$$z(\xi) = \frac{a_1 a_3 \ \mathrm{sn}^2(\mu\xi, k)}{a_3 - a_1 \ \mathrm{cn}^2(\mu\xi, k)} \ , \qquad \mu = 2\sqrt{a_3(a_3 - a_1)}, \tag{5.28}$$

and the finite solution of (5.5) does not depend on τ:

$$Q = Q_2 = Q_3 = -\sqrt{a_1 - z(\xi)} = \sqrt{\frac{a_1(a_3 - a_1)}{a_3 - a_1 \cos^2(\mu\xi)}} \ \cos(\mu \, \xi) \tag{5.29}$$

The equation for the phase function is:

$$\phi_\xi = a_1 + \frac{2a_3(a_3 - a_1)}{a_3 - a_1 \cos^2(\mu\xi)}. \tag{5.30}$$

Its solution is

$$\phi = a_1 \xi + \arctan\left[\sqrt{\frac{a_3}{a_3 - a_1}} \, \tan(\mu \xi)\right]. \tag{5.31}$$

Noting that

$$\exp(i \arctan \xi) = \frac{1 + i\xi}{\sqrt{1 + \xi^2}} \tag{5.32}$$

allows us to use the following representation:

$$e^{i\phi} = \frac{p \cos(\mu \xi) + i\sqrt{a_3} \, \sin(\mu \xi)}{\sqrt{a_3 - a_1 \cos^2(\mu \xi)}} e^{ia_1 \xi}. \tag{5.33}$$

The solution we obtain is:

$$\psi = \sqrt{a_1} \, \exp[ia_1 \xi]. \tag{5.34}$$

These solutions are c.w. and have amplitude $\sqrt{a_1}$, where a_1 is the parameter for this family. They are located along the vertical line AB in Fig. 5.2. This solution is obvious and does not need illustration.

5.5 Collision of two dark solitons

An important special case, showing the approach, collision and continued propagation of two dark solitons, can be investigated by considering $a_1 = a_2$. Then we have

$$Q_\tau^2 = (Q - Q_1)(Q - Q_2)(Q - Q_3)^2, \tag{5.35}$$

where

$$Q_1 = 2\sqrt{a_1 - z} + \sqrt{a_3 - z},$$

$$Q_2 = -2\sqrt{a_1 - z} + \sqrt{a_3 - z}, \tag{5.36}$$

$$Q_3 = Q_4 = -\sqrt{a_3 - z}.$$

Now $z(\xi)$ reduces to:

$$z(\xi) = \Delta a_1 a_3 \sinh^2(\mu \xi), \tag{5.37}$$

where $\Delta = 1/[a_3 \cosh^2(\mu \xi) - a_1]$ and $\mu = 2\sqrt{a_1(a_3 - a_1)}$. The solution of (5.35) in this case is

$$Q = \frac{Q_2(Q_1 - Q_3) - Q_1(Q_2 - Q_3) \tanh^2(p\tau)}{(Q_1 - Q_3) - (Q_2 - Q_3) \tanh^2(p\tau)}, \tag{5.38}$$

where $Q_3 \leq Q \leq Q_2$ and $p = \sqrt{a_3 - a_1}$. We can simplify this by noting that

$$Q_{1,2} = \pm 2p\sqrt{a_1 \Delta} - Q_3, \tag{5.39}$$

where $+$ is for subscript 1 and $-$ is for subscript 2, and

$$Q_3 = -p\sqrt{a_3 \Delta} \cosh(\mu \xi). \tag{5.40}$$

This allows us to rewrite Q:

$$Q = p\sqrt{\Delta} \; \frac{a_3 \cosh^2(\mu\,\xi) - 2a_1 - \sqrt{a_1 a_3}\,\cosh(\mu\,\xi)\,\cosh(2p\tau)}{\sqrt{a_3}\,\cosh(\mu\,\xi) + \sqrt{a_1}\,\cosh(2p\tau)}. \qquad (5.41)$$

The equation for the phase function is:

$$\phi_\xi = 2a_1 + a_3 - 2\Delta a_1 a_3 \sinh^2(\mu\,\xi)$$

$$= \frac{\Delta}{2}\mu^2 + a_3. \qquad (5.42)$$

Its solution is:

$$\phi(\xi) = a_3\,\xi + \arctan\left[\frac{\sqrt{a_1}}{p}\,\tanh(\mu\,\xi)\right]. \qquad (5.43)$$

Using the identity in (5.33) allows us to obtain the following representation:

$$e^{i\phi} = \sqrt{\Delta}\left[p\cosh(\mu\,\xi) + i\sqrt{a_1}\sinh(\mu\,\xi)\right]e^{ia_3\,\xi}. \qquad (5.44)$$

The field (5.2) is now found by multiplying the above phase factor (5.44) by the complex function obtained from (5.37) and (5.41) above. Thus we again obtain a simple form:

$$\psi(\tau,\xi) \qquad\qquad\qquad\qquad\qquad\qquad\qquad\qquad\qquad\qquad\qquad (5.45)$$

$$= \frac{(a_3 - 2a_1)\cosh(\mu\,\xi) - \sqrt{a_1 a_3}\cosh(2p\tau) + i\mu\sinh(\mu\,\xi)}{\sqrt{a_3}\cosh(\mu\,\xi) + \sqrt{a_1}\cosh(2p\tau)}e^{ia_3\,\xi},$$

where $a_3 > a_1$ and μ and p are given above. The modulus of this solution is shown in Fig. 5.4. This solution has also been obtained by Gagnon (1993).

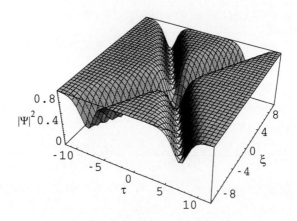

Figure 5.4 *Collision of two dark solitons ($a_3 = 1$, $a_1 = a_2 = 0.5$; point C in Fig. 5.2).*

The singular solution can be obtained from this one simply by changing the sign in front of the $\cosh(2p\tau)$ terms in the numerator and denominator of (5.45). The asymptotic background level of this function, at plus or minus infinity, is $-\sqrt{a_3}$. The solution is symmetric relative to the plane $\tau = 0$, while the mirror image in the plane $\xi = 0$ produces the complex conjugate. A two-dark-soliton solution, in a different form, was obtained by Blow and Doran (1985). After some transformations, it can be simplified to the form of (5.45).

Let us consider limiting cases. Suppose τ and ξ are large and positive. In this case we can approximate the hyperbolic functions by exponential ones:

$$\psi(\tau, \xi) \tag{5.46}$$

$$= \frac{(a_3 - 2a_1) - \sqrt{a_1 a_3}\, \exp[2p(\tau - \sqrt{a_1}\,\xi)] + 2i\sqrt{a_1(a_3 - a_1)}}{\sqrt{a_3} + \sqrt{a_1}\, \exp[2p(\tau - \sqrt{a_1}\,\xi)]}\, e^{ia_3\xi}.$$

This obviously represents a single dark temporal soliton moving with velocity $1/\sqrt{a_1}$ to the right, or a spatial soliton making an angle $\theta = \arctan(\sqrt{a_1})$ with the ξ axis. Because of symmetry, the mirror-image soliton is moving to the left:

$$\psi(\tau, \xi) \tag{5.47}$$

$$= \frac{(a_3 - 2a_1) - \sqrt{a_1 a_3}\, \exp[2p(-\tau - \sqrt{a_1}\,\xi)] + 2i\sqrt{a_1(a_3 - a_1)}}{\sqrt{a_3} + \sqrt{a_1}\, \exp[2p(-\tau - \sqrt{a_1}\,\xi)]}\, e^{ia_3\xi}.$$

Due to the initial formulation of this approach, it is convenient to represent the dark soliton on a complex plane of solutions (Fig. 5.5). In this presentation we ignore the exponential factor $\exp(ia_3\xi)$, which produces a fast rotation around the origin. The circle in this figure has radius $\sqrt{a_3}$, meaning that the background intensity is a_3. The backgrounds at infinity and in the region between the solitons can be located at any two points on this circle. In the particular case of Fig. 5.5, these are the points A and B. All the points representing the solution at any fixed time ξ are located on the straight line connecting these two points. The centre of the soliton is located at the point C closest to the origin. The distance OC is equal to $\sqrt{a_1}$. We can define the (τ, ξ) values corresponding to this point by specifying:

$$\exp[2\,p(\pm\tau - \sqrt{a_1}\,\xi)] = \sqrt{\frac{a_3}{a_1}}. \tag{5.48}$$

Differentiation of $|\psi|^2$ also shows that the intensity has a minimum when this condition holds. This minimum intensity is actually a_1. Equation (5.48) means that solitons propagate at infinity in such a way that:

$$\tau = \pm\left(\sqrt{a_1}\,\xi + \frac{1}{4p}\ln\frac{a_3}{a_1}\right) \tag{5.49}$$

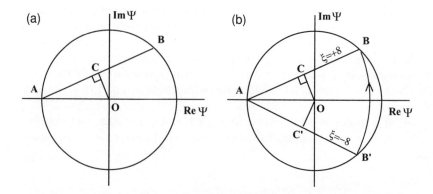

Figure 5.5 *Geometric representation of (a) dark solitons and (b) collision of dark solitons as a rotation of the line AB in the complex plane. The distance OC is equal to $\sqrt{a_1}$, while the radius OA has length $\sqrt{a_3}$. The real and imaginary parts of $\Psi = \psi \, \exp(-ia_3\xi)$ are plotted for fixed ξ, where ψ is given by (5.45).*

Hence, the temporal soliton velocities are $\pm 1/\sqrt{a_1}$ (and the corresponding angles with the ξ axis in the spatial interpretation are $\theta = \pm\arctan(\sqrt{a_1})$), and each is shifted, after the collision, by the value

$$\Delta\tau = \frac{1}{2p} \ln\frac{a_3}{a_1} \qquad (5.50)$$

relative to where it would have been if no collision had occurred. Thus, with higher values of a_1, the τ values increase more rapidly, and so the plots then require a wider range for τ. Clearly the shift $\Delta\tau$ increases as a_1 decreases. Thus, spatial solitons which are nearly parallel (θ small) suffer a large offset $\Delta\tau$, while if the approaching dark solitons have a large angle between them, then $\Delta\tau$ will be small.

From (5.45) the field value at $\xi = 0$ never changes sign if $a_3 - 2\,a_1 < \sqrt{a_1 a_3}$, i.e. if $a_1/a_3 > 0.25$. On the other hand, when $a_1/a_3 < 0.25$, the field has symmetrically placed zeros and thus a maximum in intensity at $\tau = 0$. As a_1/a_3 decreases, the central maximum increases, and approaches the background level, a_3, as $a_1 \to 0$. Physically, we can regard this as the repulsion of near–parallel dark solitons (at low a_1), rather than an actual collision (as occurs at high a_1). This transition happens gradually as a_1 is changed, and can be seen as an analogy with equally charged particles or balls coming near to each other. Particles approaching each other with a large angle between them continue almost in straight lines as they are only close enough to interact significantly for a small distance (ξ), while near-parallel ones deviate from straight lines to curved paths to repel each other, as they are close enough to interact for a long distance.

The soliton phases change in the cross-sections from point A, at minus infinity, to point B, in the region between the solitons, and back again to point A at plus infinity. Thus, the background phase outside the solitons remains constant. During the collision of the solitons, the straight line moves on the plane in such a way that point A stays on the circle but point B moves to the other side of the circle so that the solitons exchange their phase shifts. The length of the straight line is equal to

$$\Delta l = 2\sqrt{a_3 - a_1}\sqrt{\frac{\sqrt{a_3}\cosh(\mu\xi) - \sqrt{a_1}}{\sqrt{a_3}\cosh(\mu\xi) + \sqrt{a_1}}}. \tag{5.51}$$

For each soliton, $|\tan\theta|$ cannot exceed $\sqrt{a_3}$. In the limit $a_1 = a_3$, the solitons disappear. These asymptotic results can be obtained using a standard inverse scattering approach. In our case, we have the full exact solution which describes the collision in more detail at the point of impact.

This solution simplifies in the case of $a_1 = a_3/2$:

$$\psi(\tau,\xi) = \frac{q}{\sqrt{2}}\frac{-\cosh(q\tau) + i\sqrt{2}\sinh(q^2\xi/2)}{\sqrt{2}\cosh(q^2\xi/2) + \cosh(q\tau)}\exp(iq^2\xi/2), \tag{5.52}$$

where $q = \sqrt{2a_3}$. This also agrees with minus 1 times the limit $a_1 = a_2 = a_3/2$ of (5.18). Point B in Fig. 5.5 is 90° out of phase with point A in this case.

5.6 Excitation of pairs of dark solitons using symmetric initial conditions

The form of (5.45) at $\xi = 0$ is:

$$\psi(\tau,0) = \frac{(a_3 - 2a_1) - \sqrt{a_1 a_3}\cosh(2p\tau)}{\sqrt{a_3} + \sqrt{a_1}\cosh(2p\tau)}. \tag{5.53}$$

Equation (5.53) represents a symmetric function which gives an excitation of pairs of dark solitons in the Cauchy problem without any additional radiation. Any other initial condition produces some radiation along with the solitons. Hence, the initial condition (5.53) can be considered as optimal for exciting a pair of dark solitons. The critical value of a_1 is $a_1 = a_3/4$. Then

$$\psi(\tau,\xi = 0) = \sqrt{a_3}\frac{1 - \cosh(\sqrt{3a_3}\,\tau)}{2 + \cosh(\sqrt{3a_3}\,\tau)}$$

$$= -2\sqrt{a_3}\frac{\sinh^2(\frac{\sqrt{3a_3}}{2}\tau)}{1 + 2\cosh^2(\frac{\sqrt{3a_3}}{2}\tau)}. \tag{5.54}$$

In this case the function is equal to zero at the centre, $\tau = 0$, and this is the only minimum. When a_1 is below this critical value, the initial condition

(5.53) has two zeros. When a_1 is higher than $a_3/4$, the function is always positive, but still has a single minimum at $\tau = 0$.

5.7 Black soliton

In the $a_1 \to 0$ limit of the solution given by (5.45) (i.e. approaching the origin of Fig. 5.2 along the line CO), we have two black solitons, with the distance between them approaching infinity. Hence, the solution in the form of one black soliton can be obtained only if we ensure that one of these solitons is at $\tau = 0$. The same applies to (5.5). To have a solution for Q corresponding to one black soliton, solution (5.5) must be shifted along the τ axis by a quarter of the period of the elliptic functions. Then, after setting the limit $a_1 = a_2 = 0$, we obtain:

$$Q = \sqrt{a_3}\ \tanh(\sqrt{a_3}\ \tau). \tag{5.55}$$

Obviously the solution of the NLSE corresponding to this $Q(\tau)$ is the black soliton:

$$\psi = \sqrt{a_3}\ \tanh(\sqrt{a_3}\ \tau)\ \exp(ia_3\ \xi). \tag{5.56}$$

This agrees with solution (5.25), which was the limiting case of the stationary periodic solution (i.e. approaching the origin of Fig. 5.2 along the line AO), when the period has become infinite.

5.8 Relation between NLSE solutions in normal and anomalous dispersion regimes

Obviously, if we make the substitution $\tau \to i\tau$ in (5.1), we obtain the NLSE in the anomalous dispersion regime:

$$i\psi_\xi + \frac{1}{2}\psi_{\tau\tau} + |\psi|^2\psi = 0\ . \tag{5.57}$$

We can expect, then, that the solutions of (5.57), with the same transformation, $\tau \to i\tau$, will give the solutions of (5.1). For example, the bright soliton solution of (5.57), namely, $\psi(\tau,\xi) = \sqrt{2q}\,\mathrm{sech}(\sqrt{2q}\tau)\,\exp(iq\xi/2)$, on transformation gives the solution $\psi(\tau,\xi) = \sqrt{2q}\,\sec(\sqrt{2q}\tau)\,\exp(iq\xi/2)$, which is indeed a solution of (5.1). However, it is clearly a singular solution, which, up to now, has not found wide physical application. Considering this simple example, we can conclude that any solution of first order of (5.57) can be transformed into a solution of (5.1). However, to find finite solutions, we have to be careful, and in using the transformation $\tau \to i\tau$, we have to select the proper signs for square roots of the parameters a_i.

For example, this is relevant for the solution, given by (3.39) of Chapter 3, describing the process of modulation instability. Thus, in order to obtain the finite solution given by (5.45), we have to choose opposite signs in front of the terms relative to those chosen in (3.39). If this is not done, the solution obtained will again be singular. In principle, all the solutions of (5.57) can be obtained from the solutions of (5.1) using the simple transformation above, but care must be taken with signs in order to do this correctly. Also, we should remember that there are no finite solutions corresponding to the solutions of (5.57) for some values of the parameters a_i. These cases are easily seen by considering the potential $V(Q)$.

Comparing the solutions of (5.1) and (5.57), we can see that there is no overall physical equivalence of the solutions corresponding to the same values of parameters a_i. For example, the black soliton solution of (5.1) and the bright soliton solution of (5.57) are located at different points on the (a_1, a_2) plane. The solution of (5.1) describing the collision between two dark solitons corresponds to the solution of (5.57) describing the modulation instability of the c.w. solution. Hence, the simple transformation $\tau \rightarrow i\tau$ converts the solutions in such a way that they describe totally different physical processes.

The straight-line relation between real and imaginary parts is useful in obtaining, as well as in analysing, the solutions for the NLSE in the normal dispersion regime. It allows for a clear geometric interpretation of the solutions on the plane of parameters, as well as in the complex plane of solutions. Singular solutions can be classified using the techniques developed here for finite solutions.

5.9 Grey soliton phase

For the Kerr law, there is a complex solution of the NLSE which corresponds to 'grey' solitons:

$$\psi = \sqrt{q}\left[\sqrt{1-A^2} - iA\tanh[\sqrt{q}A(\sqrt{q(1-A^2)}\xi - \tau)]\right]\,\exp(iq\xi). \quad (5.58)$$

The total phase shift ϕ for grey solitons in a Kerr medium varies between 0 and π, as can be seen from Fig. 5.5. When the total phase shift is higher, then the soliton is darker (i.e. the contrast is higher). For a phase shift of π, the soliton is black ($A = 1$) and it propagates perpendicularly to the initial wavefront. The π limitation on the soliton phase shift is not just a peculiarity of the Kerr model. However, there is no restriction on the total soliton phase shift in some models of the nonlinear refractive index change (in particular, saturable nonlinearity) which are more general than the Kerr case.

Let us first consider a couple of simple examples of functions $N(I)$ describing media which are more general than the Kerr case. These examples permit exact solutions of

$$i\psi_\xi - \frac{1}{2}\psi_{\tau\tau} + \psi N(|\psi|^2) = 0, \quad (5.59)$$

enabling a full analysis of the soliton properties. We will be particularly interested in the phase properties of the dark solitons. The first model is the parabolic law model (4.6), in which the function $N(I)$ is of the form

$$N(I) = I + \nu I^2, \tag{5.60}$$

where the coefficient ν can have either positive or negative sign. When ν is positive, the refractive index changes are stronger than for a Kerr medium, so there is a 'superlinear intensity dependence'. When $\nu < 0$, the function $N(I)$ exhibits 'sublinear intensity' dependence. For $\nu = 0$ relation (5.60) describes the Kerr nonlinearity. The total phase shift of the grey soliton, for any ν, is given by

$$\phi = 2\arcsin\left(\frac{A}{A^2 + (1 - A^2)c}\right), \tag{5.61}$$

where the coefficient c is

$$c = \frac{1 + 2\nu I_0(4 - A^2)/3}{1 + 2\nu I_0(1 - A^2)/3}.$$

The second example which we consider is a threshold nonlinearity. In this case the refractive index changes as a step function if the light intensity exceeds some threshold value I_{crt}

$$N(I) = \left\{ \begin{matrix} \delta & \text{for } I \geq I_{crt} \\ 0 & \text{otherwise} \end{matrix} \right\}. \tag{5.62}$$

This model turns out to be very useful for explaining some fundamental properties of dark solitons as 'modes' that create their own waveguides (Snyder and Sheppard, 1993). The total phase shift of the dark soliton in this model is given by

$$\phi = \pi(1 - \sqrt{1 - A^2}). \tag{5.63}$$

In order to illustrate the soliton phase properties, we again use the complex plane describing the real and imaginary parts of the modified field amplitude $\psi(\tau, \xi)e^{-iq\xi}$. Any solution of (5.59) in the form of a soliton can be represented on this plane by some trajectory (a parametric curve at fixed ξ with τ as a parameter). Examples of these trajectories for Kerr, parabolic and step index saturation models are shown in Fig. 5.6.

The background of the grey soliton in each case is a solution of (5.59) in the form of a plane wave

$$\psi = q_0 \exp\left[iN(q_0^2)\xi + i\phi_0\right], \tag{5.64}$$

where ϕ_0 is the background phase. For the Kerr law, this reduces to (5.34) with $q_0 = \sqrt{a_1}$. The phase ϕ_0 is arbitrary but phases on the left and on the right of the dark soliton are related. Hence, the soliton background on the complex plane is located anywhere on a circle with radius q_0. For simplicity

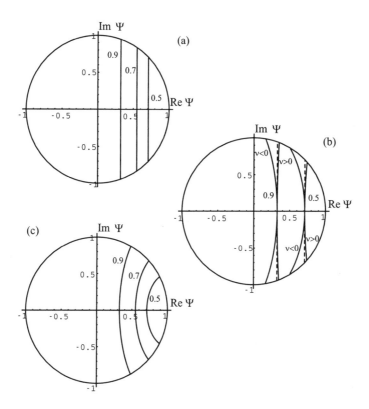

Figure 5.6 *Trajectories in the complex plane for grey solitons resulting from various nonlinear models:*
(a) Kerr-type nonlinearity, intensity contrast $A^2 = 0.5$, *0.7 and 0.9.*
(b) Parabolic nonlinearity model. Different solid curves correspond to ν *being positive or negative, while dashed lines are for* $\nu = 0$. *Intensity contrast* $A^2 = 0.5$ *and 0.9.*
(c) Threshold nonlinearity. Intensity contrast $A^2 = 0.5$, *0.7 and 0.9.*

we set $q_0 = 1$ in all our figures. The variation of the soliton amplitude and phase are displayed on the complex plane by a curve which is located inside the circle and which starts and ends on the circle. These two points correspond to the background amplitude at $\tau \to \infty$ and $\tau \to -\infty$. The total phase shift across the soliton profile is given by the angle subtended at the origin by the initial and final points. For the Kerr nonlinearity depicted in Fig. 5.6a, the soliton is represented by a straight line. The distance of closest approach to the origin corresponds to the square root of the minimum intensity. As the contrast, A^2, increases, the line moves towards the origin and the total phase shift of the soliton increases until it reaches π for a black soliton.

When the term νI^2 is added to the Kerr nonlinearity (see (5.60)) the trajectory in the complex plane is no longer a straight line (Fig. 5.6b). For the same contrast, it has only one point in common with the straight line representing the Kerr case, as the latter is a tangent to the curve. An interesting feature of this curve is that its curvature depends on the sign of the constant ν. Let us define the sign of the curvature of the curve as positive if the curve bends inwards (with respect to origin) from the tangent line and negative if it bends outwards. Thus the curve depicted in Fig. 5.6b has positive curvature when ν is less than zero and negative curvature for positive ν. Consequently, the total phase shift also depends on the sign of ν. The trajectory in general, and its curvature in particular, also depend on the contrast A^2. As the contrast approaches unity (the case of a black soliton) the curvature decreases: the curve gradually becomes a straight line and the total phase shift goes to π. Hence, π is the limiting phase shift for solitons when the nonlinearity is given by (5.60). This can be also seen from (5.61). The plot in Fig. 5.6b shows the relation between the curvature of the trajectory and the sign of ν.

The threshold nonlinearity, (5.62), can be treated similarly. In this case the trajectory in the complex plane always has negative curvature. An example of such a trajectory is shown in Fig. 5.6c. Although the trajectory has negative curvature, as the contrast approaches unity the curvature decreases to zero, so that the total phase shift again reaches π for the black soliton. The distinguishing feature of solitons in the models discussed above is that the total phase shift never exceeds π, the value usually associated with the black soliton.

5.10 'Darker than black' solitons

A question arises: is it possible to have total phase shift larger than π? Obviously, the models of nonlinearity which give negative curvature should always lead to a total phase shift lower than π. On the other hand, in cases with positive curvature there seems to be no apparent reason for a restriction on the total phase shift. In fact, as we show below, more general forms of the function $N(I)$ do lead to a total phase shift larger than π.

Let us consider the following model for a saturable nonlinearity:

$$N(I) = \frac{I_s}{2} \left[1 - \frac{1}{(1 + I/I_s)^2} \right], \qquad (5.65)$$

where I_s is a saturation parameter. For this particular form of the nonlinearity, the corresponding NLSE can be integrated implicitly (Królikowski and Luther-Davies, 1993).

The solution of (5.59), with nonlinear function (5.65), yields the following relation for the total soliton phase shift:

$$\phi = 2 \tan^{-1} \frac{\beta}{\mu} + 2\mu \tanh^{-1} \beta, \qquad (5.66)$$

where

$$\beta = A \sqrt{\frac{I_0}{I_0 + I_s}} \qquad \text{and} \qquad \mu = \sqrt{(1 - A^2)I_0/I_s},$$

while I_0 is the background intensity. The total phase shift clearly depends on the saturation parameter I_s. Some examples of trajectories for dark solitons corresponding to the function (5.65) are shown in Fig. 5.7.

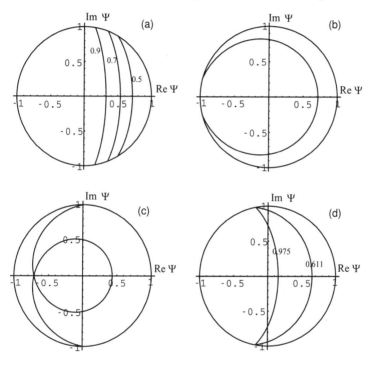

Figure 5.7 *Trajectories in the complex plane for dark solitons in a saturable nonlinear medium:*
(a) $I_0/I_s = 1.0$ *and the contrast* $A^2 = 0.5$, 0.7 *and* 0.9;
(b) $I_0/I_s = 20$, $A^2 = 0.5$;
(c) $I_0/I_s = 50$, $A^2 = 0.8$;
(d) $I_0/I_s = 5$, $A^2 = 0.611$ *and* $A^2 = 0.975$.

When the intensity is small compared with the saturation parameter, so that $I_0/I_s \ll 1$, then $N(I) \approx I$ and the behavior of the trajectory in the complex plane approaches that of Kerr model. For moderately weak saturation ($I_0/I_s < 1$), the function (5.65) can be written as a series

$$N(I) \approx I - \frac{3}{2I_s}I^2 + ... \tag{5.67}$$

which is equivalent to a sublinear intensity dependence with negative ν in (5.60). Hence, the trajectory in the complex plane is practically the same as that given by the parabolic law model. It has positive curvature, as can be seen in Fig. 5.7a, and we note that larger phase shifts in this figure correspond to darker solitons.

If the background intensity increases so that the material enters the strong saturation region ($I_0 \gg I_s$), the total phase shift drastically increases and quite easily exceeds π. As shown in Fig. 5.7b,c, solitons with a total phase shift larger than π, or even 2π, may exist. Note that when the total phase shift becomes larger than π the trajectory bends around the origin rather than crossing it. So we can expect that the energy required for the trajectory to cross the origin would be higher than that required to bend around it. Hence, these solitons do not pass through the central zero of intensity and so cannot be 'black'. On the other hand, if we stay with the standard terminology which associates black solitons with a total phase shift equal to π, then the solitons with larger phase shifts could be considered as 'darker than black'.

The π phase shift is, obviously, a special case. In a Kerr medium, π is simply the limiting value for the total phase shift. The variation of the shift with intensity contrast in a saturable medium for different values of the parameter I_0/I_s is shown in Fig. 5.8. The lowest curve in this figure corresponds to a Kerr nonlinearity. If the parameter I_0/I_s is less than 2.27 the curves are monotonic and the total phase shift is still less than π for any contrast. As soon as the parameter I_0/I_s exceeds the value 2.27 the total phase shift may become more than π in a certain range of contrasts. Moreover, there are two values of the contrast corresponding to the same value of the total phase shift $\phi > \pi$. The maximum value of the total phase shift increases with increasing I_0/I_s and is, in principle, unlimited.

There are two different dark solitons with different contrasts (and transverse velocities), but with exactly the same boundary conditions at infinity. Trajectories in the complex plane corresponding to these two solitons are shown in Fig. 5.7d. Both types of soliton are stable. The contrast of the Kerr soliton is the same as for the soliton with the higher contrast in Fig. 5.7d. The widths of the solitons in the saturable medium are much larger than in the Kerr medium.

Another special case (from the physical point of view) is $\phi = 2\pi$. In this case the background phases on both sides of the soliton can be considered as equal. Hence, in this particular case solitons can exist in the form of a perturbation of a plane wave. This means that, using initial conditions with one intensity dip on the plane wave, it is possible to excite a single

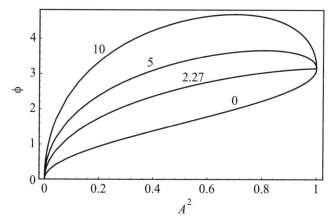

Figure 5.8 *Total phase shift ϕ versus intensity contrast A^2. The numbers above the curves quantify the parameter I_0/I_s.*

soliton rather than a pair as is the case for the Kerr medium. Special care would have to be taken to create the proper phase chirp to excite these solitons.

Grey solitons comprise a one-parameter family of solutions which depend on velocity. Hence, velocity is the parameter on which stability may depend. One of the ways to define stability is in terms of the complementary momentum and the velocity (Kivshar and Królikowski, 1995).

CHAPTER 6

Multiple-port linear devices made from solitons

6.1 General principles

The concept of 'light-guiding-light' phenomena (Snyder *et al.*, 1991a) brings the promise of new devices for all-optical information processing. Of particular interest in this area is the spatial soliton. This can be created by a strong beam which can cause a material to act as a linear waveguide with a sech-squared-type refractive index profile which can then guide a weak probe beam (De La Fuente *et al.*, 1991). This can be done both for bright (Shalaby and Barthelemy, 1991) and dark (Luther-Davies and Yang, 1992a) spatial solitons.

Furthermore, colliding solitons can serve as X-junctions (Luther-Davies, 1993; Akhmediev and Ankiewicz, 1993a), and more generally as $N \times N$ switching devices (Miller and Akhmediev, 1996a). Various types of these devices have been analysed to date (Akhmediev and Ankiewicz, 1993a; Torres-Cisneros *et al.*, 1993). The use of a probe beam with the same frequency as the pump is preferable, because these devices then have no loss in the impact area of the two solitons (Akhmediev and Ankiewicz, 1993a). Moreover, the probe beam which is sent into one of the channels does not have any reflected component in this case. It passes through the impact area of the soliton collision and is totally distributed among the output channels. This, ideally, should be one of the most important properties of any $N \times N$ switch. However, the use of a probe beam at the same frequency as the pump creates the problem of separating it from the pump. Hence, from an experimental point of view, it is more convenient to use a weak probe beam at different frequency. Then, the weak signal beam can be separated from the pump simply by using spectral filtering. However, components like the X-junction and $N \times N$ switching devices then lose their elegance because of losses in the impact area of soliton collision, and any input beam can potentially be partly reflected back into any of the input channels.

A different approach to this technique is to design linear devices using the results of soliton theory. To be specific, we suppose that an X-junction or $N \times N$ switch is initially created in a photosensitive medium by some photolithographic (e.g. ultraviolet light) process. We suppose that this switch has the same refractive index as would be induced by an (imagined) collision of N optical solitons in a Kerr medium. Such a structure can equally well be created by computerized equipment using exact analytic N-soliton solutions. Another advantage of such a device is that it can have the smallest possible size, as it is comparable with the optical wavelength, rather than with the beat length of two modes. Now, we use this $2N$-port device as an $N \times N$ switch, and consider its linear transmission properties for signals at the same frequency as that used to specify the refractive index of the structure. The transmission coefficients can be found from the transfer matrix, which in turn can be determined using the total integrability of the problem (Miller and Akhmediev, 1996a). This transfer matrix can be applied to different problems involving waveguide structures in the form of soliton collisions.

6.2 Composite waveguides made from solitons

The complex envelope, $\psi(\xi, \tau)$, needed to describe the spatial modulation of the refractive index, satisfies the cubic NLSE

$$i\psi_\xi + \frac{1}{2}\psi_{\tau\tau} + |\psi|^2\psi = 0. \tag{6.1}$$

The variable ξ is now the longitudinal coordinate and τ is the transverse coordinate. A solution, $\psi(\xi, \tau)$, of (6.1) induces an effective refractive index change Δn, proportional to $|\psi(\xi, \tau)|^2$, in the material.

Having an induced refractive index profile $\Delta n \approx |\psi(\xi, \tau)|^2$ in an isotropic medium, we consider the propagation of weak, monochromatic spatial modulations of plane light beams in the ξ direction. As we noted above, devices made in this way will have the smallest possible size. Note also that, due to the scaling transformation (2.3), the size of the device and the related maximal change of the refractive index can both vary.

Now, the weak field envelope $\phi(\xi, \tau)$ solves the *linear* Schrödinger equation

$$i\phi_\xi + \frac{1}{2}\phi_{\tau\tau} + V(\xi, \tau)\phi = 0, \tag{6.2}$$

where the potential function V is determined from the solution of (6.1):

$$V(\xi, \tau) = |\psi(\xi, \tau)|^2. \tag{6.3}$$

Of course, trivially, $\phi = \psi$ is a solution of (6.2), but, in general, waveguides made using N-soliton solutions have N linear 'modes'. Finding the remaining $N-1$ solutions of the linear problem is an involved task. To form an idea of the form of the waveguide we are considering, we may refer to Fig. 3.13. There is, however, an additional parameter which is involved in this solution, namely the phase shift between the solitons. This modifies the impact area of collision and adds an additional parameter to the refractive index profile. Hence, even in the case of the collision of two solitons of the same amplitude, there is a one-parameter family of Δn profiles (Fig. 6.1). The scaling (2.3) adds one more free parameter to this problem.

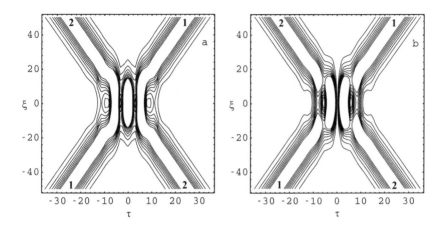

Figure 6.1 *Contour plots of the refractive index change, Δn, induced by the collision of two bright solitons (3.78) with $a = 0.25$, $b = 0.1$. Phase difference $2\Delta\varphi$ is (a) 0, (b) π. The bold numbers on each soliton give the correspondence between numbering for each input and output channel used in the text.*

The choice of N linear functions of these composite waveguides is somewhat arbitrary, because any linear superposition of 'modes' is a solution. To solve our problem, we have to choose them in a special way. Now, far from the impact area, each of the input waveguides has its single linear mode as a solution. We should choose the linear mode as that where the total power is in one input core and there is no power in the others. Linear functions constructed in this way, for an X-junction, are shown schematically in Fig. 6.2. The question is: how is the power in the input mode distributed after passing through the X-junction or $N \times N$ switching device? This question is not trivial because even for the X-junction many parameters are involved, viz. the soliton amplitudes, the angles of propagation and the relative phase between the two solitons. The problem becomes even more involved for the N-soliton collision.

Now we use result (2.43) in a form which allows us to find the answer to the above question analytically. Suppose that $\psi(\xi, \tau)$ solves the NLSE, (6.1). Let

$$R(\xi, \tau; \lambda) = \begin{pmatrix} r(\xi, \tau; \lambda) \\ s(\xi, \tau; \lambda) \end{pmatrix}$$

be any simultaneous solution of the linear problems (2.30) and (2.31), for

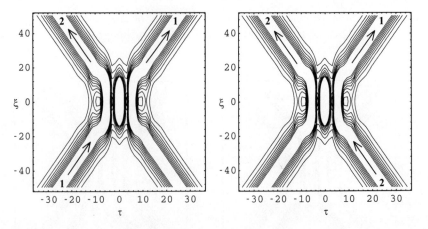

Figure 6.2 *Schematic representation of two linear modes of an X-junction formed by solitons. Arrows show the direction of propagation of linear waves. In each case the input wave is zero in one of the arms.*

any complex λ. Then the function

$$\phi(\xi,\tau|\lambda) = r(\xi,\tau;\lambda)\exp[-i(\lambda\tau + \lambda^2\xi)] \qquad (6.4)$$

is a solution of the linear Schrödinger equation (6.2).

This result connects the solution of the linear problem (6.2) with the solution of the Lax pair, (2.30) and (2.31). In the simplest case, when $\psi = 0$, the solution of the Lax pair equations (ignoring constant phase shifts) is $\exp[-i(\lambda\tau + \lambda^2\xi)]$ and the corresponding solution of the linear problem is $\exp[-2i(\lambda\tau + \lambda^2\xi)]$.

6.3 Refractive index profile of N-soliton solution

First we have to find functions $\psi(\xi,\tau)$ that solve the NLSE, (6.1). We choose the number of solitons, N, and consider vectors of the form

$$R^+(\xi,\tau;\lambda) \;=\; \begin{pmatrix} \displaystyle\sum_{n=0}^{N-1} A_n(\xi,\tau)\lambda^n \\[2ex] \displaystyle\lambda^N + \sum_{n=0}^{N-1} B_n(\xi,\tau)\lambda^n \end{pmatrix} \exp[i(\lambda\tau + \lambda^2\xi)], \qquad (6.5)$$

and

$$R^-(\xi,\tau;\lambda) = \begin{pmatrix} \lambda^N + \sum_{n=0}^{N-1} C_n(\xi,\tau)\lambda^n \\ \sum_{n=0}^{N-1} D_n(\xi,\tau)\lambda^n \end{pmatrix} \exp[-i(\lambda\tau + \lambda^2\xi)].\quad(6.6)$$

We select N complex numbers, λ_j, and N complex numbers, γ_j, $j = 1,\ldots,N$. Then, we determine the $4N$ coefficient functions, A_n, B_n, C_n and D_n, using the linear relations

$$R^+(\xi,\tau;\lambda_j) = \gamma_j R^-(\xi,\tau;\lambda_j),\qquad(6.7)$$

and

$$-\overline{\gamma}_j R^+(\xi,\tau;\overline{\lambda}_j) = R^-(\xi,\tau;\overline{\lambda}_j).\qquad(6.8)$$

If the numbers λ_j are distinct and non-real, then these relations comprise a set of $4N$ algebraic equations in the coefficient functions A_n, B_n, C_n and D_n, in which the variables ξ and τ appear as parameters. Thus the vector functions given in (6.5) and (6.6) are uniquely defined.

Now we define

$$\psi(\xi,\tau) = 2iA_{N-1}(\xi,\tau).\qquad(6.9)$$

Then, $\psi(\xi,\tau)$ solves the NLSE (6.1) and the vector functions $R^+(\xi,\tau;\lambda)$ and $R^-(\xi,\tau;\lambda)$ form a basis for the simultaneous solutions of the Lax pair (2.30) and (2.31) for all complex λ (except at the points λ_j where the basis degenerates). The solutions of (6.1) which are constructed in this way correspond to the interaction of N bright solitons, where the jth soliton, when far from the interaction region, has the form (cf. (3.77))

$$\psi_j(\xi,\tau) = \frac{2b_j \exp[-i(2a_j\tau + 2(a_j^2 - b_j^2)\xi - \theta_j)]}{\cosh(2b_j\tau + 4a_jb_j\xi - \delta_j)},\qquad(6.10)$$

where a_j, b_j, δ_j, and θ_j are all real, and $\lambda_j = a_j + ib_j$ and θ_j and δ_j are related to $\gamma_j = -\exp(i\theta_j - \delta_j)$. In all that follows, we assume, without loss of generality, that $a_j > a_k$ whenever $j > k$, and that $b_j > 0$ for all j.

When some of the numbers λ_j coincide, solutions can still be found using other techniques. An example, for the case of a two-soliton collision, is (3.80). The transmission coefficients in this case can be found with a limiting procedure. Specifically, if $a_1 = a_2$ and $b_1 = b_2$, for the X-junction, 100% of power goes to the ith channel ($i = 1,2$) if initially only the ith input is excited.

6.4 'Modes' of the composite waveguides

We now use the basis of simultaneous solutions to the Lax pair and appeal to (6.4) to provide two families of solutions, each parametrized by an arbitrary complex parameter λ, to the linear Schrödinger equation, (6.2). We define

$$
\begin{aligned}
\phi^+(\xi,\tau;\lambda) &= r^+(\xi,\tau;\lambda)\exp[-i(\lambda\tau+\lambda^2\xi)] \qquad\qquad\qquad (6.11)\\
&= \sum_{n=0}^{N-1} A_n(\xi,\tau)\lambda^n\,,
\end{aligned}
$$

$$
\begin{aligned}
\phi^-(\xi,\tau;\lambda) &= r^-(\xi,\tau;\lambda)\exp[-i(\lambda\tau+\lambda^2\xi)] \qquad\qquad\qquad (6.12)\\
&= \left(\lambda^N + \sum_{n=0}^{N-1} C_n(\xi,\tau)\lambda^n\right)\exp[-2i(\lambda\tau+\lambda^2\xi)]\,.
\end{aligned}
$$

The function $\phi^+(\xi,\tau;\lambda)$ sweeps out an N-dimensional vector space of solutions to (6.2), as the parameter λ varies. These solutions correspond to the discrete linear modes of the waveguides. On the other hand, the function $\phi^-(\xi,\tau;\lambda)$ represents the radiation modes of (6.2). For real values of λ, these solutions are linear waves in parts of the (ξ,τ) plane that are distant from the centre of mass of any of the solitons.

Here, we are only interested in the discrete modes of the composite waveguides, i.e. solutions of the linear Schrödinger equation (6.2) which are bound to the waveguide. If $\psi(\xi,\tau)$ is an N-soliton solution of the NLSE (6.1), then the family of solutions $\phi^+(\xi,\tau;\lambda)$ is N-dimensional, as can be seen from (6.11). To show that $\phi^+(\xi,\tau;\lambda)$ describes the discrete linear modes, let us analyse $\phi^+(\xi,\tau;\lambda)$ on straight lines which pass through the origin in the (ξ,τ) plane. Choose a real slope v and introduce the new variables $\chi = \tau + 2v\xi$ and $\zeta = \xi$. Let us examine the behaviour of ϕ^+ in the limits $\zeta \to \pm\infty$ for fixed χ.

Conditions (6.7) and (6.8) determine ϕ^+ and ϕ^- by giving the coefficients $A_n(\xi,\tau)$ and $C_n(\xi,\tau)$ as solutions of the linear system

$$
\sum_{n=0}^{N-1} A_n(\xi,\tau)\lambda_j^n \qquad\qquad\qquad\qquad\qquad\qquad\qquad (6.13)
$$

$$
= \gamma_j\left(\lambda_j^N + \sum_{n=0}^{N-1} C_n(\xi,\tau)\lambda_j^n\right)\exp[-2i(\lambda_j\tau+\lambda_j^2\xi)]\,,
$$

$$
\sum_{n=0}^{N-1} A_n(\xi,\tau)\overline{\lambda}_j^n \qquad\qquad\qquad\qquad\qquad\qquad\qquad (6.14)
$$

$$
= -\frac{1}{\overline{\gamma}_j}\left(\overline{\lambda}_j^N + \sum_{n=0}^{N-1} C_n(\xi,\tau)\overline{\lambda}_j^n\right)\exp[-2i(\overline{\lambda}_j\tau+\overline{\lambda}_j^2\xi)]\,,
$$

where $j = 1,\dots,N$. Taking $v > a_j$ for all j puts us in the frame of an observer moving far to the right (left) of the solitons for large negative (positive) ζ. Then, in the limit $|\zeta| \to \infty$, one obtains the relations

$$\sum_{n=0}^{N-1} A_n(\chi - 2v\zeta, \zeta)\lambda_j^n = 0, \qquad \zeta \to +\infty, \qquad j = 1, \ldots, N,$$

$$\sum_{n=0}^{N-1} A_n(\chi - 2v\zeta, \zeta)\overline{\lambda}_j^n = 0, \qquad \zeta \to -\infty, \qquad j = 1, \ldots, N.$$

$$(6.15)$$

Likewise, taking $v < a_j$ for all j yields

$$\sum_{n=0}^{N-1} A_n(\chi - 2v\zeta, \zeta)\overline{\lambda}_j^n = 0, \qquad \zeta \to +\infty, \qquad j = 1, \ldots, N$$

$$\sum_{n=0}^{N-1} A_n(\chi - 2v\zeta, \zeta)\lambda_j^n = 0, \qquad \zeta \to -\infty, \qquad j = 1, \ldots, N.$$

$$(6.16)$$

The $N \times N$ matrix having $V_{jk} = \lambda_j^{k-1}$ as its (j, k)th element, viz.

$$\mathbf{V}(\lambda_1, \ldots, \lambda_N) = \begin{bmatrix} 1 & \lambda_1 & \lambda_1^2 & \ldots & \lambda_1^{N-1} \\ 1 & \lambda_2 & \lambda_2^2 & \ldots & \lambda_2^{N-1} \\ & & \ldots & & \\ 1 & \lambda_N & \lambda_N^2 & \ldots & \lambda_N^{N-1} \end{bmatrix} \qquad (6.17)$$

is a *Vandermonde* matrix. It is invertible because the λ_j are distinct (and the same is true for $\mathbf{V}(\overline{\lambda}_1, \ldots, \overline{\lambda}_N)$). Thus the limiting values of A_n are identically zero for all $n = 0, \ldots, N - 1$. Inserting these limiting values into (6.11), we see that $\phi^+(\chi - 2v\zeta, \zeta; \lambda)$ vanishes far to the right and left of all the solitons for all values of the arbitrary parameter λ. As long as $v \neq a_j$ for $j = 1, \ldots, N$, one has $\phi^+ \to 0$ for all λ as $\zeta \to \pm\infty$. Thus, in the asymptotic limit, the solutions (6.11) of (6.2) are confined to the individual soliton waveguides and there are no losses.

6.5 Power in each waveguide

The above solutions can be characterized by the power confined in each guide as $\zeta \to \pm\infty$. Let us calculate this power. For $k = 1, \ldots, N$, let $v = a_k$ describe the frame of reference in which the soliton ψ_k, defined by (6.10), is stationary. Then, equations (6.13) and (6.14) imply that, as $\zeta \to +\infty$,

$$\sum_{n=0}^{N-1} A_n(\chi - 2v\zeta, \zeta)\lambda_j^n = 0, \qquad j < k,$$

$$\sum_{n=0}^{N-1} A_n(\chi - 2v\zeta, \zeta)\overline{\lambda}_j^n = 0, \qquad j > k,$$
(6.18)

and as $\zeta \to -\infty$,

$$\sum_{n=0}^{N-1} A_n(\chi - 2v\zeta, \zeta)\overline{\lambda}_j^n = 0, \qquad j < k,$$

$$\sum_{n=0}^{N-1} A_n(\chi - 2v\zeta, \zeta)\lambda_j^n = 0, \qquad j > k.$$
(6.19)

We would like to solve (6.18) and (6.19) for A_n, $n = 0, \ldots, N-2$, in terms of A_{N-1}. For $\zeta \to +\infty$, we find that

$$A_n(\chi - 2v\zeta, \zeta) \to A_n^{(k+)} A_{N-1}(\chi - 2v\zeta, \zeta), \qquad n = 0, \ldots, N-2, \quad (6.20)$$

where

$$A_n^{(k+)} = \mathbf{V}(\lambda_1, \ldots, \lambda_{k-1}, \overline{\lambda}_{k+1}, \ldots, \overline{\lambda}_N)^{-1} \begin{pmatrix} \lambda_1^{N-1} \\ \vdots \\ \lambda_{k-1}^{N-1} \\ \overline{\lambda}_{k+1}^{N-1} \\ \vdots \\ \overline{\lambda}_N^{N-1} \end{pmatrix}, \quad (6.21)$$

in which $\mathbf{V}^{-1}$ is the inverse of the Vandermonde matrix.
When $\zeta \to -\infty$, we find that

$$A_n(\chi - 2v\zeta, \zeta) \to A_n^{(k-)} A_{N-1}(\chi - 2v\zeta, \zeta), \qquad n = 0, \ldots, N-2, \quad (6.22)$$

where

$$A_n^{(k-)} = \mathbf{V}(\overline{\lambda}_1, \ldots, \overline{\lambda}_{k-1}, \lambda_{k+1}, \ldots, \lambda_N)^{-1} \begin{pmatrix} \overline{\lambda}_1^{N-1} \\ \vdots \\ \overline{\lambda}_{k-1}^{N-1} \\ \lambda_{k+1}^{N-1} \\ \vdots \\ \lambda_N^{N-1} \end{pmatrix}. \tag{6.23}$$

Note that $A_n^{(k-)} = \overline{A_n^{(k+)}}$. The elements of matrix $\mathbf{V}(a_1, \ldots, a_N)^{-1}$ are given by

$$(\mathbf{V}^{-1})_{ij} = (-1)^{N-i} \frac{\displaystyle\sum \prod_{m=1,\, k \neq j}^{N-i} a_k}{\displaystyle\prod_{k=1,\, k \neq j}^{N} (a_j - a_k)}, \tag{6.24}$$

where the sum includes all possible products $\prod_{m=1,\, k \neq j}^{N-i} a_k$. For example, in the bottom row ($i = N$) of the matrix $\mathbf{V}^{-1}$, the numerator of the (i,j)th term is simply 1. In the second last row ($i = N - 1$), the numerator of the (i,j)th term is $a_j - \sum_{k=1}^{N} a_k$. Thus $\mathbf{V}^{-1}$ can easily be written down. For $N = 3$, this matrix is

$$\mathbf{V}^{-1}(a_1, a_2, a_3)$$

$$= \begin{bmatrix} \dfrac{a_2 a_3}{(a_1 - a_2)(a_1 - a_3)} & \dfrac{a_1 a_3}{(a_2 - a_1)(a_2 - a_3)} & \dfrac{a_1 a_2}{(a_3 - a_1)(a_3 - a_2)} \\[2ex] -\dfrac{a_2 + a_3}{(a_1 - a_2)(a_1 - a_3)} & -\dfrac{a_1 + a_3}{(a_2 - a_1)(a_2 - a_3)} & -\dfrac{a_1 + a_2}{(a_3 - a_1)(a_3 - a_2)} \\[2ex] \dfrac{1}{(a_1 - a_2)(a_1 - a_3)} & \dfrac{1}{(a_2 - a_1)(a_2 - a_3)} & \dfrac{1}{(a_3 - a_1)(a_3 - a_2)} \end{bmatrix}.$$

6.6 Transmission coefficients

A basis of the vector space of bound state solutions of (6.2) is given by the functions $\phi_k^+(\xi, \tau) = \phi^+(\xi, \tau; \lambda_k)$. We are interested in a one-dimensional space of bound state solutions of (6.2) that vanish in all but one waveguide at $\zeta = -\infty$, say, the waveguide indexed by the soliton eigenvalue λ_j. An element of this space, $\Phi_j(\xi, \tau)$, can be expanded as:

$$
\begin{aligned}
\Phi_j(\xi, \tau) &= \sum_{k=1}^{N} f_{jk} \phi_k^+(\xi, \tau) \\
&= \sum_{k=1}^{N} f_{jk} \left(\sum_{n=0}^{N-1} A_n(\xi, \tau) \lambda_k^n \right).
\end{aligned} \tag{6.25}
$$

We determine the coefficients f_{jk} by imposing the conditions that the Φ_j vanish as $\zeta \to -\infty$ in all waveguides, m, except for $m = j$. For the latter, we shall normalize by taking Φ_j to be asymptotically equal to the soliton collision $\psi(\xi, \tau)$. For $\zeta \to -\infty$, we have, in waveguide m:

$$
\begin{aligned}
\Phi_j(\xi, \tau) &= \left[\sum_{k=1}^{N} f_{jk} \left(\lambda_k^{N-1} + \sum_{n=0}^{N-2} A_n^{(m-)} \lambda_k^n \right) \right] A_{N-1}(\xi, \tau) \\
&= \frac{1}{2i} \psi(\xi, \tau) \sum_{k=1}^{N} f_{jk} g_{km}^-,
\end{aligned} \tag{6.26}
$$

where we have defined

$$
g_{km}^- = \lambda_k^{N-1} + \sum_{n=0}^{N-2} A_n^{(m-)} \lambda_k^n. \tag{6.27}
$$

Thus, we determine the elements f_{jk} of the matrix $\mathbf{F}$ in terms of the elements g_{km}^- of the matrix $\mathbf{G}_-$ by using

$$
\mathbf{F} = 2i\mathbf{G}_-^{-1}. \tag{6.28}
$$

Now, as $\zeta \to +\infty$, we have, in waveguide m:

$$
\Phi_j(\xi, \tau) = T_{mj}\, \psi(\xi, \tau), \tag{6.29}
$$

where the *amplitude transfer matrix* is defined by

$$
\mathbf{T} = \frac{1}{2i} [\mathbf{F}\mathbf{G}_+]^T = [\mathbf{G}_-^{-1}\mathbf{G}_+]^T, \tag{6.30}
$$

and the matrix elements g_{km}^+ of the matrix $\mathbf{G}_+$ are given by

$$
g_{km}^+ = \lambda_k^{N-1} + \sum_{n=0}^{N-2} \lambda_k^n A_n^{(m+)}. \tag{6.31}
$$

Thus, an arbitrary bound beam, which, as $\xi \to -\infty$, is of the form

$$
\phi(\xi, \tau) = \sum_{j=1}^{N} \beta_j \psi_j(\xi, \tau) \tag{6.32}
$$

will be scattered by the N-soliton refractive index $|\psi(\xi,\tau)|^2$ to become

$$\phi(\xi,\tau) = \sum_{m=1}^{N} \left[\sum_{j=1}^{N} T_{mj}\beta_j \right] \psi_m(\xi,\tau), \tag{6.33}$$

as $\xi \to +\infty$.

Note that the amplitude transfer matrix depends only on the soliton eigenvalues λ_i. These contain the soliton amplitude and propagation angle information; there is no dependence whatsoever on the relative phases (contained in the parameters γ_i) of the initial solitons.

Now let us calculate the power transfer. The power of the free soliton beam $\psi_j(\xi,\tau)$ corresponding to the eigenvalue $\lambda_j = a_j + ib_j$ is defined as

$$\int_{-\infty}^{\infty} |\psi_j(\xi,\tau)|^2 d\tau = 4b_j. \tag{6.34}$$

Thus, the *power transfer matrix* is defined in terms of the amplitude transfer matrix by

$$P_{mj} = \frac{b_m}{b_j} |T_{mj}|^2. \tag{6.35}$$

We require

$$\sum_{m=1}^{N} P_{mj} = 1 \tag{6.36}$$

for power conservation. The interpretation of this matrix is that a unit of the power input into waveguide j at $\xi = -\infty$ will be split among all the waveguides at $\xi = +\infty$, with power P_{mj} in waveguide m. The power transfer matrix does not depend on the exact geometry of the N-soliton collision that forms the waveguide, but only on the sizes of the solitons and their angles of propagation. The output guide labelled m is parallel to (and almost collinear with) input waveguide m.

Now we have the response functions of $N \times N$ switching devices made from soliton collisions. The elements of these matrices depend on the amplitudes of the colliding solitons and the angles between them. Remarkably, there is no dependence on the initial relative phases of the solitons or on the overall spatial geometry of the soliton locations. Hence, we can control the switching properties of these $N \times N$ devices by specifying only the relative angles and amplitudes of the input solitons.

6.7 Example 1: soliton X-junction

Let us calculate the power transfer matrix for the special case of a soliton X-junction (Fig. 6.1). This corresponds to $N = 2$. First, we calculate $A_0^{(m\pm)}$. From (6.21) and (6.23), we obtain

$$A_0^{(1+)} = \overline{\lambda}_2, \quad A_0^{(2+)} = \lambda_1, \quad A_0^{(1-)} = \lambda_2, \quad A_0^{(2-)} = \overline{\lambda}_1. \tag{6.37}$$

Then, the coefficient matrix $\mathbf{F}$ is calculated from (6.28):

$$\mathbf{F} = 2i \left[\begin{array}{cc} \lambda_1 + \lambda_2 & \lambda_1 + \overline{\lambda}_1 \\ 2\lambda_2 & \lambda_2 + \overline{\lambda}_1 \end{array} \right]^{-1}. \tag{6.38}$$

Using (6.30), we obtain the amplitude transfer matrix

$$\begin{aligned} \mathbf{T} &= \left[\begin{array}{cc} \lambda_1 + \overline{\lambda}_2 & \lambda_2 + \overline{\lambda}_2 \\ 2\lambda_1 & \lambda_2 + \overline{\lambda}_1 \end{array} \right] \left[\begin{array}{cc} \lambda_1 + \lambda_2 & 2\lambda_2 \\ \lambda_1 + \overline{\lambda}_1 & \lambda_2 + \overline{\lambda}_1 \end{array} \right]^{-1} \\[2mm] &= \frac{1}{\overline{\lambda}_1 - \lambda_2} \left[\begin{array}{cc} \overline{\lambda}_1 - \overline{\lambda}_2 & \overline{\lambda}_2 - \lambda_2 \\ \overline{\lambda}_1 - \lambda_1 & \lambda_1 - \lambda_2 \end{array} \right]. \end{aligned} \tag{6.39}$$

Finally, using (6.35), we obtain the power transfer matrix for the soliton X-junction:

$$\mathbf{P} = \frac{1}{\Delta_{21}^2 + (b_2 + b_1)^2} \left[\begin{array}{cc} \Delta_{21}^2 + (b_2 - b_1)^2 & 4b_1 b_2 \\ 4b_1 b_2 & \Delta_{21}^2 + (b_2 - b_1)^2 \end{array} \right], \tag{6.40}$$

where $\Delta_{ij} = a_i - a_j$ determines the angle difference between the soliton waveguides. The power transfer matrix does not depend on any phase information relating to the interaction of the waveguides. There is also no dependence on each a_i – only the differences $a_i - a_j$ are involved. It is easy to see from (6.40) that for each j, the power is conserved ($P_{1j} + P_{2j} = 1$).

An important property of the transmission matrix (6.40) is its symmetry. The matrix is symmetric even when the amplitudes of the two colliding solitons are unequal. When the soliton amplitudes coincide, $\mathbf{P}$ describes transmission properties obtained numerically by Akhmediev and Ankiewicz (1993a). For unequal amplitudes, the diagonal elements of the matrix are non-zero, even when the two solitons are almost parallel ($\Delta_{21} \to 0$).

6.8 Example 2: 3×3 switch

As another example, we consider a waveguide, with three input arms and three output arms, based on the collision of three soliton solutions of (6.1). Thus, we take $N = 3$, and calculate the amplitude transfer matrix elements:

$$T_{11} = \frac{(\lambda_2 - \lambda_1)(\lambda_3 - \lambda_1)}{(\overline{\lambda}_2 - \lambda_1)(\overline{\lambda}_3 - \lambda_1)}, \tag{6.41}$$

$$T_{12} = \frac{(\lambda_3 - \lambda_1)(\overline{\lambda}_2 - \lambda_2)}{(\overline{\lambda}_2 - \lambda_1)(\overline{\lambda}_3 - \lambda_1)}, \tag{6.42}$$

$$T_{13} = \frac{\lambda_3 - \overline{\lambda}_3}{\lambda_1 - \overline{\lambda}_3}, \tag{6.43}$$

$$T_{21} = \frac{(\overline{\lambda}_1 - \lambda_1)(\lambda_3 - \lambda_1)}{(\overline{\lambda}_2 - \lambda_1)(\overline{\lambda}_3 - \lambda_1)}, \tag{6.44}$$

$$T_{22} = \frac{1}{(\overline{\lambda}_2 - \lambda_1)(\lambda_2 - \overline{\lambda}_3)(\overline{\lambda}_3 - \lambda_1)} \tag{6.45}$$
$$\times \Big[|\lambda_1|^2 \lambda_2 - \lambda_1 |\lambda_2|^2 - |\lambda_1|^2 \lambda_3 + |\lambda_2|^2 \overline{\lambda}_3 + \overline{\lambda}_1 |\lambda_3|^2 - \overline{\lambda}_2 |\lambda_3|^2$$
$$+ \lambda_1 \lambda_2 \lambda_3 - \overline{\lambda}_1 \lambda_2 \lambda_3 + \overline{\lambda}_1 \overline{\lambda}_2 \lambda_3 - \lambda_1 \lambda_2 \overline{\lambda}_3 + \lambda_1 \overline{\lambda}_2 \overline{\lambda}_3 - \overline{\lambda}_1 \overline{\lambda}_2 \overline{\lambda}_3 \Big],$$

$$T_{23} = \frac{(\overline{\lambda}_1 - \overline{\lambda}_3)(\lambda_3 - \overline{\lambda}_3)}{(\lambda_1 - \overline{\lambda}_3)(\lambda_2 - \overline{\lambda}_3)}, \tag{6.46}$$

$$T_{31} = \frac{\overline{\lambda}_1 - \lambda_1}{\overline{\lambda}_3 - \lambda_1}, \tag{6.47}$$

$$T_{32} = \frac{(\overline{\lambda}_2 - \lambda_2)(\overline{\lambda}_3 - \overline{\lambda}_1)}{(\lambda_1 - \overline{\lambda}_3)(\lambda_2 - \overline{\lambda}_3)}, \tag{6.48}$$

$$T_{33} = \frac{(\overline{\lambda}_1 - \overline{\lambda}_3)(\overline{\lambda}_2 - \overline{\lambda}_3)}{(\lambda_1 - \overline{\lambda}_3)(\lambda_2 - \overline{\lambda}_3)}. \tag{6.49}$$

The power transfer matrix elements are then

$$P_{11} = \frac{(\Delta_{21}^2 + (b_2 - b_1)^2)(\Delta_{31}^2 + (b_3 - b_1)^2)}{(\Delta_{21}^2 + (b_2 + b_1)^2)(\Delta_{31}^2 + (b_3 + b_1)^2)}, \tag{6.50}$$

$$P_{12} = P_{21} = \frac{4b_1 b_2 (\Delta_{31}^2 + (b_3 - b_1)^2)}{(\Delta_{21}^2 + (b_2 + b_1)^2)(\Delta_{31}^2 + (b_3 + b_1)^2)}, \tag{6.51}$$

$$P_{13} = P_{31} = \frac{4b_1 b_3}{\Delta_{31}^2 + (b_3 + b_1)^2}, \tag{6.52}$$

$$P_{22} = 1 - P_{12} - P_{32}, \tag{6.53}$$

$$P_{23} = P_{32} = \frac{4b_2 b_3 (\Delta_{31}^2 + (b_3 - b_1)^2)}{(\Delta_{31}^2 + (b_3 + b_1)^2)(\Delta_{32}^2 + (b_3 + b_2)^2)}, \tag{6.54}$$

$$P_{33} = \frac{(\Delta_{31}^2 + (b_3 - b_1)^2)(\Delta_{32}^2 + (b_3 - b_2)^2)}{(\Delta_{31}^2 + (b_3 + b_1)^2)(\Delta_{32}^2 + (b_3 + b_2)^2)}, \tag{6.55}$$

where $\Delta_{ij} = a_i - a_j$.

Of course, the power conservation condition (6.36) also holds for this case. Again, the power transmission is independent of the soliton phase shifts and the overall spatial geometry of the soliton locations in the material. The matrix is symmetric, as $P_{kj} = P_{jk}$. However, the power transmission matrix for this 3 × 3 switch is less symmetric than that for the soliton X-junction, in the sense that $P_{11} \neq P_{22}$. The reason for this is that the soliton corresponding to the eigenvalue λ_2 can be distinguished from the others as it is between the other two. There are five independent parameters in the above formulae (three amplitudes and two angles between the solitons). This means that it is possible to design 3×3 switches with quite diverse properties by controlling these parameters.

6.9 Deviations from optimal case

When a probe beam is used in an X-junction created by strong pump, the probe beam may 'see' a refractive index which is different from that of the strong pump. This depends on the fabrication method used to create the X-junction. The equation governing the propagation of the probe beam will then have a new coefficient for the nonlinear term. When using different frequencies, the term responsible for diffraction also has a new coefficient. Instead of (6.2), we should then use the following equation:

$$i\phi_\xi + \frac{1}{2}\alpha\phi_{\tau\tau} + \beta|\psi(\xi,\tau)|^2\phi = 0, \qquad (6.56)$$

where α is the diffraction renormalization constant and β is the renormalization constant for the linear refractive index. The X-junction then loses its elegance, in the sense that it will suffer radiation loss. Also the transmission coefficients (and radiation) can then only be calculated numerically. Such numerical simulations have been carried out by Torres-Cisneros *et al.*, (1993). When we are far from the impact zone, each individual soliton makes an angle θ with the ξ axis. In terms of the parameters of Chapter 3, in particular (3.78), and also (6.10), we have $\tan\theta = 2a_j$, $(j = 1, 2)$, where $a_2 = -a_1$ in this case. The values of transmission coefficients depend on the angle of collision and are shown in Fig. 6.3. Note that in this figure, the curve corresponding to channel 2 is numerically calculated by solving (6.56), and the curve for channel 1 is then plotted so that the total sum is equal to 1. In practice, there is radiation from the impact area and the total sum would be less then 1.

Moreover, when α and β are different from 1, the transmission depends on the relative phase of the two colliding solitons, in contrast to (6.2). The relative output probe beam power is shown in Fig. 6.4. Although varying the parameters of the soliton collision allows us to create a range of devices with diverse transmission characteristics for the X-junction, many of these will incur losses when the beam passes the junction. Hence, one can conclude that the use of an X-junction at the frequency at which it was created, and with the optimal refractive index profile, is the most preferable. Nevertheless, a complete investigation of these devices, including loss characteristics for the full range of the parameters α and β, is of great interest.

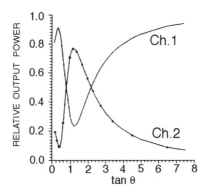

Figure 6.3 *Power transmission through soliton X-junction versus* $\tan(\theta) = 2a_1$ *for the case* $\alpha = 1$ *and* $\beta = 2$ *of (6.56). (After Torres-Cisneros et al., 1993; reproduced with permission.)*

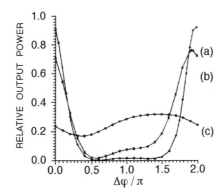

Figure 6.4 *Relative output power in channel 1 versus relative phase* $\Delta\varphi$ *between the initial solitons for the case* $\alpha = 1$ *and* $\beta = 2$ *of (6.56) with* $\tan\theta$ *being (a) 0.375, (b) 0.625 and (c) 1.25 (After Torres-Cisneros et al., 1993; reproduced with permission.)*

6.10 X-junctions based on dark soliton collisions

Up to this point in this chapter, we have been considering bright solitons. In this final section, we turn our attention to dark solitons. The solution of the NLSE, for a self-defocusing medium, which describes the collision of two dark solitons is given by (5.45). The change of refractive index which is concomitant with this field,

$$\Delta n \approx a_3 - |\psi(\tau, \xi)|^2 \tag{6.57}$$

$$= \frac{4(a_3 - a_1)[a_1 + \sqrt{a_1 a_3}\,\cosh(\mu\xi)\,\cosh(2p\tau)]}{a_3\cosh^2(\mu\xi) + a_1\cosh^2(2p\tau) + 2\sqrt{a_1 a_3}\,\cosh(\mu\xi)\,\cosh(2p\tau)},$$

where $\mu = 2\sqrt{a_1(a_3 - a_1)}$, $p = \sqrt{a_3 - a_1}$, $a_1 < a_3$, is somewhat analogous to the collision of two bright solitons, but it lacks the interference pattern (Fig. 6.5). Each dark soliton forms a high-index light-guiding channel. The linear transmission coefficients, T_{11} and T_{22}, of the coupler formed by this field are always unity (Akhmediev and Ankiewicz, 1993a,b; Miller, 1996), regardless of the angle of incidence (i.e. independent of the arbitrary parameters a_1 and a_3). This means that if the device is excited in the left channel, then the full energy of the output signal comes from the right channel. Hence, the device created by two dark solitons always behaves as a half-beat-length linear coupler. Again, the length of interaction for this coupler is the shortest possible and is thus optimal.

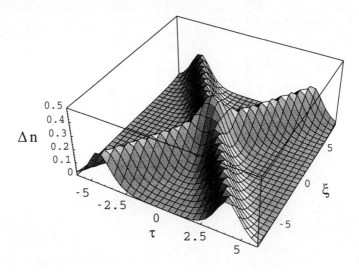

Figure 6.5 *Refractive index change (6.57) caused by the collision of two dark solitons with $a_1 = 0.25$, $a_3 = 0.5$. The interpretation of the parameters is given in Chapter 5.*

One of the possible applications for the latter device could be to make intersections of waveguides with zero cross-talk. This is important in integrated optical circuits, which could be made on one level of integration, so that it would not be necessary to construct multi-level circuits in order to keep different signals isolated. Again it should be stressed that, for total avoidance of cross-talk and the minimization of loss, matching wavelengths should be used.

CHAPTER 7

Nonlinear pulses in birefringent media

Analysis of the propagation of plane waves in Kerr media was initiated by Maker et al. (1964). Detailed investigations of polarization evolution in an isotropic Kerr medium have also been made (Winful, 1985; Gregori and Wabnitz, 1986 and Tratnik and Sipe, 1988). The generalization to anisotropic media is due to David et al. (1990). Spatial solitons in Kerr media have been considered by Snyder et al. (1994).

Propagation of soliton-like pulses in birefringent nonlinear fibers has attracted much attention in recent years (e.g. Menyuk, 1987; Evangelides, 1992; Cao and and Meyerhofer, 1994; Islam et al., 1987; Blow, 1987; Wright, 1989; Dowling, 1990; Kivshar, 1990; Romagnoli, 1992; Akhmediev and Soto-Crespo, 1994b; and Soto-Crespo et al., 1995b). The equations which describe pulse propagation in these fibers were derived by Menyuk (1989). In a reference frame traveling along the ξ axis with the average group velocity, this set of equations is of the form:

$$iU_\xi + i\delta U_\tau + \beta U + \tfrac{1}{2}U_{\tau\tau} + (|U|^2 + A|V|^2)U + BV^2U^* = 0,$$

$$iV_\xi - i\delta V_\tau - \beta V + \tfrac{1}{2}V_{\tau\tau} + (A|U|^2 + |V|^2)V + BU^2V^* = 0,$$

$$\tag{7.1}$$

where U and V are the slowly varying envelopes of the two linearly polarized components of the field along the x and y axes, δ is half of the inverse group velocity difference, β is half the difference between the propagation constants, A is the ratio of the cross-phase modulation term to the self-phase modulation term, B is the ratio of the four-wave mixing term to the self-phase modulation term, ξ is the normalized longitudinal coordinate, τ is the normalized retarded time and the asterisk denotes the complex conjugate. The nonlinear terms in (7.1) are related as follows. In cubic anisotropic media, $B = A/2$. In isotropic media like fused silica (far from resonances), $A = 2/3$ and $B = 1/3$. The latter is a particular case of a more general relation, $B = 1 - A$, valid for isotropic media (Maker et al., 1964). In this Chapter, we use $B = 1 - A$ in the general analysis, and $A = 2/3$ and $B = 1/3$ in examples of numerical simulations.

These equations can only be solved analytically for certain specific cases. Two main cases have been studied in depth, viz. high- and low-birefringence fibers, and separate approximations have been developed for each. The case of high birefringence has been studied in detail by Menyuk (1989); Evangelides *et al.* (1992); and Cao and Meyerhofer (1994). In this regime, one considers the two linearly polarized components of the field as having different phase velocities and different group velocities. Due to the nonlinearity, the pulses in these two components can capture each other, but their central frequencies become different, so as to make their group velocities equal. As a result of averaging, the fast oscillatory terms which relate the phases of the two components can be ignored, and usually only trapping effects are considered in this approach (Ueda and Kath (1990); Cao and and Meyerhofer, 1994). This trapping has been observed experimentally and it forms the basis of ultrafast optical logic gates (Islam, 1992).

The approximation of low birefringence takes into account the difference in phase velocities between the two linearly polarized components, but neglects their difference in group velocities, as this is assumed to be a higher-order effect. The two components of the soliton travel with the same group velocity, and phase locking of these two components can occur. This approach has been considered in the numerical work of Blow *et al.*, (1987). In particular, polarization instabilities were first found by Blow *et al.* (1987), and studied in more detail by Wright *et al.* (1989). The full polarization dynamics of solitons in polarization-preserving fibers, in the approximation of low birefringence, has been considered by Akhmediev and Soto-Crespo (1994b).

Birefringence in optical fibers can be induced deliberately, or it can be a residual effect due to imperfections in the drawing process. In the case of linear pulses, the output can be completely depolarized because different parts of the spectrum behave differently. On the other hand, even fibers with no linear birefringence influence the polarization properties of nonlinear signals. Hence, birefringence must be taken into account for pulse propagation along the fiber. However, solitons have different properties. It was shown by Evangelides *et al.* (1992) that the whole soliton has the same state of polarization. Hence, we have to know how this state of polarization evolves upon propagation.

The Stokes parameter formalism is the most appropriate method for studying this evolution. In the case of pulse propagation, the main difficulty is that the dynamical system has an infinite number of degrees of freedom, and the full analysis has to be done in an infinite-dimensional phase space. However, in the case of soliton-like pulses, we can reduce the system to a finite-dimensional one, as an 'approximation of average profile' can be applied to the problem. In this special approximation, equations for differential soliton Stokes parameters can be reduced to equations for integrated Stokes parameters. This allows us to consider solutions of the system as trajectories on the Poincaré sphere.

Another important effect is the radiation of small-amplitude waves by 'dynamic' solitons during propagation along the fiber. Radiation emission is a general property of nonintegrable systems, and it accompanies soliton-like pulse propagation. It appears due to coupling between the soliton and linear waves. In the case of birefringent fibers, the reason for radiation is the oscillatory behaviour of the soliton-like pulse itself. The intensity of radiation depends on the amplitude of the oscillations. Therefore, there can be loss of energy from the soliton. This effect influences the propagation dynamics of the pulse in general, and its state of polarization in particular.

7.1 Symmetries and conserved quantities

The set of equations (7.1) has at least three integrals of motion, the action (i.e. the total energy of the pulse),

$$Q = \int_{-\infty}^{\infty} (|U|^2 + |V|^2)d\tau; \tag{7.2}$$

the 'momentum'

$$M = -i \int_{-\infty}^{\infty} (UU_\tau^* - U^*U_\tau + VV_\tau^* - V^*V_\tau)d\tau; \tag{7.3}$$

and the Hamiltonian,

$$H = \int_{-\infty}^{\infty} \Big[\frac{1}{2}(|U_\tau|^2 + |V_\tau|^2) - \beta(|U|^2 - |V|^2) - \frac{1}{2}(|U|^4 + |V|^4)$$

$$- \frac{i\delta}{2}(UU_\tau^* - U^*U_\tau + VV_\tau^* - V^*V_\tau)$$

$$- A|U|^2|V|^2 - \frac{1}{2}(1-A)(U^2V^{*2} + U^{*2}V^2)\Big]d\tau. \tag{7.4}$$

Due to Noether's (1918) theorem, each conserved quantity corresponds to a symmetry of the system. Thus, conservation of energy is a result of the translational invariance of (7.1) relative to phase shifts $U, V \rightarrow U, V \exp(i\phi)$. Conservation of the momentum is a consequence of the translational invariance in τ, and the conservation of the Hamiltonian is a consequence of the translational invariance in ξ. If $\delta = 0$, then equations (7.1) are symmetric relative to the Galilean transformations:

$$U(\xi,\tau) = U(\xi,\tau - v\xi) \exp(iv\tau + iv^2\xi/2),$$
$$V(\xi,\tau) = V(\xi,\tau - v\xi) \exp(iv\tau + iv^2\xi/2). \tag{7.5}$$

This symmetry produces a linear equation for the motion of the 'centre of mass' of the pulse:

$$\int_{-\infty}^{\infty} \tau(|U|^2 + |V|^2)\,d\tau = -M(\xi - \xi_0)/2. \tag{7.6}$$

Equations (7.1) can be written in a canonical form (Faddeev and Takhtad-jan, 1987):

$$iU_\xi = \frac{\delta H}{\delta U^*}, \qquad iV_\xi = \frac{\delta H}{\delta V^*}. \tag{7.7}$$

Equations (7.4) and (7.7) define a Hamiltonian dynamical system on an infinite-dimensional phase space of two complex functions U, V which decrease to zero at infinity and can be analysed using the theory of Hamiltonian systems. This means that the behaviour of the solutions is defined, to a large extent, by the singular points of the system (i.e. stationary solutions of (7.1)), and depends on the nature of these points (as determined by the stability of its stationary solutions).

7.2 Approximation of low birefringence

In this approximation, in (7.1) we ignore the difference between the group velocities of each component and we thus obtain:

$$iU_\xi + \beta U + \tfrac{1}{2}U_{\tau\tau} + (|U|^2 + A|V|^2)U + (1 - A)V^2U^* = 0,$$
$$\tag{7.8}$$
$$iV_\xi - \beta V + \tfrac{1}{2}V_{\tau\tau} + (A|U|^2 + |V|^2)V + (1 - A)U^2V^* = 0.$$

This can be done for optical pulse durations in the picosecond range and for relatively small linear birefringences (Menyuk, 1987). The particular case with $A = 1$ and $\beta = 0$ is integrable using the inverse scattering method (Manakov, 1974).

7.3 Transformation to circularly polarized components

Equations (7.8) can be written in a different form if we use the transformation (Winful, 1985; Blow *et al.*, 1987):

$$P = U + iV, \quad G = U - iV. \tag{7.9}$$

Equations (7.8) become:

$$iP_\xi + KG + \tfrac{1}{2}P_{\tau\tau} + \left(\tfrac{A}{2}|P|^2 + \tfrac{2-A}{2}|G|^2\right)P = 0,$$
$$\tag{7.10}$$
$$iG_\xi + KP + \tfrac{1}{2}G_{\tau\tau} + \left(\tfrac{2-A}{2}|P|^2 + \tfrac{A}{2}|G|^2\right)G = 0,$$

where $K = \beta$. The variables P and G in this set correspond to the representation of the field in terms of its right- and left-hand circularly polarized components, respectively. The main differences between (7.10) and (7.8) are the appearance in (7.10) of an explicit linear coupling term and the disappearance of the terms containing complex conjugate variables that coupled the phases of the variables in (7.8). Each coupling term represents an energy exchange between the components. However, these terms are physically different because the energy is transferred between different types of component. If $\beta = 0$, for example, there is no energy exchange between the circularly polarized components P and G. However, there is energy transfer between the linearly polarized components U and V.

Another important difference is that the cross-phase modulation coefficient has the rescaled value $\frac{2-A}{A}$. When A is in the range $0 < A < 1$, the new coefficient varies in the range $\infty > \frac{2-A}{A} > 1$. Which equations are chosen is a matter of convenience. Both result in the same physics, if analysed correctly. We consider below mainly linearly polarized components, because the stationary solutions then have the simplest forms.

7.4 Stationary solutions (linearly polarized solitons)

For Hamiltonian systems, the stationary solutions play a pivotal role in the dynamics. To find them, we represent the field components in the form:

$$U = u(\xi, \tau, q)e^{iq\xi}, \quad V = v(\xi, \tau, q)e^{iq\xi}, \tag{7.11}$$

where q (the nonlinearly induced shift to the wavenumber) is the parameter of this family of solutions, and u and v are real functions of their parameters.

Now the equation for finding stationary solutions, in the variational formulation, can be written as:

$$\delta(H - qQ) = 0. \tag{7.12}$$

This variational formulation of the problem also defines the stability of stationary states. That is to say, for any fixed Q, the stationary state is stable if the corresponding H has a local minimum, with q being a Lagrange multiplier. The differential equations for finding stationary solutions obtained from (7.8) are the following:

$$\frac{1}{2}u_{\tau\tau} - (q - \beta)u + (|u|^2 + A|v|^2)u + (1 - A)v^2u^* = 0,$$
$$\frac{1}{2}v_{\tau\tau} - (q + \beta)v + (A|u|^2 + |v|^2)v + (1 - A)u^2v^* = 0. \tag{7.13}$$

Equations (7.13) have two simple stationary solutions, viz. linearly polarized soliton waves along the slow axis,

$$u = \frac{\sqrt{2(q - \beta)}}{\cosh(\sqrt{2(q - \beta)}\tau)}, \quad v = 0; \tag{7.14}$$

and linearly polarized soliton waves along the fast axis,

$$u = 0; \quad v = \frac{\sqrt{2(q+\beta)}}{\cosh(\sqrt{2(q+\beta)}\tau)}. \tag{7.15}$$

In the absence of linear birefringence ($\beta = 0$), the pulses (7.14) and (7.15) degenerate into a soliton of a single NLSE:

$$\sqrt{u^2 + v^2} = \frac{\sqrt{2q}}{\cosh(\sqrt{2q}\tau)}. \tag{7.16}$$

This solution can be linearly polarized along any direction in the (u, v) plane.

7.5 Elliptically polarized solitons

Apart from the linearly polarized solitons, there are stationary solutions of (7.8) with two components locked in phase. These are elliptically polarized solitons with u and v in quadrature (Akhmediev *et al.*, 1994a). If we take u real and v imaginary, i.e. $v \to iv$, (7.13) can be written in the form

$$\frac{1}{2}u_{\tau\tau} - (q - \beta)u + [u^2 + (2A - 1)v^2]u = 0,$$
$$\frac{1}{2}v_{\tau\tau} - (q + \beta)v + [(2A - 1)u^2 + v^2]v = 0. \tag{7.17}$$

To study the bifurcation from the fast soliton states, (7.15), we perturb these solutions, so that

$$u = \epsilon G, \qquad v = \sqrt{2(q+\beta)} \; \text{sech} \left[\sqrt{2(q+\beta)} \; \tau \right] + \epsilon^2 F, \tag{7.18}$$

where ϵ is a small parameter, and F and G are perturbation functions. Substituting (7.18) into (7.17), and linearizing with respect to the small parameter ϵ, we find

$$\ddot{G} - \mu^2\alpha^2 G + \frac{4(2A-1)\alpha^2}{\cosh^2(\sqrt{2(q-\beta)}\tau)}G = 0, \tag{7.19}$$

where $\mu = \sqrt{(q-\beta)/(q+\beta)}$ and $\alpha = \sqrt{2(q+\beta)}$. This equation has a lowest-order bounded solution when

$$4(2A - 1) = \mu(\mu + 1). \tag{7.20}$$

This is the only solution when $1/2 < A < 1$.

The point of bifurcation follows from this equation: $q_{cr}/\beta = (3\sqrt{33} - 5)/8 \approx 1.529$. The soliton energy at this point is $Q_{cr} = \sqrt{3(\sqrt{33}+1)\beta} \approx 4.498\sqrt{\beta}$ (assuming $A = 2/3$). Stationary elliptically polarized solitons do not have exact analytical forms at higher values of q. The two components of these soliton states have different amplitudes. These solutions form a one-parameter family with q being the parameter. The component with the smaller amplitude is zero at the point of bifurcation, but becomes equal to the other component in the limit $q \to \infty$. Hence, at high energies, these states converge to circularly polarized solitons. The curves for u and v, as functions of τ, are shown in Fig. 7.1 for a particular value of q. The curve for energy versus q for new soliton states, calculated numerically, is shown in Fig. 7.2. There are also multi-soliton stationary states represented by the upper curve; we consider them later.

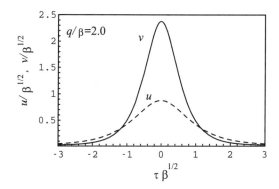

Figure 7.1 *Profiles u and v for stationary solutions with elliptic polarizations* $(q/\beta = 2)$.

7.6 Energy–dispersion diagram

All stationary solutions can be represented on an energy–dispersion diagram (Fig. 7.2). The energy Q on this diagram is given by (7.2). The lowest solid curve corresponds to the family of slow solitons (7.14). Its energy is given by $Q = 2\sqrt{2(q - \beta)}$. The solid line above it corresponds to the family of fast solitons, and its energy is given by $Q = 2\sqrt{2(q + \beta)}$. The solid line which splits off from it at the point M corresponds to the family of elliptically polarized solitons, where M is the point of bifurcation. Twice the energy of one NLSE soliton (7.16) is shown by the dashed line. A single-soliton input pulse, whose propagation is governed by the coupled NLSE, evolves in such a way that its energy remains in the intervals defined by the region between the curve for elliptically polarized solitons and fast solitons, and that between the curves for fast and slow solitons in Fig. 7.2, provided that the initial conditions are soliton-like pulses. These two regions form the area of allowed motion for these pulses in this dynamical system. Any given soliton-like pulse can be shown on this diagram by its *representative point*, which specifies its energy and average q parameter. The dynamics is different in each of the two regions in Fig. 7.2. The phases of the two components are locked in the upper region and unlocked in the lower strip.

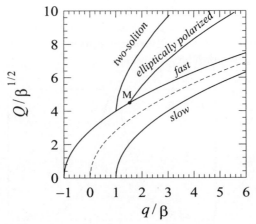

Figure 7.2 *Energy–dispersion diagram for the stationary soliton states.*

7.7 Hamiltonian versus energy diagram

These notions can be expressed more accurately if we present the data in a Hamiltonian versus energy diagram (Fig. 7.3). For any particular family of stationary solutions, there is a one-to-one correspondence between the q-parameter, the energy and the Hamiltonian. The explicit expression for the Hamiltonian in terms of energy has the form

$$\hat{H} = -\hat{Q}^3/24 - \hat{Q} \tag{7.21}$$

for slow solitons and the form

$$\hat{H} = -\hat{Q}^3/24 + \hat{Q} \tag{7.22}$$

for fast solitons, where $\hat{H} = H/\beta^{3/2}$ and $\hat{Q} = Q/\sqrt{\beta}$. For elliptically polarized solitons, this dependence has to be calculated numerically. The H–Q curve for them splits off from the curve for fast solitons at the point M. The H–Q diagram can be useful in several aspects.

Firstly, the stability criterion discussed in section 7.4 is based on the Hamiltonian. Hence, we can expect the lowest of the stationary solutions at any fixed Q to be stable, so we can find stable branches on the H–Q diagrams and transpose them to the energy–dispersion diagram.

Secondly, the values H and Q can be calculated for any initial pulse (in contrast to the q parameter). Hence, the representative point on the H–Q diagram corresponding to an arbitrary soliton-like initial condition is rigorously defined. If this point is located inside the strips between the stationary solutions, or at least close to them, it will stay in the strip, evolve in some complicated way, and eventually will converge to a stable stationary solution below and to the left of the initial point, after emitting some amount of radiation. Although the Hamiltonian and energy are conserved quantities for the whole solution, they can be calculated for the soliton and for radiation separately, so that H and Q for the main pulse can move on the (H,Q) plane, thus making the transition to stationary stable states visible. The trajectories marked from a to f in Fig. 7.3 show some examples of evolution found numerically.

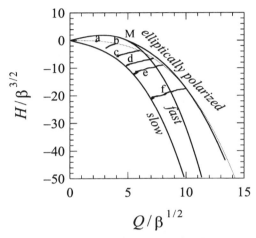

Figure 7.3 *Hamiltonian versus energy diagram for the stationary soliton states. Trajectories (a)–(f) show examples of soliton evolution emitting radiation.*

7.8 Stokes parameters

A convenient way to analyse the polarization dynamics and also to solve equations (7.8) is to use the Stokes parameters, which are defined by:

$$s_0 = |u|^2 + |v|^2,$$

$$s_1 = |u|^2 - |v|^2,$$

$$s_2 = u^*v + uv^*,$$

$$s_3 = -i(u^*v - uv^*).$$

(7.23)

These four parameters are all real functions of ξ and τ. For general solution, they vary along the fiber as well as across the pulses, so we call them 'differential Stokes parameters'. Using them, and taking into account the fact that the fields for pulses decay to zero at infinity, (7.13) can be written in the form:

$$\frac{d}{d\xi} \int_{-\infty}^{\infty} s_0 d\tau \;=\; 0,$$

$$\frac{d}{d\xi} \int_{-\infty}^{\infty} s_1 d\tau \;=\; -2(1-A) \int_{-\infty}^{\infty} s_2 s_3 d\tau,$$

$$\frac{d}{d\xi} \int_{-\infty}^{\infty} s_2 d\tau \;=\; 2\beta \int_{-\infty}^{\infty} s_3 d\tau + 2(1-A) \int_{-\infty}^{\infty} s_1 s_3 d\tau,$$

$$\frac{d}{d\xi} \int_{-\infty}^{\infty} s_3 d\tau \;=\; -2\beta \int_{-\infty}^{\infty} s_2 d\tau,$$

(7.24)

where the dependence on τ has been eliminated due to the integration, and all the magnitudes are real.

Equations (7.24) are invariant relative to the transformation $1 - A \rightarrow -(1-A)$ and $s_i \rightarrow -s_i$. This means that if $1 - A$ is negative (i.e. $A > 1$), then all solutions of (7.24) can be obtained from the solutions for positive $1 - A$ (i.e. $A < 1$) by changing the signs of the s_i. The corresponding solutions can be obtained by inversion of the solutions and the whole phase space for s_i with respect to the origin. This allows us to simplify the analysis by considering only the case $A < 1$. This is true, however, only in problems with no radiation. The inversion of the Stokes parameters is not followed by the inversion of ξ in (7.24). Hence, if loss is involved, then the transformation given above does not give new solutions.

Equations (7.24) are integro-differential equations, which can be even more difficult to solve than the original set (7.8). However, if we are interested in integrated values, $S_i = \int_{-\infty}^{\infty} s_i d\tau$, then these equations can be simplified. Moreover, if we are interested in soliton propagation, then the integrated values provide most of the information we need in the problem. The first equation, for example, is nothing other than the conservation of the total energy of the solitary pulse. The third and the fourth equations in (7.24) show that the integrated values $\int_{-\infty}^{\infty} s_2 d\tau$ and $\int_{-\infty}^{\infty} s_3 d\tau$ rotate around the axis s_1 with frequency 2β. However, the integrals of the products $s_2 s_3$ or $s_1 s_3$ on the right-hand side of these equations make the problem quite complicated to solve. It can be done using an approximation which is suggested by the numerical solution of (7.8).

7.9 Dynamic solitons

Obviously, the behaviour of the whole pulse depends on the initial conditions. Two conserved quantities, namely, the energy and the Hamiltonian, are unique parameters of the pulse pair, as they define the whole dynamic behaviour of the pulse. These quantities can be calculated for any pulse-like initial condition. They define a point on a plane (H, Q). To form a single-soliton state, the point must be in the strip between the curves for stationary solutions in Fig. 7.3. If the corresponding point is much higher than the upper curve, corresponding to the elliptically polarized states, then the pulse can subdivide into two or more soliton-like pulses. If the point is much lower than the lowest curve (slow), then the pulse will be dispersed as radiation. Otherwise, the pulse will remain as a single pulse with periodic energy exchange between the components U and V. Eventually, due to radiation of small-amplitude waves, it will converge to one of the stationary solutions. This pulse, which behaves periodically on propagation, can be called a *dynamic soliton*. (Dynamic solitons in birefringent media have been considered by Snyder *et al.* (1994).)

7.10 Approximation of the average profile

We restrict ourselves to soliton-like pulse dynamics. These pulses form a very special class of solutions, which is the most important one in practice. For soliton-like pulses, the shapes of the two components of the pulses should change only slightly upon propagation, and the phase chirp across each pulse should be negligible. This idea has been expressed clearly by Evangelides *et al.* (1992) (where the whole soliton has the same state of polarization) and can be viewed as a generalization of the 'soliton phase model' (Blow *et al.*, 1992) for two-component complex fields. Only the component amplitudes should change with distance ξ, because of the energy transfer between the two field components. Obviously, the actual shape will oscillate around the calculated average profile. However, these oscillations should be small – in non-integrable dynamical systems, any field oscillation generates small-amplitude waves around the soliton, and these are radiated away. Their intensity is proportional to the amplitude of the oscillations. In this case, the energy of the soliton will decrease with distance ξ.

To a first approximation, we assume that the solution is separable in the following way:

$$u = X(\xi)f(\tau), \quad v = Y(\xi)f(\tau), \qquad (7.25)$$

where $f(\tau)$ is a real function defining the common profiles, and $X(\xi)$ and $Y(\xi)$ are complex amplitudes. Equations (7.24) become:

$$\frac{d}{d\xi}S_0 = 0,$$

$$\frac{d}{d\xi}S_1 = -2gS_2S_3,$$

$$\frac{d}{d\xi}S_2 = 2\beta S_3 + 2gS_1S_3, \qquad (7.26)$$

$$\frac{d}{d\xi}S_3 = -2\beta S_2,$$

where the normalized integrated Stokes parameters, $S_i = \int\limits_{-\infty}^{\infty} s_i d\tau / \int\limits_{-\infty}^{\infty} f^2 d\tau$, are given by

$$
\begin{aligned}
S_0 &= |X|^2 + |Y|^2, \\[2mm]
S_1 &= |X|^2 - |Y|^2, \\[2mm]
S_2 &= X^*Y + XY^*, \\[2mm]
S_3 &= -i(X^*Y - XY^*);
\end{aligned}
\tag{7.27}
$$

and g, the nonlinear birefringence coefficient, is defined by

$$
g = (1 - A)\frac{\int\limits_{-\infty}^{\infty} f^4 d\tau}{\int\limits_{-\infty}^{\infty} f^2 d\tau}.
\tag{7.28}
$$

Clearly, the value of g depends on A, as well as on the soliton shape. The value of g can be calculated exactly on the two edges of the strip of allowed motion in Fig. 7.3. For the fast solitons, $g = \frac{4}{3}(q + \beta)(1 - A)$, and for the slow solitons, $g = \frac{4}{3}(q - \beta)(1 - A)$. Note that g changes sign at $A = 1$.

7.11 Comparison with c.w. beams

Apart from the factor g, equations (7.26) are the same as those for continuous waves (Gregori and Wabnitz, 1986). Hence, qualitatively, the state of polarization of soliton-like pulses evolves along the fiber in the same way as for c.w. beams. We stress here that this is true for solitons alone, as only solitons have a fixed state of polarization. The governing parameter is the total energy, Q, of the pulse for solitons, while it is the power in the c.w. case. This is clear from (7.28), which is expressed in terms of integrals over the whole pulse. Other differences are discussed below.

7.12 Linear and nonlinear beat lengths

Equations (7.26) can be written in a vector form:

$$
\frac{d}{d\xi}\mathbf{S} = 2\beta[\mathbf{e}_1 \times \mathbf{S}] + 2gS_3[\mathbf{e}_3 \times \mathbf{S}],
\tag{7.29}
$$

where $\mathbf{S} = (S_1, S_2, S_3)$ is the Stokes vector in a three-dimensional space, $\mathbf{e}_1$ and $\mathbf{e}_3$ are unit vectors along axes 1 and 3 respectively, and $\times$ indicates vector product. Equation (7.29) describes a double rotation around axes 1 and 3, respectively. The rotation around the S_1 axis is linear and does not depend on the amplitude of the pulse, while the rotation around the S_3 axis is nonlinear. Hence, we can introduce two beat lengths. If the pulse energy is small, the relative phase of the two components oscillates with a certain period in ξ, viz. π/β. This period is called the *linear beat length*.

On the other hand, if we put $\beta = 0$ in (7.29) (i.e. we remove the linear oscillations), then the period of the remaining phase oscillations will depend on the pulse shape and the total energy of the pulse. In this case the two components exchange their energy. The beat length L_{nl} related to this process is clearly inversely proportional to the coefficient g: $L_{nl} \approx \pi/g$. The value of g depends on the soliton parameter q as well as on A. It can be called the *nonlinear beat length* because it depends on the soliton's peak intensity. To estimate roughly the dependence on q, we calculate g using the average pulse shape. In this case $g = \frac{4}{3}q(1 - A)$. The deviation of this value from the average of that corresponding to the slow and fast linearly polarized solitons is $\Delta g = \frac{4}{3}\beta(1 - A)$, which is small at small β. Then g is, roughly speaking, proportional to q, and changes its sign at $A=1$; thus g becomes zero when $A = 1$. The nonlinear beat length is 4.5 times higher than the *soliton period* which is usually defined as $T_{sol} = \pi/2q$.

In general, the soliton oscillations are complicated and are not purely harmonic. They include phase oscillations, as well as energy exchange occurring at the same time. The fundamental period of these oscillations can be written in terms of the two beat lengths by using elliptic integrals. The dynamic behaviour of the soliton is mainly defined by the relation between these two parameters. For small amplitudes, A, when the nonlinear beat length is much longer than the linear, the pulse behaviour is defined mainly by the linear beat length. For the opposite relation between the two parameters, the soliton components only exchange energy. When the two are comparable, bifurcations occur and new solutions can appear.

7.13 Analysis of the system

Equations (7.26) have a constant of motion,

$$S_1^2 + S_2^2 + S_3^2 = S_0^2, \tag{7.30}$$

which is a consequence of energy conservation (7.2), and indicates that, within the approximation we are making, the evolution of any solution can be analysed qualitatively as a trajectory of the Stokes vector **S** on the Poincaré sphere.

Equations (7.26) have a second invariant:

$$W = \frac{g}{2\beta}S_3^2 - S_1, \tag{7.31}$$

which is a consequence of the conservation of the Hamiltonian (7.4), as can easily be verified. In our approximation

$$W = \frac{H}{\beta I} + \frac{gS_0^2}{2\beta(1 - A)} - \frac{\rho S_0}{2\beta},$$

$$\rho = \frac{\int\limits_{-\infty}^{\infty} f_\tau^2 d\tau}{\int\limits_{-\infty}^{\infty} f^2 d\tau}, \qquad I = \int\limits_{-\infty}^{\infty} f^2 d\tau. \qquad (7.32)$$

Different values of W, the *evolution parameter*, correspond to different regimes of soliton propagation (and thus different trajectories on the Poincaré sphere). The convenience of the evolution parameter (in contrast to the Hamiltonian) is that it is constant along the energy-dispersion curves for fast and slow solitons in Fig. 7.2. We can conclude, preliminarily, that a given value of W corresponds to a fixed type of solution as q varies.

Let us choose, as an initial condition, one of the linearly polarized solitons given by (7.14) or (7.15). We can now consider that initially $X(\xi = 0) = 1, Y(\xi = 0) = 0$, or vice versa, and that therefore the integrated Stokes parameters in (7.27) are normalized so that $S_0(\xi = 0) = 1$. In principle, the value of S_0 is conserved, as the first equation in (7.27) proves. However, if the pulse changes its shape, and if it radiates energy as a result of this reshaping, then S_0 can, and usually does, decrease during propagation. For small perturbations of the NLSE, these changes are slow ($\frac{dS_0}{d\xi} \ll S_0$) and can be considered adiabatic. The Hamiltonian also decreases if radiation takes place, as does W. These processes are slow, and can be ignored in a first approximation.

7.13.1 Particular case: (a) $\beta = 0$

In (7.13), β is responsible for the linear birefringence. When $\beta = 0$ there is no linear birefringence, the medium is isotropic and the pulse evolution is determined completely by the nonlinear terms. In this case, the equations (7.26) take a simple form given by:

$$\frac{d}{d\xi} S_0 = 0,$$

$$\frac{d}{d\xi} S_1 = 2gS_2 S_3,$$

$$\frac{d}{d\xi} S_2 = -2gS_1 S_3, \qquad (7.33)$$

$$\frac{d}{d\xi} S_3 = 0.$$

S_0 and S_3 are therefore conserved and the pulse evolution appears as a rotation of the Stokes vector around the $\mathbf{e}_3$ axis onto the Poincaré sphere.

The solution to (7.33) is:

$$S_1 = S_0 \cos\theta \cos(\omega\xi + \phi/2)$$

$$S_2 = -S_0 \cos\theta \sin(\omega\xi + \phi/2) \qquad (7.34)$$

$$S_3 = S_0 \sin\theta$$

where the frequency $\omega = 2gS_3 = 2gS_0 \sin\theta$, while θ is the constant angle formed between $\mathbf{S}$ and the (S_1, S_2) plane , and ϕ defines the initial phase of this rotation. The direction of the rotation depends on the sign of S_3, and is different in each hemisphere. If $\beta = 0$, then the two linearly polarized solitons given by (7.14) and (7.15) coincide. This means that the 'approximation of constant profile' is a good one in this case. The rotation around the $\mathbf{e}_3$ axis takes place along one of the circles parallel to the equator on the Poincaré sphere. The circle depends on the initial conditions. The state of polarization always remains elliptic. The field components along the axes u and v oscillate periodically. The phase difference between the two components oscillates around $\pi/2$ (for $S_3 > 0$) or around $-\pi/2$ (for $S_3 < 0$). Equation (7.34) can be described as a *solution with oscillating phase*.

The phase only oscillates in the frame of reference associated with the u and v axes. In a frame rotating with the angular frequency $\omega/2 = gS_0 \sin\theta$ relative to these axes, the state of polarization is fixed (elliptic). Hence, the nonlinearity causes a rotation of the polarization ellipse. The ellipticity is defined by the angle θ (given by the initial conditions). The polarization changes from linear at $\theta = 0$ to circular at $\theta = \pm\pi/2$. The angular frequency of rotation at a given ellipticity is defined by g, i.e. by the energy of the soliton. The direction of rotation is clockwise for right-hand elliptic polarization (Fig. 7.4a) and counterclockwise for left-hand polarization (Fig. 7.4b). The polarization ellipse does not rotate when $\theta = \pi/2$ (linear polarization) and/or g is small, as should be the case in the limit of small intensities.

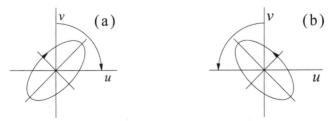

Figure 7.4 *Rotation of the polarization ellipse in an isotropic ($\beta = 0$) nonlinear fiber: (a) $0 < \theta < \pi/2$; (b) $-\pi/2 < \theta < 0$.*

The solutions to (7.13) can only be found in analytic form for particular cases. Equations (7.34), for $\theta = \pi/2$ or $-\pi/2$, correspond to stationary points $\mathbf{S} = (0,0,\pm S_0)$ on the Poincaré sphere, where the circularly polarized input wave does not change its state of polarization, with the two components u and v being $\pi/2$ out of phase. In terms of u and v, these solutions can be written:

$$u = \frac{\sqrt{q}}{\sqrt{A}\,\cosh(\sqrt{2q}\tau)}, \quad v = +iu \quad \text{at} \quad \theta = \pi/2, \tag{7.35}$$

$$u = \frac{\sqrt{q}}{\sqrt{A}\,\cosh(\sqrt{2q}\tau)}, \quad v = -iu \quad \text{at} \quad \theta = -\pi/2. \tag{7.36}$$

Another set of stationary points is located on the equator of the Poincaré sphere ($\theta = 0$) at $\mathbf{S} = (S_1, S_2, 0)$. The solution for this case is:

$$u = \frac{\sqrt{2q}}{\cosh(\sqrt{2q}\tau)}\cos\frac{\phi}{2}, \quad v = \frac{\sqrt{2q}}{\cosh(\sqrt{2q}\tau)}\sin\frac{\phi}{2}, \tag{7.37}$$

where ϕ is an arbitrary phase. This solution corresponds to a wave linearly polarized in a direction which forms an angle $\phi/2$ with the fast axis.

7.13.2 Particular case: (b) $A = 1$

The parameter A is related to the nonlinear birefringence (or ratio between the self- and cross-phase modulation terms). This ratio is equal to one when $A = 1$; this is a Manakov model with the two components having different phase velocities. In this case equations (7.26) become:

$$\frac{d}{d\xi}S_0 = 0,$$
$$\frac{d}{d\xi}S_1 = 0,$$
$$\frac{d}{d\xi}S_2 = 2\beta S_3, \tag{7.38}$$
$$\frac{d}{d\xi}S_3 = -2\beta S_2.$$

Therefore S_1 is conserved, in addition to S_0. The Stokes vector $\mathbf{S}$ now rotates around the $\mathbf{e}_3$ axis with frequency 2β. The solution to (7.38) is:

$$S_1 = S_0\cos\theta',$$
$$S_2 = S_0\sin\theta'\sin(2\beta\xi + 2\phi_0), \tag{7.39}$$
$$S_3 = S_0\sin\theta'\cos(2\beta\xi + 2\phi_0),$$

where θ' is the angle formed between $\mathbf{S}$ and $\mathbf{e}_1$, and ϕ_0 is another constant to be determined from the initial conditions. The solution for u and v is given by:

$$u = \frac{\sqrt{2q}\,\cos{(\theta'/2)}}{\cosh(\sqrt{2q}\tau)}\,\exp(i\beta\xi + i\phi_0),$$

(7.40)

$$v = \frac{\sqrt{2q}\,\sin{(\theta'/2)}}{\cosh(\sqrt{2q}\tau)}\,\exp(-i\beta\xi - i\phi_0),$$

where θ' defines the relative values of the two components, and $2\phi_0$ denotes the phase difference between them. The solution oscillates with frequency 2β. The state of polarization changes from linear (when $\beta\xi + \phi_0 = N\pi/2$, N being an integer) to elliptical (when $\beta\xi + \phi_0 = (N + 1/2)\pi/2$), and vice versa. There is no energy transfer between the components. Hence, the amplitudes of the two components do not change in this case, but their phase difference increases linearly. This means that the phase of the whole solution rotates. Accordingly, this solution can be called a *solution with rotating phase*.

The two examples given above show, qualitatively, the role of each parameter in (7.13) in the evolution of soliton-like pulses. These two cases correspond to two different regimes of propagation of soliton-like pulses. The qualitative peculiarities of these two regimes are preserved in more complicated cases which we consider in the next section. In general, these two regimes of propagation are defined not only by the parameters of the problem, but also by the initial conditions, i.e. by the initial state of polarization.

7.13.3 General case

If neither β nor $(1 - A)$ is zero, then the total motion consists of a combination of two rotations, as (7.26) or (7.29) show. To analyse this complicated motion, we first find its singular points on the Poincaré sphere. When $S_0 < \frac{\beta}{g}$, equations (7.26) have only two stationary points (Fig. 7.5a):

$$S_1 = +S_0, \quad S_2 = 0, \quad S_3 = 0,$$

(7.41)

and

$$S_1 = -S_0, \quad S_2 = 0, \quad S_3 = 0.$$

(7.42)

The point $\mathbf{S} = (+S_0, 0, 0)$ is always stable. It corresponds to the slow linearly polarized pulse of (7.14). The second point $\mathbf{S} = (-S_0, 0, 0)$ is stable if $\frac{\beta}{g} > S_0$. It corresponds to the fast linearly polarized soliton given by (7.15).

If $\frac{\beta}{g} < S_0$ then, in addition to (7.41) and (7.42), we find two more stationary points (Fig. 7.5b):

$$S_1 = -\frac{\beta}{g}, \quad S_2 = 0, \quad S_3 = \pm\sqrt{S_0^2 - \frac{\beta^2}{g^2}},$$

(7.43)

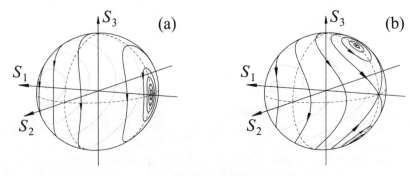

Figure 7.5 *Trajectories on the Poincaré sphere for the periodic solutions for (a)* $\beta > gS_0$ *and (b)* $\beta < gS_0$. *These trajectories are solutions of (7.26).*

which are always stable. These two stationary solutions correspond to elliptically polarized solitons with

$$|X|^2 = \frac{1}{2}(S_0 - \beta/g), \quad |Y|^2 = \frac{1}{2}(S_0 + \beta/g), \qquad (7.44)$$

and ellipticity $|X|^2 - |Y|^2 = -\frac{\beta}{g}$. The phase difference between the two components u and v is $\pi/2$ or $-\pi/2$. These two points correspond to right- and left-hand elliptically polarized pulses which change neither their state of polarization nor the amplitudes of their components during the propagation. With the appearance of these two points, the stationary point corresponding to the fast soliton (7.42) loses its stability. However, the Hamiltonian, H, for these stationary solutions is higher than that corresponding to the fast solitons (Fig. 7.3). This indicates that they can also be unstable.

7.13.4 Solutions with oscillating phase

When points (7.43) exist, there are solutions oscillating around them. These solutions exist only when $\frac{\beta}{g} < S_0$. They correspond to closed loops inside two separatrices in Fig. 7.5b. The solutions can be written as:

$$S_1 = \frac{A_1^2 k^2}{4\beta g}\left[2\operatorname{cn}^2(A_1\xi, k) - 1\right] - \frac{\beta}{g},$$

$$S_2 = \pm \frac{A_1^2 k^2}{2\beta g}\operatorname{sn}(A_1\xi, k)\operatorname{cn}(A_1\xi, k), \qquad (7.45)$$

$$S_3 = \pm \frac{A_1}{g}\operatorname{dn}(A_1\xi, k),$$

where $A_1 = \frac{2\beta}{k^2}\sqrt{\sqrt{4(1 - k^2) + \frac{g^2 k^4}{\beta^2}S_0^2} - (2 - k^2)}$, and sn, cn and dn are Jacobian elliptic functions with amplitude k. Plus and minus signs correspond to loops at $S_3 > 0$ and $S_3 < 0$, respectively. These solutions exist only when $\frac{\beta}{g} < S_0$. The value of the second integral is $W = \beta/g + A_1^2 (2 - k^2)/(4\beta g)$.

The three parameters k, A_1 and W are related to each other. In principle, k and W can be expressed in terms of A_1 (or k and A_1 in terms of W). However, we find that the above formulae are the most convenient way to relate them. The best choice is to consider k as independent parameter. Different values of k correspond to different trajectories in Fig. 7.5b. At $k \to 0$ these solutions tend to the stationary solution given by (7.43). Note that $A_1 \to \sqrt{g^2 S_0^2 - \beta^2}$ as $k \to 0$. Hence, in this limit, the solution (7.45) oscillates with frequency $2\sqrt{g^2 S_0^2 - \beta^2}$. The frequency decreases to zero when k increases to 1. At $k \to 1$, these solutions reduce to the separatrix solutions.

The state of polarization is elliptic, with both components oscillating along the principal axes, and with the phase difference between them oscillating around $\pi/2$ or $-\pi/2$. This regime of propagation is analogous to the regime with oscillating phase considered before. Moreover, as $\beta \to 0$, these solutions have the limit presented earlier. Solutions with oscillating phase can be excited if the initial state of polarization is elliptic, with the major axis of the ellipse being directed along the principal axis of the fiber ($S_2 = 0$), and with the amplitudes satisfying the following inequality: $S_1 < S_0 - 2\beta/g$ (or $|X|^2 < |Y|^2 + S_0 - 2\beta/g$). When $\beta \neq 0$, a second type of solution appears for the same values of material parameters. It characterizes a different regime of propagation.

7.13.5 Solutions with rotating phase

For these solutions, the phase difference between the two components increases monotonically. These solutions exist for both $\frac{\beta}{g} < S_0$ and $\frac{\beta}{g} > S_0$. When $\frac{\beta}{g} < S_0$, they correspond to closed loops outside the two separatrices in Fig. 7.5b. They are given by:

$$S_1 = \frac{A_2^2}{4\beta g} [2 \operatorname{dn}^2(A_2\xi, k_1) - 1] - \frac{\beta}{g},$$

$$S_2 = \frac{k_1 A_2^2}{2\beta g} \operatorname{sn}(A_2\xi, k_1) \operatorname{dn}(A_2\xi, k_1), \qquad (7.46)$$

$$S_3 = \frac{k_1 A_2}{g} \operatorname{cn}(A_2\xi, k_1),$$

where $A_2 = 2\beta\sqrt{(1 - 2k_1^2) + \sqrt{\frac{g^2}{\beta^2} S_0^2 - 4k_1^2(1 - k_1^2)}}$, and modulus k_1 satisfies $0 < k_1 < 1$. The second integral is given by $W = \frac{\beta}{g} + \frac{A_2^2}{2\beta g} (k_1^2 - \frac{1}{2})$. The corresponding trajectories rotate around the $\mathbf{e}_1$ axis, with a frequency which goes from zero (when $k_1 = 1$) to $2\beta\sqrt{\frac{g}{\beta} S_0 + 1} > 2\beta$ (when $k_1 = 0$).

When $\frac{\beta}{g} > S_0$, only the solutions with rotating phase exist (Fig. 7.5a). These solutions are also given by equations (7.46), but with

$$A_2 = 2\beta \sqrt{(1 - 2k_1^2) \pm \sqrt{\frac{g^2}{\beta^2} S_0^2 - 4k_1^2(1 - k_1^2)}}. \qquad (7.47)$$

The modulus k_1 satisfies $0 < k_1^2 < \frac{1}{2} - \sqrt{\frac{1}{4} - \frac{g^2 S_0^2}{4\beta^2}}$, so that its upper limit (viz. $1/\sqrt{2}$) is now smaller then before (where it could reach unity). There are two values for A_2, corresponding to plus and minus signs in (7.47) for each value of k_1. They coincide when $k_1^2 = \frac{1}{2} - \sqrt{\frac{1}{4} - \frac{g^2 S_0^2}{4\beta^2}}$. The function $A_2/2\beta$ is plotted against k_1 in Fig. 7.6. The upper ($\sqrt{1 + gS_0/\beta}$) and the lower ($\sqrt{1 - gS_0/\beta}$) values of $A_2/2\beta$ at $k_1 = 0$ correspond to the fast and slow stationary solutions, respectively. The frequency of the oscillations ranges from $2\beta\sqrt{1 - gS_0/\beta}$ to $2\beta\sqrt{1 + gS_0/\beta}$. Therefore, there is some intermediate point (given by the intersection of the curve with the dashed line in Fig. 7.6) where the difference in propagation constants is exactly 2β.

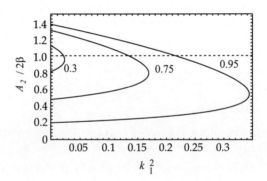

Figure 7.6 *The function $A_2/2\beta$ versus k_1^2; each curve is labelled with its value of the parameter gS_0/β.*

When $\frac{\beta}{g} \gg S_0$, the modulus k_1 is close to zero in the whole interval, because the upper limit of k_1 goes to zero. Then, solution (7.46) can be approximated by :

$$S_1 = \frac{A_2^2 - 4\beta^2}{4\beta g},$$

$$S_2 = \frac{k_1 A_2^2}{2\beta g} \sin A_2\xi, \qquad (7.48)$$

$$S_3 = \frac{k_1 A_2}{g} \cos A_2\xi.$$

In this limit, the Stokes vector evolution is similar to that in the linear birefringent medium, except that the rotation frequency depends on the pulse energy. There is no energy exchange between the components (S_1 is constant). The rotation frequency at $S_1 = 0$ is 2β, but it differs from this linear limit for other trajectories.

The state of polarization for these solutions evolves from linear to elliptically polarized and then back to linear. Solutions with rotating phase can be excited in an optical fiber in different ways. The simplest uses the fact that any linearly polarized soliton ($S_3 = 0$) (except that which coincides with the fast soliton) can be used as an input. This is clearly seen in Fig. 7.5.

7.13.6 Separatrix solution

Solution (7.45), when $k \to 1$, and solution (7.46), when $k_1 \to 1$, have a common limit in the form of the separatrix solution:

$$S_1 = \frac{\delta^2}{4\beta g} \left[\frac{2}{\cosh^2(\delta\xi)} - 1 \right] - \frac{\beta}{g},$$

$$S_2 = \pm \frac{\delta^2}{2\beta g} \frac{\sinh(\delta\xi)}{\cosh^2(\delta\xi)}, \qquad (7.49)$$

$$S_3 = \pm \frac{\delta}{g \cosh(\delta\xi)},$$

where $\delta = \sqrt{4\beta[gS_0 - \beta]}$. The evolution parameter W equals S_0 for this solution. At $\xi \to -\infty$, the trajectory corresponding to this solution tends to the point $\mathbf{S} = (-S_0, 0, 0)$. The Stokes vector moves away from this point exponentially when ξ increases. Hence, in the vicinity of this point, the solution can be approximated by:

$$S_1 = \frac{\delta^2}{4\beta g} - \frac{\beta}{g}, \quad S_2 = \pm \frac{\delta^2}{\beta g} \exp(\delta\xi), \quad S_3 = \pm \frac{2\delta}{g} \exp(\delta\xi). \qquad (7.50)$$

The Stokes vector moves exponentially towards the same point when $\xi \to \infty$. Hence, close to this point, the solution can be approximated by:

$$S_1 = \frac{\delta^2}{4\beta g} - \frac{\beta}{g}, \quad S_2 = \mp \frac{\delta^2}{\beta g} \exp(-\delta\xi), \quad S_3 = \pm \frac{2\delta}{g} \exp(-\delta\xi). \qquad (7.51)$$

Two trajectories start at the point $\mathbf{S} = (-S_0, 0, 0)$, and two trajectories terminate at this point, so it is a saddle-type point. The separatrices are shown in Fig. 7.5b as two closed loops starting and finishing at the point $\mathbf{S} = (-S_0, 0, 0)$.

7.14 Instability of the fast soliton

The linearly polarized solitons along the slow axis (lower branch in the H–Q diagram) are stable for $0 < A < 1$. This is to be expected because, when $0 < A < 1$, there are no other stationary soliton states below this curve to which they could converge during propagation. The linearly polarized solitons along the fast axis (upper branch in the H–Q diagram) are unstable in a certain range of parameters. This instability was first observed in numerical simulations by Blow *et al.* (1987). The perturbation growth rates at $A = 2/3$, which were found numerically by Akhmediev *et al.* (1995), are shown in Fig. 7.7. The growth rates are complex in general. Only the real parts of the growth rates are shown in the figure.

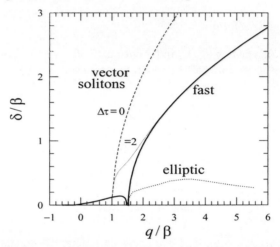

Figure 7.7 *Instability growth rates for the fast soliton (solid curves), for the elliptically polarized soliton (lower dashed curve), and for vector solitons with $\Delta\tau = 0$ (leftmost dashed curve) at $\Delta\tau = 2$ (dotted curve).*

The growth rate for the fast soliton consists of two parts. The growth rate is purely real when $q/\beta > 1.53$. This part of the growth rate curve can be described within the framework of the approximation of average profile. Expressions (7.50) and (7.51) show that, when $\beta/g < S_0$, the point $\mathbf{S} = (-S_0, 0, 0)$ is of saddle type and that fast solitons are unstable. Obviously, δ is the growth rate of this instability. Let us estimate it by calculating the value of g. The point $\mathbf{S} = (-S_0, 0, 0)$ corresponds to the fast linearly polarized soliton (7.15). If it is used as an initial condition, then $X(\xi) = 0$ and $Y(\xi) = 1$. For a silica fiber, $A = 2/3$, and

$$g = \frac{4}{9}(q + \beta). \tag{7.52}$$

Hence, the point $\mathbf{S} = (-S_0, 0, 0)$ is unstable if $g < \beta$ or $q > \frac{5}{4}\beta$. In the alternative case, $g > \beta$ or $q < \frac{5}{4}\beta$, it is stable. The instability growth rate is equal to:

$$\delta = \frac{4}{3}\sqrt{\beta\left(q - \frac{5}{4}\beta\right)}. \tag{7.53}$$

This expression qualitatively describes the exact curve plotted in Fig. 7.7. The slight shift in the edge of the stability region ($q/\beta = 1.53$ rather than 1.25 given by (7.53)) is due to the approximation we have made.

In addition to this instability, an instability with a complex growth rate exists in the interval $-0.2 < q/\beta < 1.475$. Only the real part of the growth rate is shown in Fig. 7.7. This instability is related to the oscillations of the pulse shape during propagation, and radiation due to these oscillations. It cannot be described using the above simple approximation. For values of q/β outside of these two intervals, the fast waves are stable, or at least their perturbation growth rates are very small. For example, in the small range $1.475 < q/\beta < 1.53$, the frequency of the oscillations is very close to zero. The radiation is small in this case and the fast soliton is stable. All elliptically polarized solitons are also weakly unstable. The growth rate for them is shown by the dotted line in Fig. 7.7. Their instability is caused purely by radiation.

7.15 Radiation of energy from the soliton

Analytical solutions are obtained in the approximation of constant shape. Of course, the shape is not fixed but oscillates slightly. This implies that a small amount of radiation is emitted in each period of these oscillations. Thus, the values of Q and H for the main pulse have to decrease. We assume that the values of Q and H decrease adiabatically with ξ. Hence, in Fig. 7.3, the representative point moves down and to the left. Examples of evolution are given by the lines with arrows in Fig. 7.3. The rate of this process depends on $\beta/|g|$, and is negligible in two cases: $\beta/|g| \gg 1$ (or $q/\beta \to 0$) and $\beta/|g| \ll 1$ (or $q/\beta \to \infty$). It becomes faster when β and g are comparable, and the representative point is close to the point of bifurcation.

When radiation is taken into account, the solutions behave qualitatively as described above, but the paths traced out in Fig. 7.5 change slowly from one closed loop to another with a slightly smaller value of H. For an infinite length of propagation, any initial condition will converge to a slow soliton, because this state has the lowest H at a given Q. The second invariant, W, is equal to $-S_0$ for the slow soliton and to $+S_0$ for the fast soliton. The value of S_0 also decreases adiabatically, but this process does not change the qualitative features of the trajectories.

After a while, an initial condition in the form of a soliton-like pulse will separate into a main pulse and dispersed radiation (Fig. 7.8). Due to this radiation, Q_{sol} and H_{sol} for the main pulse change with propagation. These

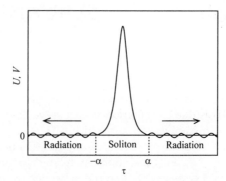

Figure 7.8 *The full solution consists of a soliton plus radiation.*

values are defined as in (7.2) and (7.4), but with a finite range of integration which includes the main pulse. Let us calculate these changes. The overall values of Q and H are conserved, so that the rates of change of Q_{sol} and H_{sol} are related to the energy and Hamiltonian flows at the boundaries of the main pulse. We divide the t axis into three parts, where the central part contains the main pulse and the two other parts contain the radiation. The energy and Hamiltonian flows are calculated at the points $t = -\alpha$ and $t = \alpha$.

To find the explicit relation, we obtain two differential conservation laws (cf. section 2.5) which can be derived from the initial equations (7.8):

$$\frac{\partial}{\partial \xi} \rho_Q = -\frac{\partial}{\partial \tau} j_Q,$$

$$\frac{\partial}{\partial \xi} \rho_H = -\frac{\partial}{\partial \tau} j_H, \tag{7.54}$$

where

$$\rho_Q = |U|^2 + |V|^2$$

and

$$\rho_H = \frac{1}{2}(|U_\tau|^2 + |V_\tau|^2) - \frac{1}{2}(|U|^4 + |V|^4) - (UV^* + U^*V)$$

are the energy and Hamiltonian densities, respectively, and

$$j_Q = -(i/2)(U_\tau U^* - U_\tau^* U + V_\tau V^* - V_\tau^* V)$$

and

$$j_H = -\frac{1}{2}(U_\tau U_\xi^* + U_\tau^* U_\xi + V_\tau V_\xi^* + V_\tau^* V_\xi)$$

are, respectively, the energy and Hamiltonian flows. Therefore, (7.54) are generalized one-dimensional continuity equations. Integration of (7.54) over the interval $(-\alpha, \alpha)$ gives

$$\frac{\partial}{\partial \xi} Q_{sol} = -[j_Q(\alpha) - j_Q(-\alpha)],$$

$$\frac{\partial}{\partial \xi} H_{sol} = -[j_H(\alpha) - j_H(-\alpha)]. \tag{7.55}$$

Equations (7.55) determine the rate of change of H_{sol} and Q_{sol} in terms of the energy and Hamiltonian flows outside the main pulse. These flows are determined by radiation from the main pulse.

The radiation can be considered as a Fourier series of small-amplitude linear waves. The linearized equations (7.8) give two possible types of linear wave (symmetric and anti-symmetric) for each frequency, ω_n, of radiation:

$$u = a_n \exp(i\omega_n \xi + i\Omega_n \tau),$$

$$v = \pm a_n \exp(i\omega_n \xi + i\Omega_n \tau). \tag{7.56}$$

The dispersion relation for these waves has the form

$$-2\omega_n \pm 2 = \Omega_n^2. \tag{7.57}$$

Now, substituting (7.56) into (7.55), one finds

$$\frac{\partial Q_{rad}}{\partial \xi} = \sum_n S_n, \quad \frac{\partial H_{rad}}{\partial \xi} = \sum_n \omega_n S_n, \tag{7.58}$$

where S_n is the total energy flow for the Fourier component at the frequency $\omega = \omega_n$. The values of ω_n are defined by oscillation frequencies of the soliton. Now we can simplify the problem, making the following assumptions:

(i) The largest contribution to the radiation comes from the soliton amplitude oscillations at the fundamental (lowest) frequency, ω_0. The contributions from all higher harmonics of ω_0 are small.

(ii) The radiation wave has the same symmetry (i.e. is anti-symmetric) as the amplitude oscillations of the two main pulse components.

These assumptions allow us to take into account only the ω_0 term in each of the equations in (7.55), and thus obtain the result

$$\frac{\partial H_{sol}}{\partial Q_{sol}} = -\omega_0(H_{sol}, Q_{sol}). \tag{7.59}$$

In this equation, the frequency of oscillations, ω_0, depends on the energy and Hamiltonian of the main pulse. This frequency is always negative due to the dispersion relation (7.57). Thus, we can write

$$\frac{\partial H_{sol}}{\partial Q_{sol}} = |\omega_0(H_{sol}, Q_{sol})|. \tag{7.60}$$

Expression (7.60) is a dynamical equation, based on the two main conservation laws of the initial system, (7.8). The frequency ω_0 can be expressed in terms of complete elliptic integrals. Then (7.60) gives the slope of the trajectory at any point on the H–Q diagram. H can only decrease as Q decreases, so that the direction of evolution can only be to the left and down in the H–Q diagram. Several numerical examples (a to f) of this transformation, which start from various initial conditions, are indicated by arrows in Fig. 7.3.

7.16 Numerical examples

When q/β is below the bifurcation point, only solutions with rotating phase can exist. Figure 7.9 shows the evolution of the Stokes parameters for initial conditions corresponding to a fast soliton with $q/\beta = 0.5$. In this region, the instability of the fast soliton is of radiative type. In terms of the Stokes parameters, we start from the point $\mathbf{S} = (-S_0, 0, 0)$. This point is unstable inside the intervals we discussed before, and any perturbation, such as the addition of a small value $u = \mu v$ ($\mu \ll 1$) to the stationary solution, makes it diverge from the initial point. The main pulse decreases its energy on propagation, and therefore S_0 decreases and the Poincaré sphere shrinks in volume. The trajectory rotates around the S_1 axis along the curves with constant W, while W decreases adiabatically. This decrease is negligible initially, as the amplitude of the oscillations is small and radiation is also minimal. The decrease becomes faster after some time, when the amplitude of the oscillations increases. The decrease becomes noticeably slower later on ($S_1 > 0$), when the oscillations of the amplitudes of the fast and the slow components become smaller. This is the reason why the trajectory lines are denser to the left of the plane, $S_1 = 0$.

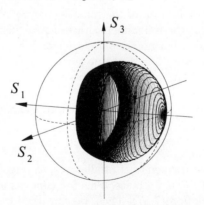

Figure 7.9 *Evolution of the integrated Stokes parameters when the initial condition has the form of a fast soliton with $q/\beta = 0.5$.*

Figure 7.10a shows the evolution of the Stokes parameters when the initial condition is a fast soliton with a small perturbation, and we have set

q/β (= 100) to be well above the bifurcation point. The trajectory is close to the separatrix. The perturbation is chosen in such a way that the solution with rotating phase is excited. After moving along the first separatrix (with $S_3 > 0$) and returning to the saddle-type point, it follows the second separatrix (with $S_3 < 0$). The trajectory could follow the same separatrix for a different perturbation. The deviation of the differential Stokes parameter $s_0(\tau = 0)$ at $\tau = 0$ from its initial value (at $\xi = 0$) is less than 2% during this evolution. This indicates that the pulse shape changes very slightly, and that the approximation of average profile is good enough to describe this simulation. Radiation emission is also very small, and consequently the trajectory is predicted to high accuracy by this approximation.

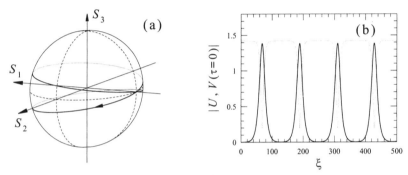

Figure 7.10 *(a) Evolution of the integrated Stokes parameters for $q/\beta = 100$. The initial condition is a fast linearly polarized wave (7.15). A periodic solution close to the separatrix is excited. (b) Evolution of pulse amplitudes $|u|$ and $|v|$ at the centre of pulses ($\tau = 0$).*

Figure 7.10b shows the evolution of the field amplitudes at the centres of the two components. This figure shows that, initially, the energy is concentrated largely in the fast mode and that it is certainly unstable, so that, after some distance of propagation, the energy switches to the slow mode. Because of the recurrency which takes place for separatrices, the energy switches back to the fast mode, where it stays for a longer distance. The behaviour is almost periodic. The periodicity becomes weaker for smaller values of q/β.

When the initial condition is closer to the bifurcation point but is still above it, the radiation become appreciable. Figure 7.11 shows this type of fast transformation. The trajectory starts very close to a stationary point corresponding to an elliptically polarized soliton, makes several loops corresponding to a solution with oscillating phase, then crosses the separatrix and transforms into a solution with rotating phase. This trajectory ends up converging to a slow soliton. Its radius in the Poincaré sphere decreases, so that a large amount of radiation is emitted in this process. In Fig. 7.11, only three parts of the full trajectory are plotted.

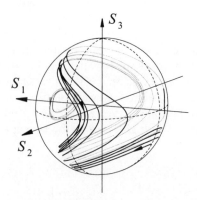

Figure 7.11 *Evolution of the integrated Stokes parameters. The initial condition is chosen close to an elliptically polarized soliton with $q/\beta = 10$.*

7.17 Approximation for long soliton period

If the soliton period is longer than the linear beat length, then the soliton energy Q is well below the bifurcation point. The energy exchange between the two components U and V is small, and S_1 in Fig. 7.5a changes only by a small amount. This means that the energy exchange terms in (7.8) can be ignored. This has been done in numerical simulations by Islam *et al.* (1990). The width of the soliton is greater than the time offset accumulated between the two components in the fiber. Hence, we can still ignore the difference in group velocities. Then (7.8) can be written in the form

$$iU_\xi + \beta U + \tfrac{1}{2}U_{\tau\tau} + (|U|^2 + A|V|^2)U = 0,$$

$$iV_\xi - \beta V + \tfrac{1}{2}V_{\tau\tau} + (A|U|^2 + |V|^2)V = 0. \tag{7.61}$$

The set (7.61) conserves energies separately in each component:

$$Q_1 = \int\limits_{-\infty}^{\infty} |U|^2 d\tau = \text{const.},$$

$$Q_2 = \int\limits_{-\infty}^{\infty} |V|^2 d\tau = \text{const.} \tag{7.62}$$

Thus, the Stokes parameter, S_1, is constant. If two components are equal, then the nonlinear terms in equations (7.61) produce equal changes in the phase delay of solitons. Thus an input soliton which is linearly polarized at 45° relative to the principal axes of the birefringent fiber will not suffer any nonlinear phase shift between its components. Any small deviation of a linearly polarized soliton relative to 45° gives an additional nonlinear shift in phase difference between its components. This can be registered by a half-wavelength plate and polarizer at the fiber output. This description and the one given by (7.48) are equivalent. In practice, for small soliton amplitudes, we need long fibers to accumulate the nonlinear intensity-dependent phase shift. This would correspond to many rotations of the Stokes vector around the S_1-axis.

7.18 Transformation to rotating frame

By representing the field components in the form

$$U = U'(\xi, \tau, q)e^{i\beta\xi}, \quad V = V'(\xi, \tau, q)e^{-i\beta\xi}, \tag{7.63}$$

equations (7.8) become:

$$iU'_\xi + \tfrac{1}{2}U'_{\tau\tau} + (|U'|^2 + A|V'|^2)U' + (1 - A)V'^2U'^*e^{-i4\beta\xi} = 0,$$
$$iV'_\xi + \tfrac{1}{2}V'_{\tau\tau} + (A|U'|^2 + |V'|^2)V' + (1 - A)U'^2V'^*e^{i4\beta\xi} = 0. \tag{7.64}$$

Obviously, transformation (7.63) does not change the moduli of the field components, but it removes the oscillatory factors from the solution at low intensities. However, these oscillations are the cause of birefringence in the linear limit. Hence, to compare the results with experiment, it is more appropriate to use the variables U and V.

The Stokes parameters s''_i, defined for the U' and V' variables, are related to the Stokes parameters s_i, defined above, by:

$$s''_0 = (|U'|^2 + |V'|^2) = s_0,$$

$$s''_1 = (|U'|^2 - |V'|^2) = s_1,$$

$$s''_2 = (U'^*V' + U'V'^*) = s_2\cos(2\beta\xi) - s_3\sin(2\beta\xi), \tag{7.65}$$

$$s''_3 = -i(U'^*V' - U'V'^*) = s_2\sin(2\beta\xi) + s_3\cos(2\beta\xi).$$

These equations show that the Stokes parameters for the original variables evolve as the Stokes parameters for the new variables but rotate with angular velocity 2β around the common s_1 axis. Hence, any solution obtained for one set of variables can be transformed into a solution for the other set simply by using this rotation.

This transformation allows the terms proportional to βU and βV to be removed from the system. This can be useful in analysing the nonlinear phase shift between the components. However, the state of polarization is clear only in the original frame and is lost in this new frame.

7.19 Multi-soliton solutions

In addition to one-soliton solutions of (7.8), there are multi-soliton solutions. If both components, u and v have the same phase, then (7.13) can be rewritten as

$$\frac{1}{2}u_{\tau\tau} - (q - \beta)u + (u^2 + v^2)u = 0,$$
$$\frac{1}{2}v_{\tau\tau} - (q + \beta)v + (u^2 + v^2)v = 0, \tag{7.66}$$

where we consider u and v to be real. A two-soliton solution of (7.66) is the following (Akhmediev *et al.*, 1989):

$$u = -\frac{\lambda_2\sqrt{\lambda_1^2 - \lambda_2^2}\,\sinh(\lambda_1\tau_1)}{\left(\lambda_1\cosh(\lambda_1\tau_1)\,\cosh(\lambda_2\tau_2) - \lambda_2\sinh(\lambda_1\tau_1)\,\sinh(\lambda_2\tau_2)\right)}, \tag{7.67}$$

$$v = \frac{\lambda_1\sqrt{\lambda_1^2 - \lambda_2^2}\,\cosh(\lambda_2\tau_2)}{\left(\lambda_1\cosh(\lambda_1\tau_1)\,\cosh(\lambda_2\tau_2) - \lambda_2\sinh(\lambda_1\tau_1)\,\sinh(\lambda_2\tau_2)\right)}, \tag{7.68}$$

where $\lambda_1 = \sqrt{2(q + \beta)}$ and $\lambda_2 = \sqrt{2(q - \beta)}$, $\tau_1 = \tau - \tau_{01}$, $\tau_2 = \tau - \tau_{02}$, and τ_{01} and τ_{02} are arbitrary real constants. The two components of the solution have the same phase. Hence, at any τ, the state of polarization for this solution is linear. The angle of polarization changes along the τ axis.

The solution defined by (7.67) and (7.68) is a three-parameter family of solutions with arbitrary parameters q, τ_{01} and τ_{02}. One of them corresponds to trivial translations along the τ axis. Then the parameter $\Delta\tau = \tau_1 - \tau_2$ corresponds to the distance between the two solitons involved in the nonlinear superposition. For large $\Delta\tau$, it splits into two separate solitons, one fast and one slow (Fig. 7.12b). When $\Delta\tau$ is small, the solution has a more complicated shape. The particular case of the solution (7.67)-(7.68) having $\Delta\tau = [(\lambda_2 - \lambda_1)/(2\lambda_2\lambda_1)]\,\ln[(\lambda_2 + \lambda_1)/(\lambda_2 - \lambda_1)]$ has been obtained by Christodoulides and Joseph (1988a).

Solution (7.67)-(7.68) exists when $q > \beta$, and, in the limit $q = \beta$, it converges to a polarized solution, (7.15). It can be shown that the energy Q of the two-soliton solution (7.67)-(7.68) is the direct sum of energies of its two constituents (the fast and slow solitons) for each value of q. This shows the integrability of the reduced problem when U and V have the same phase. Hence, this family of solutions can be represented on a Q versus q diagram (Fig. 7.2) as a curve above the curve for fast solitons. Each point on this curve is a one-parameter family of solutions with variable $\Delta\tau$.

The two-soliton solution is unstable on propagation for any q and any $\Delta\tau$. The growth rate of instability is shown in Fig. 7.7 by the dashed and dotted curves for two different $\Delta\tau$. The growth rate is highest (dashed curve) when $\Delta\tau = 0$, and it decreases for non-zero $\Delta\tau$. When $\Delta\tau$ is much higher than the width of either soliton, the growth rate practically coincides with that of the fast soliton. This is expected, because one of the components

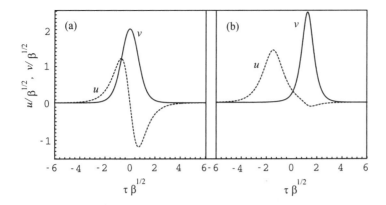

Figure 7.12 *Profiles of the two-soliton solution with $q/\beta = 2$ and $\Delta\tau\sqrt{\beta}$ equal to: (a) 0, (b) 2.*

of the solution in this case is the unstable fast soliton. At small $\Delta\tau$, the two-soliton solution splits, on propagation, into two separate one-soliton solutions moving away from each other. The process of splitting is shown in Fig. 7.13. The behaviour of each soliton after splitting can be described in the approximation of average profile. In particular, the two resulting solitons in Fig. 7.13 initially oscillate around the elliptically polarized soliton states.

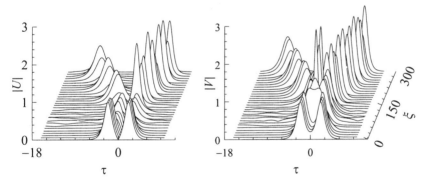

Figure 7.13 *Perspective plot of propagation of two-soliton solution with a small initial perturbation. Parameters of the simulation are $A = 2/3$, $\Delta\tau = 0$ and $q/\beta = 100$.*

Solutions of (7.8) with more than two solitons also exist, but analytical expressions for them are quite complicated. A pulse of general form, with zero boundary conditions, which is not necessarily an exact stationary solution, may contain several solitons. In general, if the initial condition is such a pulse, it will split into several one-soliton pulses and radiation. After complete separation, when the radiation has dispersed, the evolution of each soliton can be described on the basis of a finite-dimensional dynamical system.

7.20 The role of the difference in group velocities

If the birefringence is high, then we have to take into account not only the difference in phase velocities, but also differences in group velocities. Then we have to retain all the terms in equations (7.1):

$$iU_\xi + i\delta U_\tau + \beta U + \tfrac{1}{2}U_{\tau\tau} + (|U|^2 + A|V|^2)U + (1-A)V^2U^* = 0,$$

$$iV_\xi - i\delta V_\tau - \beta V + \tfrac{1}{2}V_{\tau\tau} + (A|U|^2 + |V|^2)V + (1-A)U^2V^* = 0.$$
$$(7.69)$$

As the birefringence is high, the trajectories corresponding to soliton solutions rotate rapidly around the S_1 axis on the Poincaré sphere. This rotation can be removed if we use a rotating frame (section 7.18).

This approach has been developed by Menyuk (1987). Transformation of the components to a rotating frame removes the linear terms proportional to β:

$$U = U'(\xi, \tau, q)e^{i\beta\xi}, \quad V = V'(\xi, \tau, q)e^{-i\beta\xi}. \tag{7.70}$$

Equations (7.69) become:

$$iU'_\xi + i\delta U'_\tau + \tfrac{1}{2}U'_{\tau\tau} + (|U'|^2 + A|V'|^2)U' + (1-A)V'^2U'^*e^{-4i\beta\xi} = 0,$$

$$iV'_\xi - i\delta V'_\tau + \tfrac{1}{2}V'_{\tau\tau} + (A|U'|^2 + |V'|^2)V' + (1-A)U'^2V'^*e^{4i\beta\xi} = 0.$$
$$(7.71)$$

The last terms are oscillatory, and if β is large, then they are fast and can be averaged to zero. We thus obtain:

$$iU'_\xi + i\delta U'_\tau + \tfrac{1}{2}U'_{\tau\tau} + (|U'|^2 + A|V'|^2)U' = 0,$$
$$iV'_\xi - i\delta V'_\tau + \tfrac{1}{2}V'_{\tau\tau} + (A|U'|^2 + |V'|^2)V' = 0. \tag{7.72}$$

Removing these last terms means that there is no longer any energy exchange between the components, so they always maintain their initial energies. This gives the physics, and it can easily be interpreted. Each component creates a potential well through the cross-phase modulation (CPM) term, and this attracts the other component. As a result, the two components can move together with the same group velocity or oscillate around each other if their relative velocity does not exceed a certain threshhold. When the relative velocity is higher, the two components cannot stay in the potential well and the coupled state breaks down. The detailed dynamics of these oscillations has been studied by Ueda and Kath (1990). The stability of the coupled states relative to various perturbations has been considered by Menyuk (1987).

7.21 Transformation to different frequencies

Equations (7.72) can be simplified further using the transformation (Menyuk, 1989; Ueda and Kath, 1990)

$$U' = U'' \exp\left(-i\frac{\delta^2}{2}\xi + i\delta\tau\right),$$

$$V' = V'' \exp\left(-i\frac{\delta^2}{2}\xi - i\delta\tau\right),$$

which gives

$$iU''_\xi + \tfrac{1}{2}U''_{\tau\tau} + (|U''|^2 + A|V''|^2)U'' = 0,$$

$$iV''_\xi + \tfrac{1}{2}V''_{\tau\tau} + (A|U''|^2 + |V''|^2)V'' = 0.$$

(7.73)

Physically, this transformation corresponds to separating the central frequencies of the two components. If the pulse energy is high enough, the nonlinearity locks the group velocities of the two components. Equalizing the group velocities, however, causes them to have different frequencies, so that the group velocities at these frequencies become equal.

Equations (7.73) admit stationary solutions with two even non-zero components. Exact analytical expressions for them are unknown but they can be found numerically (Inoue, 1976) or by using a variational technique (Ueda and Kath, 1990; Afanasjev, 1995c). The amplitude ratio between the two components depends on the initial conditions and can be arbitrary.

Equations (7.73) can be reduced to a coupled ordinary differential equation using the ansatz:

$$U'' = X(\tau)\exp(iq_1\xi),$$

$$V'' = Y(\tau)\exp(iq_2\xi),$$

where the propagation constants q_1 and q_2 can take two different values. For stationary solutions, the functions X and Y must be real. Hence, the set (7.73) is reduced to a coupled set of ordinary differential equations, viz.

$$\tfrac{1}{2}X_{\tau\tau} - q_1 X + (|X|^2 + A|Y|^2)X = 0,$$

$$\tfrac{1}{2}Y_{\tau\tau} - q_2 Y + (A|X|^2 + |Y|^2)Y = 0,$$

(7.74)

which is integrable only for certain values q_i and A (Eleonskii *et al.*, 1991). The problem is equivalent to that for the two-dimensional motion of a particle moving in the potential well

$$W(X,Y) = \frac{1}{4}(X^4 + 2AX^2Y^2 + Y^4) - \frac{1}{2}q_1 X^2 - \frac{1}{2}q_2 Y^2.$$

For solitons, the condition $W(X, Y) = 0$ defines the boundary of allowed motion. This boundary is shown in Fig. 7.14 for three different sets of the parameters q_1 and q_2.

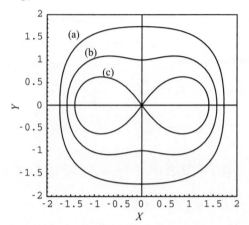

Figure 7.14 *Boundary of allowed motion in the (X, Y) plane for different values of q_1 and q_2: (a) $q_1 = 1.5$, $q_2 = 1.5$, (b) $q_1 = 1.25$, $q_2 = 1.0$, and (c) $q_1 = 1.0$, $q_2 = -0.5$.*

The solution with even components can be approximated by sech functions
$$X = \sqrt{I} \cos \alpha \operatorname{sech}(\tau), \qquad Y = \sqrt{I} \sin \alpha \operatorname{sech}(\tau),$$
where α is the angle of polarization. The dependence of the intensity on the angle of polarization can be found using a variational technique (Afanasjev, 1995c):
$$I(\alpha) = \frac{4}{(3 + A) + (1 - A) \cos 4\alpha}.$$
This function, for three different values of A, is shown in Fig. 7.15.

The approximate values of q_1 and q_2 can be expressed, in this case, in terms of the same parameter α:
$$q_1 = 2\frac{(1 + A) + (1 - A) \cos 2\alpha}{3 + A + (1 - A) \cos 4\alpha} - \frac{1}{2}$$
$$q_2 = 2\frac{(1 + A) - (1 - A) \cos 2\alpha}{3 + A + (1 - A) \cos 4\alpha} - \frac{1}{2}$$
They become equal when $\alpha = 45°$.

Equations (7.73) are integrable when $A = 3$ (Eleonskii et al., 1991). A particular solution in this case (using $q_1 = q_2$) can be written in the form
$$X = \frac{3}{\sqrt{2}}[\operatorname{sech}(\tau + \tau_0) + \operatorname{sech}(\tau - \tau_0)],$$

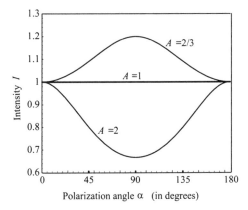

Figure 7.15 *Soliton intensity I versus polarization angle α for $A = 2/3$, $A = 1$ and $A = 2$.*

$$Y = \frac{3}{\sqrt{2}}[\text{sech}(\tau + \tau_0) - \text{sech}(\tau - \tau_0)],$$

where τ_0 is an arbitrary parameter. The solution is asymmetric in τ and unstable in ξ. Other examples of exact solutions can be found in the work of Eleonskii *et al.* (1991). Unfortunately, they require different values of A in each of the equations in (7.73).

Equations (7.73) are similar in form to those in (7.61), and they can be made identical using a suitable transformation of (7.61). However, they are related to different physical quantities. Equations (7.73) describe the evolution of two fields at two different frequencies. Hence, they cannot be represented on the same Poincaré sphere as the field components U and V.

Equations (7.73) are the most frequently used in studies of the dynamics of two–component fields. For any particular case, it is important to refer to the physical meaning of variables involved in (7.73), in order to draw physically meaningful conclusions.

7.22 Stationary solutions in the presence of group velocity delay

Averaging out the fast oscillatory terms is an approximation which describes the phenomenon with great accuracy. What will happen, however, if we aim for even greater accuracy? Let us retain the last terms in equations (7.69), to allow for exchange of energy between the components, and solve the problem numerically. The energy exchange makes the physics more complicated. Exact analytic solutions do not exist, but numerical analysis allows us to observe the most important features of the solution.

Let us start this analysis with known stationary solutions and their stability. Equations (7.69) again have two simple stationary solutions, viz. linearly polarized soliton waves along the slow axis,

$$u = \frac{\sqrt{2(q-\beta)}}{\cosh(\sqrt{2(q-\beta)}(\tau - \delta\xi))}, \qquad v = 0; \qquad (7.75)$$

and linearly polarized soliton waves along the fast axis,

$$u = 0, \qquad v = \frac{\sqrt{2(q+\beta)}}{\cosh(\sqrt{2(q+\beta)}(\tau + \delta\xi))}. \qquad (7.76)$$

Each pulse has its own group velocity. However, each of these solutions has zero 'momentum' ($M = 0$), in spite of the fact that they are moving in the chosen frame of reference. The value of the 'momentum' is not necessarily related to the motion of the peak amplitude of the solution.

7.23 Soliton states with locked phase and group velocities

The stability analysis of stationary solutions (7.75) and (7.76) differs from that of solutions with zero δ. For a state which is strictly stationary in a moving frame of reference, the stability analysis must be done in that moving frame. For this reason, it is convenient to make a change of co-ordinates to obtain a new frame of reference where these solutions remain fixed:

$$\tau = \tau' \pm \delta\xi, \quad \xi = \xi. \qquad (7.77)$$

Equations (7.69) then become:

$$iu_\xi + i2\delta u_{\tau'} - (q-\beta)u + \tfrac{1}{2}u_{\tau'\tau'} + (|u|^2 + A|v|^2)u + (1-A)v^2 u^* = 0,$$

$$iv_\xi - (q+\beta)v + \tfrac{1}{2}v_{\tau'\tau'} + (A|u|^2 + |v|^2)v + (1-A)u^2 v^* = 0; \qquad (7.78)$$

or

$$iu_\xi - (q-\beta)u + \tfrac{1}{2}u_{\tau'\tau'} + (|u|^2 + A|v|^2)u + (1-A)v^2 u^* = 0,$$

$$iv_\xi - i2\delta v_{\tau'} - (q+\beta)v + \tfrac{1}{2}v_{\tau'\tau'} + (A|u|^2 + |v|^2)v + (1-A)u^2 v^* = 0. \qquad (7.79)$$

The perturbation growth rates for these solutions have been calculated by Soto-Crespo *et al.* (1995b). For small δ, namely $\delta < 1$, the results are similar to those obtained for $\delta = 0$, in that slow solitons are stable while fast solitons are unstable. The regions of instability are those where the eigenvalues have non-zero real parts. The instability growth rates for non-zero δ are shown in Fig. 7.16b. In this case, the fast mode is unstable, as before, but the onsets of the two instability regions shift to higher values of q.

At non-zero δ, the slow solitons can also be unstable, but they only

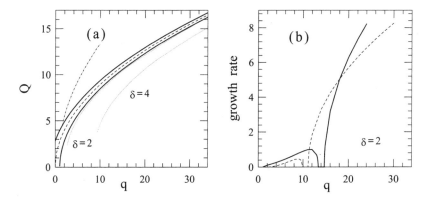

Figure 7.16 *(a) Energy–dispersion diagram for the fast and slow linearly polarized solitons and coupled soliton states (upper and lower solid lines, respectively). The dashed line is for a single linearly polarized soliton in the absence of bire-fringence. The dot-dashed line is for elliptically polarized soliton states at $\delta = 0$. (b) Perturbation growth rates for the fast and slow linearly polarized solitons for $\delta = 2$. The solid line shows the growth rate for the fast soliton, while the dotted line is for the slow soliton.*

become unstable at a certain threshold value, viz. $\delta \approx 1.1$. Figure 7.16b shows the perturbation growth rates of the slow (dashed line) and fast (solid line) modes as a function of q for $\delta = 2$. These curves consist of two segments, corresponding to zones where the perturbation with the largest growth rate has either complex (low q) or purely real eigenvalues (high q). The curves for higher values of δ have the same features as these two curves: when q is higher, the perturbation growth rates are higher, and the region with real perturbation eigenmodes is broader. The instability of slow solitons means that other soliton states, located below the curve for slow solitons in Fig. 7.16a, exist at values of $\delta > 1.1$.

As an example, Fig. 7.17a represents the propagation of a slow soliton, clearly showing that it is unstable. When the slow soliton propagates a sufficient distance for the perturbation to develop, it produces a coupled soliton state plus radiation. On the other hand, Fig. 7.17b shows the propagation of a fast soliton. Now the final state consists of a stable coupled soliton state propagating forwards (having the mean group velocity) and a slow soliton which has smaller q ($q = 2$), and is therefore stable (Fig. 7.16).

The cases shown in Fig. 7.17 are just two explicit examples which indicate that, when $\delta > 1.1$, the linearly polarized states along the principal axes become unstable, but stable coupled soliton states can exist. After propagating a certain distance, the pulses reach a stationary profile (Fig. 7.17) which moves as a unit. The figure shows that, upon convergence to it, each component oscillates with a small amplitude around the stationary state,

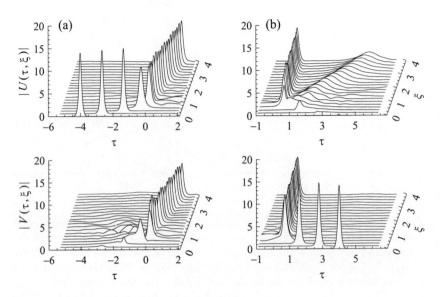

Figure 7.17 *Perspective plots showing the field envelopes* $|U|$ *and* $|V|$ *for propagation of (a) the slow soliton and (b) the fast soliton for* $q = 100$, $\delta = 6.4$. *Because both are unstable, the slow soliton transforms into a coupled soliton state after radiating its excess energy, whereas the fast soliton (which has higher energy) splits into a coupled soliton state and a stable low-q ($q = 2$) slow soliton.*

which therefore must be stable. The amplitudes of the two components are unequal, and their ratio defines the direction of propagation.

When $\delta = 0$, the energy Q and the velocity of the soliton are independent. Soliton solutions which travel with a given velocity can be obtained from the solutions which are at rest in the initial frame by using a Galilean transformation. They have the same stability properties as solitons which do not move in the chosen frame. In the present case, equations (7.8) are not symmetric relative to Galilean transformations. The group velocity of the soliton is defined by the ratio of the soliton components.

These stable solutions are stationary in a suitable frame of reference. The coupled soliton states are a two-parameter family of solutions. For a given q, they can have various values of the inverse group velocity relative to the frame of reference where v_{gr}^{-1} is zero, within the range $-\delta \leq v_{gr}^{-1} \leq \delta$. The corresponding Q values are very similar for the family of solutions with fixed q. In the two limiting cases, when v_{gr}^{-1} coincides with $-\delta$ or $+\delta$ (slow and fast modes, respectively), one of the two components is zero. For a given q, the energy Q of the coupled states is lower than the corresponding energy of the slow soliton. We assume that the state with the lowest energy is stable. The energy-dispersion diagram for these states, found numerically, is given in Fig. 7.16a for $\delta = 2$ and $\delta = 4$ (dotted lines). These curves are located below the curve for the slow linearly polarized soliton. As δ increases, the 'coupled states' curve in this diagram moves down. The coupled states with smallest energy appear at those q values where both the fast and slow modes are unstable.

The soliton states for high values of δ are similar to those found by Aceves and Wabnitz (1989) and Christodoulides and Joseph (1989). For each q, they comprise a one-parameter family of solutions, where the inverse group velocity is the parameter of the family. Although these solutions have the same qualitative properties as those found by Aceves and Wabnitz (1989), the equations which we consider are more complicated. They include second-derivative terms which are responsible for dispersion. As a result, the amplitudes of the two components are not equal when $v_{gr}^{-1} = 0$, while they were equal in the work of Aceves and Wabnitz (1989) and Christodoulides (1989). No analytic solutions have been found so far.

To summarize, the above solutions have a non-trivial dependence on two variables, the parameter q and the velocity v. The Hamiltonian then depends on the energy Q and the momentum M. This is a surface in the three-dimensional space of these variables. The lowest branch of this surface defines the stable solutions.

CHAPTER 8

Pulses in nonlinear couplers

Wave propagation in couplers (Fig. 1.4) at relatively high field intensities is described by coupled nonlinear equations (Jensen, 1982; Maier, 1984). The physics of nonlinear coupler response to c.w. has been presented by Snyder *et al.* (1991b). The nonlinear coupler response to solitons has been described by Paré and Florjanczyk (1990), using a variational approach. This description gives all the basic properties of the simple switching phenomenon. In the more complicated case of long twin-core fibers (Vallée and Essambre, 1994), and sometimes even for the switching operation, we have to use more elaborate approaches (Ankiewicz *et al.*, 1993; Akhmediev and Soto-Crespo, 1994a). An analysis, based on Hamiltonian dynamics, has been presented by Romagnoli *et al.* (1992). To understand the operation of devices based on the coupling of two nonlinear optical fibers, we use an approach which is related to the qualitative analysis of systems having an infinite number of degrees of freedom, as described in earlier chapters. First, we consider the stationary states of our system. Then we consider their stability, and, as a final step, we study their dynamics.

The switching behaviour of the coupler follows from this analysis as a particular case. It is the special transmission response of a coupler with a certain length to a certain input. The 'ideal' transmission would be one which is a step function in terms of the input energy. To approach this behaviour, the input pulse must be a soliton of an individual core. The transmission characteristic of the device in this case is similar to that when c.w. input is used, as it is close to a step function. Different types of couplers, e.g. those with saturable nonlinearity or with unequal cores, can have special properties.

8.1 Couplers with Kerr-type nonlinearity

First, we establish an analogy with birefringent fibers. Moreover, we will show that the equations for pulses in a coupler can be obtained from the equations for pulses in birefringent fiber using simple transformations. In fact, in equations (7.10), if we interpret P and G as the slowly varying pulse amplitudes of the fields in two cores, and $K = \beta$ as the normalized coupling coefficient between the cores, and set $A = 2$, then equations (7.10) describe pulse propagation in dual-core fibers, neglecting cross-phase modulation effects. In terms of the variables P and G defined by (7.9), the equations for pulses in the Kerr-law coupler are

$$iP_\xi + \tfrac{1}{2}P_{\tau\tau} + |P|^2 P + KG = 0,$$
$$iG_\xi + \tfrac{1}{2}G_{\tau\tau} + |G|^2 G + KP = 0. \tag{8.1}$$

The Hamiltonian in this case is

$$H_c = \int\limits_{-\infty}^{\infty} \Big[\frac{1}{2}(|P_\tau|^2 + |G_\tau|^2) - \frac{1}{2}(|P|^4 + |G|^4)$$
$$-K(PG^* + P^*G)\Big]d\tau, \tag{8.2}$$

and the energy Q_c is

$$Q_c = \int\limits_{-\infty}^{\infty} (|P|^2 + |G|^2)d\tau. \tag{8.3}$$

Their associated differential Stokes parameters can be expressed as follows:

$$2s_0 = |P|^2 + |G|^2 = s_0',$$
$$2s_1 = PG^* + P^*G = s_2',$$
$$-2s_2 = -i(P^*G - PG^*) = s_3', \tag{8.4}$$
$$-2s_3 = |P|^2 - |G|^2 = s_1'.$$

Hence, if we use the variables P and G, the analysis is similar to that in Chapter 7, except for the subscript notation and signs. Moreover, equations (8.4) show that the Stokes parameters associated with birefringence in fibers are related to those associated with the nonlinear fiber coupler by a simple permutation of axes and a renormalization.

Equations (8.4) allow us to consider soliton-like pulse propagation for a variety of physical systems within a unified theory. They show that the same trajectories will represent the solutions of (7.10) when the axes are permuted. Physically, this gives different functions in terms of U and V, of course. We should also bear in mind that $A = 2/3$ for a fiber, while $A = 2$ for the coupler.

8.2 Stationary soliton states

Solution (7.14) is transformed to the symmetric soliton state of the coupler

$$P = G = \frac{\sqrt{2(q-K)}}{\cosh(\sqrt{2(q-K)}\tau)} e^{iq\xi}, \tag{8.5}$$

while solution (7.15) is transformed to the anti-symmetric soliton state of the coupler

$$P = -G = i\frac{\sqrt{2(q + K)}}{\cosh(\sqrt{2(q + K)}\tau)} e^{iq\xi}. \tag{8.6}$$

The normalized energies of the symmetric (lower signs) and anti-symmetric (upper signs) states are thus

$$\hat{Q} = \frac{Q_c}{\sqrt{K}} = 4\sqrt{2a} = 4\sqrt{2(\bar{q} \pm 1)} = 4\sqrt{2\left(\frac{q}{K} \pm 1\right)}, \tag{8.7}$$

where $\bar{q} \equiv q/K$. If the coupling is absent (i.e. $K = 0$), the pulses (8.5) and (8.6) degenerate into solitons of a single NLSE:

$$|P| = |G| = \sqrt{u^2 + v^2} = \frac{\sqrt{2q}}{\cosh(\sqrt{2q}\tau)}. \tag{8.8}$$

This solution represents two independent solitons propagating in the cores of the coupler.

8.3 Asymmetric states

Apart from the solutions considered above, there are also soliton states which lack symmetry with respect to permutation $P \leftrightarrow G$, i.e. states where $|P| \neq |G|$. These are the states where the pulse shape in the first core differs from that in the second. As we shall see, they play a pivotal role in switching. Consideration of the coupled equations shows (Akhmediev and Ankiewicz, 1993a) that these states exist over a wide range of spatial frequency offsets, q. In Kerr-law media, symmetric states exist whenever $q/K > 1$, but there is a bifurcation point at $q/K = 5/3$ (where energy $= 8/\sqrt{3}$), and an asymmetric (A-type) branch starts at that point. Near this point, on the asymmetric branch, one component is higher and the other is lower; the eigenfunction representing this difference is $\text{sech}^2\tau$. The two components of these states have different shapes and different amplitudes (Fig. 8.1). The amplitudes become closer to each other as q approaches the point of bifurcation. One of the components approaches zero when $q \to \infty$.

The anti-symmetric states exist when $q/K > -1$, but there is a bifurcation point at $q/K = 1$ and another asymmetric (B-type) branch starts. The two components take on the same shape but have different signs at the point of bifurcation. The two components of these states have quite complicated shapes when q is above the point of bifurcation (Fig. 8.2). These states degenerate into three solitons (two in one core and one in the other) in the limit $q \to \infty$. The B-type states have the highest energy among the states considered for a given value of q. They become important when a pulse has more energy than just one soliton. If the input pulse energy is close to the energy of a single-fiber soliton

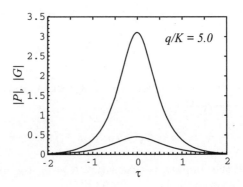

Figure 8.1 *The two components of an A-type state of the Kerr-law nonlinear coupler.*

(at given q), then only the lowest soliton states can influence the total soliton dynamics. Hence, the A-type states are quite important in switching.

Bifurcations for the coupled set of equations (8.1) were found first by Akhmediev et al. (1989a). An approximate model for the appearance of asymmetric states, using a variational approach, has been considered by Paré and Florjanczyk (1990). Transmission characteristics of the coupler have been calculated by Uzunov et al. (1995). Improved trial functions for the A-type states, which give a qualitatively correct ('subcritical') bifurcation, have been given by Malomed et al. (1996).

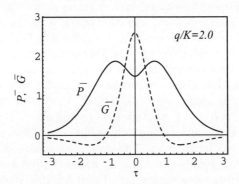

Figure 8.2 *Components of the B-type state of the Kerr-law nonlinear coupler for $q/K = 2$. Here $\bar{P}(\tau)$ and $\bar{G}(\tau)$ are $P(\tau,\xi)$ and $G(\tau,\xi)$ respectively, with the ξ phase variation factored out.*

8.4 The energy–dispersion diagram

All solutions considered above are one-parameter families of solutions. They
have one free parameter, viz. q, which is determined by the energy of the
input pulse. Hence, we can again plot the energy Q of these solutions against
the free parameter q (Fig. 8.3). Each type of solution has its own curve on
the diagram. The two parabolic curves in this diagram, defined by (8.7),
correspond to symmetric and anti-symmetric states, while the lowest curve
is for the asymmetric states (A-type branch). The upper curve (dotted
line) is for B-type states. The plot shows the point of bifurcation (point
M) where the curve for asymmetric solitons splits off from the curve for
symmetric states. The B-type state bifurcation occurs at the point N on
this diagram.

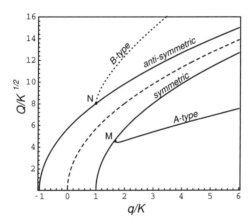

Figure 8.3 *Energy-dispersion diagram for the stationary soliton states in a coupler.*

Transformations (7.9) establish the overall analogy between the solutions
for the birefringent fiber and the nonlinear coupler. However, there are
essential differences due to different values of A. The symmetric $(P = G)$
soliton states of the coupler are mathematically equivalent to the slow
linearly polarized solitons of the birefringent fiber. But if $A > 1$, novel A-
type asymmetric soliton states appear below the lower curve in Fig. 8.3,
and the symmetric pairs lose their stability at the point of bifurcation. If
$A < 1$, then the bifurcation occurs on the upper curve (fast solitons) in
Fig. 7.2. The main difference between the elliptically polarized stationary
soliton states and the A-type asymmetric soliton states is that the curves
for these special solutions split off from the lower curve in Fig. 8.3 but from
the upper curve in Fig. 7.2. This difference arises because the parameter
$A(= 2) > 1$ for the coupler, while $A(= 2/3) < 1$ for the birefringent fiber.

Physically, this occurs because the energy exchange term changes sign at $A = 1$.

8.5 The Hamiltonian versus energy diagram

There is a one-to-one correspondence between the parameter q, the energy and the Hamiltonian. The plot of Hamiltonian versus energy shows all the important physical properties of the system in a natural way. H characterizes the behaviour of any solution (P, G) of this dynamical system with just a single scalar number. Extrema of the Hamiltonian for fixed energy correspond to soliton solutions. Secondly, it shows how close any particular pulse is to a soliton-like pulse. Using this knowledge, we can also predict, to some extent, the evolution of any initial condition in the system. For the symmetric and anti-symmetric states, the dependence of H on Q is given by:

$$\hat{H} = \frac{H_c}{K^{3/2}} = \hat{Q}\left(\pm 1 - \frac{\hat{Q}^2}{96}\right). \tag{8.9}$$

The upper sign is for anti-symmetric states, while the lower is for symmetric states. For asymmetric states, the curves can be found numerically.

The normalized values of H_c and Q_c are plotted in Fig. 8.4 for the Kerr-law coupler. Only the three lowest curves, corresponding to symmetric, anti-symmetric and asymmetric states, are shown. There would be many curves, corresponding to B-type states and coupled states of multi-soliton pulses, on this plot above these ones, but they are not shown. The analogue of Q_c for c.w. inputs is the energy flow, i.e. the power. Note that there is no analogue of H in the c.w. case.

The difference between a birefringent fiber and a coupler has important consequences. In Figs 7.2 and 7.3, the energy and Hamiltonian for elliptically polarized solutions are higher than the energy and Hamiltonian for other stationary states. Hence, these states may be unstable even if they are elliptic fixed points, rather than saddle points, in analysis which uses a reduced number of degrees of freedom. On the other hand, the asymmetric states in Figs 8.3 and 8.4 are absolutely stable (above the point with minimum energy) because they have the lowest energy and Hamiltonian of all the stationary solutions.

Another consequence is the following. The dotted curve in Fig. 8.4 shows the energy of a soliton of an isolated core. It is close to an A-type state towards the right of the diagram. This occurs because, at high energy, the asymmetric A-type state is close (in shape and consequently in functional space) to an individual core soliton. Almost all of its energy is in one core. This means that solitons of an individual core will almost exactly excite the A-type states at high energies (higher than at the point of bifurcation). This is the root of the switching phenomenon in the coupler.

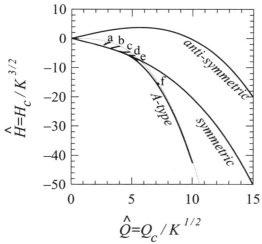

Figure 8.4 *Hamiltonian versus energy diagram for stationary solutions of a non-linear Kerr-law fiber coupler. The dotted line corresponds to a soliton of an individual core. The letters (a – f) and the short lines associated with them correspond to the evolution of input pulses, with the shapes of single core solitons (8.14), towards final symmetric or A-type states. Their input energies and trajectories are given in Fig.8.6.*

8.6 Stokes parameter formalism

Integrated Stokes parameters for the P and G functions can be obtained by permutation of the axes of the birefringent fiber case, as was done for the differential Stokes parameters:

$$2\,S_0 \;=\; S_0', \qquad 2\,S_1 = S_2',$$
$$2\,S_2 \;=\; -S_3', \qquad 2\,S_3 = -S_1'. \tag{8.10}$$

All of the above equations can easily be written in terms of these parameters. For example, we have, instead of (7.26),

$$\tfrac{d}{d\xi}S_0' \;=\; 0,$$
$$\tfrac{d}{d\xi}S_2' \;=\; -2g'S_1'S_3',$$
$$\tfrac{d}{d\xi}S_3' \;=\; 2KS_1' + 2g'S_1'S_2', \tag{8.11}$$
$$\tfrac{d}{d\xi}S_1' \;=\; -2KS_3',$$

where

$$g' = \frac{(1-A)}{2} \frac{\int\limits_{-\infty}^{\infty} f^4 d\tau}{\int\limits_{-\infty}^{\infty} f^2 d\tau} = \frac{g}{2}. \tag{8.12}$$

Trajectories for the $\mathbf{S}'$ vector, describing soliton propagation in a nonlinear coupler, are shown in Fig. 8.5. These trajectories are obtained from Fig. 7.5, using transformations (8.10). There are two stationary (elliptic type) points when $|g|/\beta > S_0$ (Fig.8.5a). They correspond to the stationary solutions (8.5) and (8.6). There are four stationary points when $|g|/\beta < S_0$ (Fig. 8.5b). Two additional points represent asymmetric soliton states. Solutions which move around the stationary elliptic points are periodic. There are two separatrix trajectories connected by a saddle-type point.

We recall that the transformations of (8.10) are a consequence of the transformation $g \rightarrow -g$ with $S_i \rightarrow -S_i$, which, in turn, depends on the inversion symmetry of (7.24). This effectively means that, when changing the sign of g, the correspondence between the fixed points on the Poincaré sphere and the type of solutions also changes. In the case of the birefringent fiber, the elliptic point $\mathbf{S} = (S_0, 0, 0)$ corresponds to the lower (slow) branch of the energy-dispersion diagram, while, for the nonlinear coupler, the equivalent point, $\mathbf{S}' = (0, -S_0', 0)$, corresponds to the upper branch (anti-symmetric soliton state). Both the Hamiltonian and the energy decrease when we move over the Poincaré sphere from the point corresponding to elliptically polarized solitons to that for slow solitons in Fig. 7.5b, but these values increase when we move from points corresponding to asymmetric states to those for anti-symmetric solitons in Fig. 8.5b. This has important consequences when considering pulse evolution, due to the radiation effects.

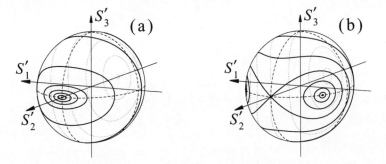

Figure 8.5 *Trajectories on the Poincaré sphere for the nonlinear coupler: (a)* $K > |g| S_0$ *and (b)* $K < |g| S_0$.

The second important difference is related to the value of the parameter

g (because A is different in each case). We note that $|g'|$ for the nonlinear coupler is 1.5 times as large as g for the birefringent fiber. Hence, the birefringent fiber can be viewed as being closer to the integrable case than is the nonlinear coupler. The result is that, at some values of the soliton energy which are slightly below the energy at the bifurcation point (M in Fig. 8.3), both the symmetric and A-type asymmetric states are stable, and an additional unstable A-type state exists. This means that two more fixed points (saddles) can be present on the Poincaré sphere. This is not taken into account in our finite-dimensional approximation. Only two (elliptic) fixed points split off from the point corresponding to the symmetric state (Fig. 8.5b). This qualitative difference can be observed in numerical simulations.

8.7 Stability of soliton states

Although all stationary solutions of (8.1) are pairs of pulses which can propagate without any change in shape or intensity along the coupler, this propagation can be stable or unstable relative to a small perturbation of the soliton profile. The two conserved quantities, Q and H, determine the stability of the stationary solutions. The stationary state is stable if, at fixed energy, the Hamiltonian has a local minimum in function space. We can suppose, using this criterion, that the state with lowest H at any fixed Q in Fig. 8.4 is stable. The states with higher values of Hamiltonian need special consideration.

The stability can be studied numerically using linear stability analysis, and this has been done by Soto-Crespo and Akhmediev (1993). That study confirms the above stability criterion for the states with the lowest Hamiltonian. The symmetric state is stable from zero energy up to the point of bifurcation. The A-type state is stable whenever its slope, dQ/dq, is positive. Note that the criterion $dQ/dq > 0$ is a consequence of that expressed in terms of the Hamiltonian. Hence, the A-type asymmetric state is stable for all q values greater than the point where it has minimum energy. The asymmetric branch thus has a small unstable section close to the point of bifurcation. (It is too small to observe on the scale of Fig. 8.4.) Numerical results of Soto-Crespo and Akhmediev (1993) show that the anti-symmetric state is unstable for the whole range of energies. However, the instability growth rate is different for each q, and can be very small at small energies. Note that the stability criterion ($dQ/dq > 0$), which is valid for the lowest states, is the same as that for nonlinear surface waves governed by a single NLSE. (These will be analysed in Chapter 12).

8.8 Radiation of small-amplitude waves by solitons

Generally speaking, input pulses are not stationary solutions of (8.15). This means that the pulse profile will be reshaped during pulse propagation in the coupler. One consequence of reshaping is that a part of the soliton energy is radiated away from the soliton in the form of small-amplitude waves (Akhmediev and Soto-Crespo, 1994a). The switching properties of these waves are clearly different from the switching properties of solitons. Hence, the radiation can, to some extent, degrade the desirable transmission properties of the coupler. The radiation is small for short (half-beat-length) couplers, but becomes noticeable for long (at least one-beat-length) couplers. Solitons radiate small-amplitude waves at the base frequency (and its harmonics) of energy exchange between the components of the soliton. This frequency is defined by the linear and the nonlinear beat lengths. Hence, the energy of the soliton decreases and the soliton itself converges to one of the stable soliton states. The path of convergence and the amount of radiation depend on the initial energy and Hamiltonian of the pulse. The analysis of this process is similar to that given in Chapter 7.

8.9 Linear and nonlinear beat lengths

If the energy of the pulse is small, i.e. the nonlinear term in (8.1) is negligible, then the cores will swap energy with a certain period in ξ which is equal to π/K. This period is called the *linear beat length*. On the other hand, if we put $K = 0$ in (8.1) (i.e. we remove linear oscillations), then the period of the remaining oscillations due to the phase factor $\exp(iq\xi)$ will depend on the pulse shape and the total energy of the pulse. In this case the cores do not exchange energy, but independent solitons in each core (with equal amplitudes) have their own periodic phase oscillations. Their period is given by

$$L_{NL} = \frac{2\pi}{q} = \frac{4\pi}{A^2}, \tag{8.13}$$

where A is the amplitude of the soliton. It can be called the *nonlinear beat length* because it depends on the soliton's peak intensity. The dynamic behaviour of the soliton is mainly defined by the relation between these two parameters. For small amplitudes, A, when the nonlinear beat length is much longer than the linear, the pulse behaviour is defined mainly by the linear beat length. For the opposite relation between the two parameters, the soliton behaves like a soliton of an individual core, in that the energy exchange is very weak. When the two are comparable, bifurcations occur and new (asymmetric) solutions can appear. The combined beat length in this case can even be infinite.

8.10 Numerical examples

The main difference from the birefringent fiber is that the A-type asymmetric state has minimal H_c at given Q_c, while the anti-symmetric state has maximal H_c at given Q_c. Radiation causes transitions from one closed loop to another in the reverse direction to that which occurs in the birefringent fiber. Thus, depending on the initial condition, the trajectory can converge to a symmetric state or an asymmetric A-type state. Figure 8.6 shows six examples of the evolution of the Stokes parameters for a nonlinear coupler with $K = 1$. The initial condition is a soliton of the unperturbed system, with different values of Q being chosen, i.e.

$$P = 0, \qquad G = \frac{Q}{2} \operatorname{sech}\left(\frac{Q}{2}\tau\right). \qquad (8.14)$$

On the Poincaré sphere, this initial condition corresponds to the point $\mathbf{S'} = (-S_0', 0, 0)$. At small energies ($Q = 3.2$ and $Q = 4.0$), the trajectory rotates around the S_2' axis. Radiation is relatively large (5% of the soliton energy) during the first half-period of this rotation, but then becomes smaller. This shows that the initial condition, even if it is in the form (8.14), is not an exact solution. It has to adjust to a soliton-like pulse before it follows the solution defined by our simple analytic approach. At higher values of Q (e.g. $Q = 4.8$), when the initial condition is closer to the bifurcation point (Fig. 8.6c), the radiation in the initial states of the process is greater, and then the trajectory slowly converges to a stationary symmetric state.

The initial radiation becomes even larger when $Q = 5.2$. For energies very close to this, we have a special situation, and the behaviour is qualitatively outside the domain of our simple model. There are six singular points on the Poincaré sphere, rather than just four. Four of them correspond to A-type states (two stable and two unstable). The trajectory quickly converges to a closed loop enclosing four A-type states and a symmetric state (Fig. 8.6d). This can occur because the energy-dispersion curve for A-type states has a minimum, and so three soliton states can have the same energy. The appearance of the two additional (unstable) A-type states at the same energy is not predicted by our simple approximation of the average profile. This energy, $Q = 5.2$, is about 10% above the bifurcation point energy ($Q_M \approx 4.62$). The excess energy is radiated away during the first part of the process. If the energy of the initial soliton is higher, e.g. $Q = 5.6$, then the solution converges to an A-type state (Fig. 8.6e). At higher values of energy (e.g. $Q = 7.2$), the elliptic-type point on the sphere is closer to the initial point $\mathbf{S'} = (-S_0', 0, 0)$, and convergence to an A-type state occurs very quickly (Fig. 8.6f).

In Fig. 8.6e,f, the initial conditions are chosen in such a way that the trajectory follows the separatrix in the initial stages of the process. Hence,

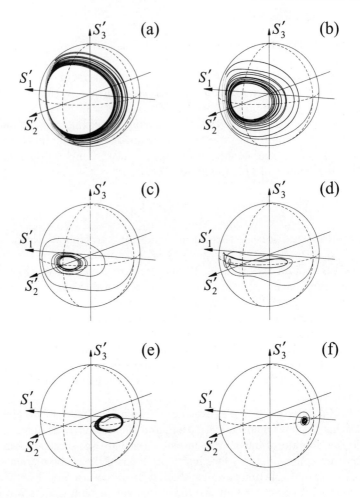

Figure 8.6 *Evolution of the integrated Stokes parameters for pulses in a nonlinear coupler. The initial condition is a soliton of an individual core as given by (8.14) with Q being: (a) 3.2, (b) 4.0, (c) 4.8, (d) 5.2, (e) 5.6 and (f) 7.2.*

initially the radiation is still large. However, after a half beat length, the trajectory starts to spiral around the stationary A-type state and the amount of radiation becomes noticeably smaller.

8.11 Switching

The use of a coupler for switching purposes implies that the input is a soliton-like pulse in a single core. For an input pulse with a given energy

Q, the system will tend to evolve to the stable state of lowest energy. In doing this, some energy is lost due to radiation as the light traverses the coupler. The process can be understood if we look at Fig. 8.4. The dashed line in this figure corresponds to a single-core soliton input. If the energy of the pulse is much less than that at the bifurcation point, $\frac{Q_c}{\sqrt{K}} = \frac{8}{\sqrt{3}}$, then the coupler is in an almost linear regime, and the energy will swap back and forth from one core to the other. The spatial frequency of these oscillations is the linear beat length, and not much energy is radiated. For short couplers (where the length equals half the linear beat length) and low energies, only a half-period of these oscillations occurs. The energy is transferred to the second core completely.

If the energy is closer to the point of bifurcation, but the length is the same, the pulse will evolve differently, gradually losing energy and approaching a symmetric state which is still stable, as we are below the point of bifurcation. Hence, the initial pulse converges to a stable symmetric state with an equal amount of energy in each core. Above the point of bifurcation, the pulse converges to an asymmetric A-type state (which is stable). For most of its range, this state has by far the greater part of its energy in one core, and so resembles a single-core input. Thus a high-energy input pulse in core 1 just changes slightly to assume this shape, and the observed effect is that the input energy remains in the input core. The two latter processes are indicated by the short lines in Fig. 8.4. Thus the core from which the pulse emerges depends on the input energy and the pulse shape. This forms the basis of an intensity-dependent switch, and shows how the dynamical behaviour is understood in terms of the exact stationary states.

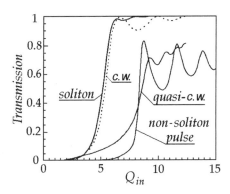

Figure 8.7 *Transmission characteristics of (half-beat-length) nonlinear coupler showing response to different input pulses, as labelled, as a function of energy input Q_{in}. In the case of c.w. input (dotted curve), the horizontal axis shows normalized energy flow rather than energy. Hence, the c.w. curve is plotted in such a way as to equalize the switching thresholds. 'Quasi-c.w.' means Gaussian pulses whose widths are many times the soliton width.*

Transmission curves, giving the fraction of the total energy of the pulse emerging from the first core when the input is launched into the same core, are shown in Fig. 8.7. The solid line labelled 'soliton' in this diagram corresponds to an input in the form of a soliton of an individual core. This curve illustrates what was said above. The transmission is zero at low energies, but it increases sharply around the point of bifurcation, and is very close to 1 at energies much higher than that at the point of bifurcation. If the input pulse has a shape different from that of a soliton of an individual core, then the pulse Hamiltonian will be above the dotted line in Fig. 8.4. The pulse has to adjust its shape to a soliton-like profile before oscillating and converging to one of the stable curves. The transmission curve then loses its step-like character, and so is far from ideal. An example is shown in Fig. 8.7.

Pulses which are very long in comparison with the soliton length (i.e. quasi-c.w. pulses) separate into subpulses, and then each subpulse switches into the second core. The amplitudes of subpulses differ from pulse to pulse and the averaged transmission curve also becomes quite different from the ideal curve (Fig. 8.7). The dotted line in Fig. 8.7 corresponds to the switching of a c.w. input (i.e. the second derivatives are ignored in (8.15)). Instead of the total energy of the pulse, we plot along the horizontal axis the intensity of continuous radiation (energy flow) renormalized in such a way that the thresholds of the two curves coincide.

8.12 Arbitrary initial conditions

An arbitrary initial input in a coupler will separate into one or more solitons plus radiation. This process depends upon the location, on the (H, Q) plane, of the point corresponding to the initial condition. The same can be said for c.w. or quasi-c.w. radiation. Due to the effects of modulation instability, c.w. in couplers with anomalous dispersion will be transformed into a train of pulses. Each pulse in the train can be considered as a soliton, and can be treated separately if these subpulses are far enough apart so that they do not interact. A pulse of finite energy will separate into a finite number of soliton-like pulses. This effect can be used in different applications, such as amplitude discrimination. In this chapter, we consider only the evolution of separate soliton-like pulses.

For a soliton-like pulse, the amplitude and the width of the pulse must be related. Thus, for a single-core soliton, the product of the amplitude and full width at half-maximum intensity is always 1.76, and the area under the amplitude curve is π. If the initial profile of the pulse is not a soliton, the transmission properties of the coupler can change drastically. Numerical simulations showing the importance of initial conditions have been carried out by Peng and Ankiewicz (1992).

8.13 Non-Kerr-law anomalous dispersion couplers

When the material of the couplers has a saturable nonlinearity, the stationary states, stability and switching properties of the coupler are changed. The following equations describe light pulse propagation in a nonlinear coupler where the index change with intensity I is of the form $N(I)$:

$$iP_\xi + \tfrac{1}{2}P_{\tau\tau} + PN(|P|^2) + KG = 0,$$

$$iG_\xi + \tfrac{1}{2}G_{\tau\tau} + GN(|G|^2) + KP = 0.$$

(8.15)

In 'Kerr-law' fibres, $N(I) = I$, where $I = |P|^2$ in the first core, and $I = |G|^2$ in the second core, and so we recover (8.1). However, if the refractive index of the fibres saturates at high light intensities, then we may have

$$N(I) = \frac{I}{1+\gamma I},$$

(8.16)

as in (4.16). Another saturating case (Ankiewicz and Peng, 1993),

$$N(I) = \frac{1}{\gamma}[1 - \exp(-\gamma I)],$$

(8.17)

was also introduced in Chapter 4. Both forms, (8.16) and (4.19), reduce to the Kerr law when the intensity is low, and asymptotically approach a maximum value $(1/\gamma)$ when the intensity is high. The 'parabolic' law, (4.6), is given by:

$$N(I) = I + \nu I^2.$$

(8.18)

In experimental determinations of N, it can be useful to express N as a polynomial (using about three terms) in I. *Single-core* propagation of solitons has been considered for a double-doped saturable fiber (Gatz and Herrmann, 1992a), where N then takes the form:

$$N(I) = \frac{I + \nu I^2}{1 + \gamma I}.$$

(8.19)

8.13.1 Stationary soliton states

To find stationary states, we set $P = \sqrt{K}\, h\, \exp(iq\xi)$ and $G = \sqrt{K}\, w\, \exp(iq\xi)$ in (8.15). We arrive at:

$$\tfrac{1}{2}h_{tt} - \tfrac{q}{K}h + \tfrac{N(Kh^2)}{K}h + w = 0,$$

$$\tfrac{1}{2}w_{tt} - \tfrac{q}{K}w + \tfrac{N(Kw^2)}{K}w + h = 0,$$

(8.20)

where $t = \tau\sqrt{K}$. As in Chapter 4, we define the function $\bar{F}(I)$ so that the nonlinearity function N is related to $\bar{F}$ by $N = d\bar{F}/dI$, and thus (4.5) still applies.

We now note that (h, w) can be regarded as the position of a particle moving in a plane, where the motion is due to the potential R, such that

$$R(h, w) = 2hw - \bar{q}(h^2 + w^2) + K^{-2}[\bar{F}(Kh^2) + \bar{F}(Kw^2)], \quad (8.21)$$

with $\bar{q} \equiv q/K$. For states which are symmetric $(h = w)$ or anti-symmetric $(h = -w)$, the solutions are found by solving a single differential equation:

$$\frac{1}{2}h_{tt} - ah + \frac{N(Kh^2)}{K}h = 0, \quad (8.22)$$

where

$$a = \frac{q}{K} \pm 1. \quad (8.23)$$

Here, and for the rest of this chapter, the upper signs are for anti-symmetric states, and the lower are for symmetric states. If an analytic form is available for the single-core solitons for a given nonlinearity N, then we may easily obtain the symmetric and anti-symmetric soliton pairs for that nonlinearity.

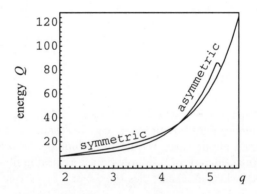

Figure 8.8 *Energy–dispersion diagram for saturable coupler with K=1 for saturation parameter (8.16) of $\gamma = 0.16$. The energies are similar (and the curves actually cross in the middle), although the component pulse shapes differ greatly.*

8.13.2 Couplers with saturable nonlinearity

The energy-dispersion curves for the saturable nonlinearity of (8.16) are shown in Fig. 8.8. Firstly, the saturation effect means that the bifurcation energy increases (Ankiewicz et al., 1995b). However, the most important qualitative difference from the Kerr nonlinearity case is that the asymmetric A-type states join the curve for the symmetric states at a second bifurcation point, located higher than the first one (Ankiewicz et al., 1995a). The ratio of peak component values, w_0/h_0, for the asymmetric state, equals 1 at the lower bifurcation point, then decreases and reaches a minimum at higher energies and finally increases to 1 again at the second bifurcation point, as the asymmetric state rejoins the symmetric branch. This effect is clearly apparent in Fig. 8.9. The asymmetric branch is stable for virtually its whole range of existence, viz. for $Q_l(\gamma) < Q < Q_u(\gamma)$, where Q_l and Q_u are, respectively, the lower and upper limits on its energy, Q, for a given γ.

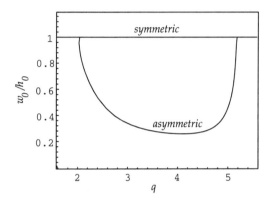

Figure 8.9 *Ratio of peak amplitudes, w_0/h_0, for saturable coupler $(K = 1)$ with $\gamma = 0.16$. This ratio is obviously equal to 1 for the symmetric states. The diagram clearly shows that the asymmetric branch joins the symmetric branch in two places.*

We plot Q against γK in Fig. 8.10, and this shows that these limits exist whenever $\gamma K < 0.233$. Of course, when γ is very low, Q_u is very high, and would not be commonly reached, but at moderate saturation values, Q_u is only a few times the Kerr-law material value of Q_l. The symmetric states become stable again above this point. Consequently, the switching behaviour below the lower bifurcation point and above the upper bifurcation point can be very similar. This provides the physical explanation for the unusual saturation switching results obtained numerically by Peng et al. (1994). Transmission characteristics (Ankiewicz et al., 1995a) show these peculiar properties of the coupler with saturable nonlinearity (Fig. 8.11). The energy switches to the second core at low and high energies but stays in the input core at medium energies. For high saturation values $(\gamma K > 0.233)$, there is no asymmetric state, and the coupler could not then be an effective switch. For a practical switch, the components of the asymmetric state should be significantly different, so we should not operate too close to the bifurcation points. This would mean that γK should be less than about 0.18.

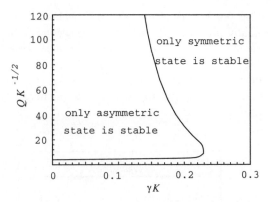

Figure 8.10 *Energy Q versus γK for saturable coupler.*

This behaviour is generic in the sense that similar nonlinearities (e.g. the exponential one of (8.17)) would also produce the same coupling behaviour.

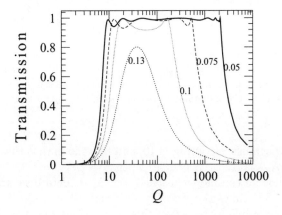

Figure 8.11 *The transmission characteristics for a half-beat-length saturable coupler with N given by (8.16). The transmission is the fraction of energy emerging from core 1 when the input energy launched into core 1 is Q. Each curve is labelled with its value of γ.*

8.13.3 Parabolic nonlinearity couplers

For these couplers (Ankiewicz and Akhmediev, 1996), the parabolic law (8.18) gives the potential

$$R = h^2 \left(h^2 - 2a - \frac{2}{3}vh^4 \right), \qquad (8.24)$$

where $v = -K\nu$.

For the parabolic law (8.18), if $v > 0$, then the potential exhibits a maximum (at $h = h_m$, say), and then becomes negative for large h. In order to have a bound state with $h = \mp w$, we need $R(h = h_m) \geq 0$. Now $h_m^2 = [1 + \sqrt{1 - 4va}]/2v$, so this condition means that we must have $K\nu a \geq -3/16$. For this parabolic law (8.18), the general solution is

$$h = \mp w = 2\sqrt{a}\,[1 + \sqrt{1 + s}\,\cosh(\sqrt{8a}\,t)]^{-1/2}, \qquad (8.25)$$

where $s = 16K\nu a/3$. Thus $-1 < s < \infty$.

It is easily verified that R returns to zero at the maximum value of h, viz. $h(0)$. If $\nu > 0$, then clearly there is an index increase for each part of the pulse. When $\nu < 0$, it is plain that the maximum intensity appearing in these states is $4aK$; this is less than $3/(4|\nu|)$, due to the condition on s. Thus each part of each soliton experiences an increase in index, as the peak intensity is less than $1/|\nu|$. If we define the width as the full width of intensity at half the maximum intensity, then the amplitude–width product is $r(s)$, where

$$r(s) = \sqrt{2}\,\frac{\operatorname{arccosh}[2 + (1 + s)^{-1/2}]}{\sqrt{1 + \sqrt{1 + s}}} \qquad (8.26)$$

The limit of this product, as $s \to 0$, is $\operatorname{arccosh}(3) \approx 1.7627$, which is the standard result for a Kerr-law soliton. The function $r(s)$ here increases monotonically as s decreases, indicating that these solitons become larger as s decreases. In fact $r(s) \to \infty$ as $s \to -1$.

Here for $\nu < 0$, the normalized energies (still defined by (8.3)) are

$$\hat{Q} = Q_c/\sqrt{K} = \frac{4}{\sqrt{-s}}\,\sqrt{2a}\,\operatorname{arctanh}(\sqrt{-s}) = \sqrt{\frac{-6}{K\nu}}\,\operatorname{arctanh}(\sqrt{-s}). \quad (8.27)$$

If v is small, we have

$$\hat{Q} = 4\sqrt{2a}\left[1 - \frac{16}{9}K\nu a + \cdots\right], \qquad (8.28)$$

and the bifurcation occurs at

$$\bar{q} \approx \frac{1}{3}\left(5 - \frac{128}{45}K\nu\right). \qquad (8.29)$$

The eigenfunction near the beginning of the asymmetric branch is

$$\left[c - \frac{1}{4}\operatorname{sech}^2(T) + \frac{1}{5}\log\,\cosh(T)\right]\operatorname{sech}^2(T), \qquad (8.30)$$

where $T = \sqrt{2a}\,t$.

When $v \to 0$, (8.25) to (8.29) reduce to results for the Kerr-type nonlinearity. When v is below 0.1102, there are two values of $\bar{q}$, while above it there are no eigenvalues. Thus the asymmetric state again exists only over a finite range of energies. The upper eigenvalues occur at comparatively low energies, so it may be feasible to make couplers with three switching regimes, where the behaviour changes occur at energies lower than those in the saturable coupler case (Ankiewicz and Akhmediev, 1996). For other nonlinearities, these states may not have explicit forms, but they are none the less easily found numerically (e.g. Ankiewicz et al., 1995b).

We may summarize this subsection by noting that we have provided analytic results for nonlinear couplers formed from parabolic nonlinearity materials. At low energies these are similar to saturable couplers, allowing verification of numerical studies on saturable cases. However, significant physical effects, such as the appearance of an additional bifurcation point, occur at much lower energies in the parabolic case.

8.13.4 Hamiltonian as function of energy

The Hamiltonian for a general nonlinearity coupler can be written in the form

$$
H_c = \int_{-\infty}^{\infty} \left[\frac{1}{2}|P_\tau|^2 + \frac{1}{2}|G_\tau|^2 - \bar{F}(|P|^2) - \bar{F}(|G|^2) - K(PG^* + P^*G) \right] d\tau.
$$
(8.31)

Thus, for the Kerr-law we have $\bar{F}(I) = I^2/2$ as before, while the saturation law of (8.16) gives

$$
\bar{F}(I) = \frac{1}{\gamma} \left[I - \frac{1}{\gamma} \ln(1 + \gamma I) \right].
$$
(8.32)

For (8.17), we find that

$$
\bar{F}(I) = \frac{1}{\gamma} \left[I + \frac{1}{\gamma}(e^{-\gamma I} - 1) \right].
$$
(8.33)

Both of these forms reduce to $I^2/2$ when $\gamma \to 0$. The parabolic form of (8.18) leads to

$$
\bar{F}(I) = \frac{I^2}{2} + \frac{\nu}{3}I^3,
$$
(8.34)

and, in this case, the normalized Hamiltonian can be found analytically for symmetric and anti-symmetric states. For $\nu < 0$, we have

$$
\hat{H} = \frac{H_c}{K^{3/2}} = \sqrt{\frac{-6}{K\nu}} \left[\frac{-3\sqrt{-s}}{16K\nu} + \left(\frac{3}{16K\nu} \pm 1 \right) \operatorname{arctanh}(\sqrt{-s}) \right],
$$
(8.35)

with s given after (8.25). We can now use (8.27) to relate $\hat{H}$ directly to $\hat{Q}$:

$$
\hat{H} = \frac{3\sqrt{6}}{16(-K\nu)^{3/2}} \tanh\left(\hat{Q}\sqrt{-\frac{K\nu}{6}} \right) + \left(\frac{3}{16K\nu} \pm 1 \right) \hat{Q}.
$$
(8.36)

Analogous results for $\nu > 0$ involve the function tan instead of tanh. As the parabolic coefficient ν approaches zero, we have $s \to 0$, and (8.35) reduces to the correct Kerr-law coupler form, (8.9). The Hamiltonian for the asymmetric state has to be calculated numerically in each particular case. The results above reduce to the single-fiber parabolic law result of Chapter 4 when the coupling $K \to 0$, i.e. when $t\sqrt{a} \to \tau\sqrt{q}$ and $aK = q \pm K \to q$. Thus $Q_c = 2Q$ and $H_c = 2H$, where Q and H are the energy and Hamiltonian for a soliton in a single core, so that (8.25) reduces to (4.26) and (8.36) reduces to (4.30).

8.14 Dissimilar cores

The symmetry of the original equations is lost when the dispersion values differ, or the core propagation constants, β_1 and β_2, are unequal (due to differing radii or indices). The physical effects of these symmetry-breaking perturbations are similar. We consider the latter case, where the cores are unequal, and define:

$$\delta = \frac{1}{2}(\beta_2 - \beta_1). \tag{8.37}$$

The system is then described by:

$$iP_\xi + \tfrac{1}{2}P_{\tau\tau} + |P|^2 P + KG\,\exp(2i\delta\xi) = 0,$$

$$iG_\xi + \tfrac{1}{2}G_{\tau\tau} + \sigma|G|^2 G + KP\,\exp(-2i\delta\xi) = 0, \tag{8.38}$$

where σ represents the ratio of the nonlinear coefficient of the second core to that of the first. If we set the time derivative terms to zero, then we recover the c.w. coupling equations for unequal cores. On the other hand, if the cores are equal, then $\sigma = 1$ and $\delta = 0$ and the above set reduces to the Kerr-law case of (8.15).

We now let $P = u\,\exp[i(q+\delta)\xi]$ and $G = v\,\exp[i(q-\delta)\xi]$, with $t = \tau\sqrt{K}$, $u = \sqrt{K}f$ and $v = \sqrt{K}g$. The set of equations is then:

$$\tfrac{1}{2}f_{tt} - (\bar{q}+\bar{\delta})f + g + f^3 = 0,$$

$$\tfrac{1}{2}g_{tt} - (\bar{q}-\bar{\delta})g + f + \sigma g^3 = 0, \tag{8.39}$$

where $\bar{q} = q/K$, $\bar{\delta} = \delta/K$.

Using typical fiber coupler values, we obtain $\sigma = 1 + 0.1\bar{\delta}$, and this value is used with $\bar{\delta} = 0.01$ in the numerical analysis. The net result of this symmetry breaking is that the A-type curve separates into two curves. One of these joins the lower near-symmetric branch, while the other joins the upper near-symmetric branch. The loss of symmetry has thus caused the disappearance of the bifurcation point (Fig. 8.12). This effect has also been investigated using a variational approach (Malomed *et al.*, 1996).

This behaviour is reminiscent of the splitting of (degenerate) asymmetric

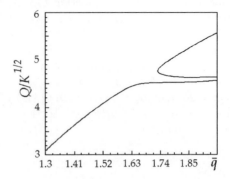

Figure 8.12 *Energy–dispersion diagram for a coupler with unequal cores.*

states in the planar nonlinear guide (Moloney *et al.*, 1986), with the energy of the soliton pair replacing the power of the spatial field of the planar guide, and frequency offset q replacing the propagation constant. Using this analogy, one would expect the whole lower branch in Fig. 8.12 to be stable, and that only the part of the upper branch corresponding to strongly asymmetric states (this is located close to the lower branch) would be stable. Even for a symmetric planar guide, the part of the lower branch having negative slope disappears at low values of the waveguide parameter V, so, in this range, the lower branch is always stable (Fig. 6 of Tran and Ankiewicz, 1992). The soliton pairs energy–dispersion diagram here (Fig. 8.12) is thus analogous to the low-V power–propagation constant diagram.

8.15 Solitons with time offsets

Viewing soliton interaction in terms of the underlying stationary states also provides us with a clear understanding of earlier numerical work on various aspects. One of these is the investigation of the interaction when one soliton is launched into each input core when there is a time offset between the two. For 0 or π phase difference, the pair was called a *bisoliton* by Abdullaev *et al.* (1989). For small time offsets, the in-phase pair (which we now describe as being close to a symmetric state) maintains its integrity, as it propagates with only a slight wobble, while the out-of-phase pair (anti-symmetric state) does not, since the solitons move apart and dissipate upon propagation (Ankiewicz and Peng, 1991a). Now, we can plainly see that this behaviour occurs because the symmetric state is stable (at low energies) while the anti-symmetric state is not. When the phase difference is $\pi/2$, no bisoliton is formed, and the energy swaps back and forth between the cores (Fig. 5 of Ankiewicz and Peng, 1991a). It can now be seen that the system is oscillating around the symmetric state, while slowly losing energy. It would eventually reach this stable state, in a similar manner to the low-energy, *single-input* case (section 8.11). Of course, large offsets are not perturbations, and the solitons then couple independently.

CHAPTER 9

Multi-core nonlinear fiber arrays

Nonlinear effects in three-core (Chen *et al.*, 1990a, Finlayson and Stegeman, 1990, Mitchell *et al.*, 1990), and more generally in *n*-core (Chen *et al.*, 1990b; Christodoulides and Joseph, 1988; Schmidt-Hattenberger *et al.*, 1991), nonlinear couplers have attracted much attention in recent years. The switching of solitons in these devices is of special interest (Soto-Crespo and Wright, 1991) because of the possibility of very short switching times (down to the femtosecond range).

The theory of three-core (and more generally *n*-core) couplers can be constructed by analogy with the twin-core fiber. First, we consider soliton states and their bifurcations (Akhmediev and Buryak, 1994). The lowest-order soliton states and bifurcations have features similar to those in two-core fibers. Stable stationary solutions, and solutions close to them, may appear as the final states of pulse evolution in a fiber. Hence, knowledge of the set of all possible stationary soliton states roughly answers the question: what type of output signals can we expect for a given input pulse energy? However, the general dynamics for an arbitrary initial pulse is a complicated matter and has not been considered in detail in the existing literature. Switching is a particular form of the general dynamics: it is the process of energy redistribution between the cores for a given input. The problem of switching is rather involved, but it can be solved when the stability of the soliton states is known. Examples of numerical simulations of short-pulse propagation in multi-core fibers have been given by Soto-Crespo and Wright (1991).

9.1 *n*-core nonlinear fiber arrays

The propagation of pulses in an array of *n* coupled nonlinear fibers with circular symmetry can be described in terms of *n* linearly coupled NLSEs. An example of this geometry is shown in Fig. 1.5c. In normalized form, this set of NLSEs is given by:

$$i\frac{\partial U_1}{\partial \xi} + \frac{1}{2}\frac{\partial^2 U_1}{\partial \tau^2} + |U_1|^2 U_1 + K(U_n + U_2) = 0,$$

$$i\frac{\partial U_2}{\partial \xi} + \frac{1}{2}\frac{\partial^2 U_2}{\partial \tau^2} + |U_2|^2 U_2 + K(U_1 + U_3) = 0, \qquad (9.1)$$

$$\cdots$$

$$i\frac{\partial U_n}{\partial \xi} + \frac{1}{2}\frac{\partial^2 U_n}{\partial \tau^2} + |U_n|^2 U_n + K(U_{n-1} + U_1) = 0,$$

where $U_i(\xi, \tau)$ is the envelope in the ith core, ξ is the normalized distance along the fiber, τ is the normalized retarded time and K is the coupling coefficient between any pair of neighbouring cores.

We can use the same approach as in the previous Chapter to solve this problem. That is, we can find stationary pulses, study their stability and finally consider the dynamics for arbitrary initial conditions. The presence of many variables in the problem makes it quite complicated. However, the existence of certain symmetries allows us to make simplifications. We now list the symmetries in this situation. Equations (9.1) are invariant with respect to four major symmetry transformations:

(a) rotation symmetry ($U_i \rightarrow U_{i+1}$, for $1 \leq i \leq n-1$ and $U_n \rightarrow U_1$);

(b) mirror symmetry ($U_1 \rightarrow U_1$ and $U_{1+i} \rightarrow U_{n+1-i}$ for $1 \leq i \leq n-1$) (for the case of even n we also have the other type of mirror symmetry, $U_{n-i} \rightarrow U_{1+i}$ for $0 \leq i \leq n-1$);

(c) sign inversion symmetry ($U_i \rightarrow -U_i$, for $1 \leq i \leq n$); and

(d) time-reversal symmetry ($U_i(\tau) \rightarrow U_i(-\tau)$ for $1 \leq i \leq n$).

The internal symmetries of (9.1) result in the existence of a simple solution, which exists for every value of $n\,(n > 3)$, and can be expressed in an analytical form. This is the fully symmetric solution where the shape and the amplitude are the same in each core:

$$U_i = \sqrt{2(q-2)}\,\text{sech}(\sqrt{2(q-2)}\tau)\,\exp(iq\xi) \qquad (9.2)$$

for all i values ($1 \leq i \leq n$). We shall call this the symmetric solution. Other solutions of high symmetry, which are expressible in analytic form, exist for some particular values of n, but only the fully symmetric solution exists for every n.

9.2 Soliton states in three-core couplers

In what follows, we concentrate on the three-core coupler. The propagation of pulses in a nonlinear triple-core directional coupler, with the cores in an equilateral triangle arrangement, is described in terms of three linearly coupled NLSEs (Soto-Crespo and Wright, 1991). In normalized form, this set of coupled NLSE's is given by:

$$iU_\xi + \tfrac{1}{2}U_{\tau\tau} + |U|^2 U + K(V + W) = 0,$$

$$iV_\xi + \tfrac{1}{2}V_{\tau\tau} + |V|^2 V + K(U + W) = 0, \qquad (9.3)$$

$$iW_\xi + \tfrac{1}{2}W_{\tau\tau} + |W|^2 W + K(V + U) = 0,$$

where $U(\xi, \tau)$, $V(\xi, \tau)$ and $W(\xi, \tau)$ are the dimensionless envelope functions in the three cores and K is the normalized coupling coefficient between each pair of cores. K, which we assume is positive, is the only physical constant in the problem.

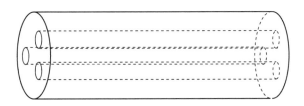

Figure 9.1 *Triple-core directional coupler with the cores in an equilateral triangle arrangement.*

Stationary pulse-like solutions can be represented in the form:

$$U(\xi,\tau) = u(\xi,\tau)\ \exp(iq\xi),$$
$$V(\xi,\tau) = v(\xi,\tau)\ \exp(iq\xi), \qquad (9.4)$$
$$W(\xi,\tau) = w(\xi,\tau)\ \exp(iq\xi),$$

where q is the parameter of a soliton state family of solutions and $u(\xi,\tau)$, $v(\xi,\tau)$, $w(\xi,\tau)$ are real functions which decay to zero at infinity. Using (9.4), the set of partial differential equations (9.3) can be reduced to the following set of three real ordinary differential equations:

$$\tfrac{1}{2}u_{\tau\tau} - qu + u^3 + K(v+w) = 0,$$

$$\tfrac{1}{2}v_{\tau\tau} - qv + v^3 + K(u+w) = 0, \qquad (9.5)$$

$$\tfrac{1}{2}w_{\tau\tau} - qw + w^3 + K(v+u) = 0.$$

Rescaling the functions and variables in (9.5) in such a way that

$$u = \sqrt{K}f, \qquad v = \sqrt{K}g, \qquad w = \sqrt{K}h, \qquad \tau = t/\sqrt{K}, \qquad q' = q/K,$$

we transform the set (9.5) into the set

$$\tfrac{1}{2}f_{tt} - q'f + f^3 + g + h = 0,$$

$$\tfrac{1}{2}g_{tt} - q'g + g^3 + f + h = 0, \qquad (9.6)$$

$$\tfrac{1}{2}h_{tt} - q'h + h^3 + g + f = 0,$$

which now has only one combined parameter, $q' = q/K$. If the initial value of K is fixed, then this combined parameter is the only parameter of the soliton state family of solutions.

Equations (9.6) are invariant with respect to the symmetry transformations outlined in section 9.1. Specifically they are:

(i) permutational symmetry $(f \to g,\ g \to h,\ h \to f)$;
(ii) mirror symmetry $(g \to h,\ h \to g,\ f \to f)$;
(iii) inversion symmetry $(f \to -f,\ g \to -g,\ h \to -h)$;
(iv) time-reversal symmetry $(t \to -t)$.
Different combinations of the listed symmetries are also possible, such as inversion mirror symmetry $(g \to -h,\ h \to -g,\ f \to -f)$.

One of the consequences of the symmetry relations is that the following relatively simple types of solutions can exist:
(1) $f = g = h$ (totally symmetric solution);
(2) $f = -g,\ h = 0$ (anti-symmetric solution);
(3) $f = g \neq h$ (partly symmetric solution).

These considerations allow us to write analytic expressions for some of the simple solutions. The totally symmetric solution, which we simply call the 'symmetric solution', is given by

$$f = g = h = \sqrt{2(q'-2)}\operatorname{sech}(\sqrt{2(q'-2)}t).$$

The anti-symmetric solution is given by

$$f = -g = \sqrt{2(q'+1)}\operatorname{sech}(\sqrt{2(q'+1)}t) \quad \text{and} \quad h = 0.$$

Partly symmetric solutions also exist as a solution of the set (9.6), but explicit analytic expressions for them cannot be found easily. These solutions (as well as any other ones) are triply degenerate because of the permutational symmetry of the problem $(f \to g,\ g \to h,\ h \to f)$. Each of the symmetries (ii)–(iv) also doubles the degeneracy.

The symmetry relations show that it is more convenient to change variables so that the three types of simple solution listed above can be expressed using a smaller number of components. Due to the permutational symmetry of the problem, we can do this in many ways. For example, we can use the orthogonal transformations:

$$x = (g-h)/\sqrt{2}, \qquad y = (2f - g - h)/\sqrt{6}, \qquad z = (f + g + h)/\sqrt{3}.$$

Equations (9.6) then take the form:

$$\ddot{x} - 2(q'+1)x + x^3 + x(y^2 + 2z^2) - 2\sqrt{2}xyz = 0,$$
$$\ddot{y} - 2(q'+1)y + y^3 + y(x^2 + 2z^2) - \sqrt{2}z(x^2 - y^2) = 0, \qquad (9.7)$$
$$\ddot{z} - 2(q'-2)z + \frac{2}{3}z^3 + 2z(x^2 + y^2) - \sqrt{2}y\left(x^2 - \frac{1}{3}y^2\right) = 0,$$

where each overdot denotes a derivative with respect to t. The advantage of this form is that, in the linear limit, equations (9.7) describe independent harmonic oscillations along the three axes.

The set (9.7) can be considered as governing the motion of a particle in a three-dimensional potential well. The Hamiltonian for this motion is:

$$H = \frac{\dot{x}^2}{2} + \frac{\dot{y}^2}{2} + \frac{\dot{z}^2}{2} - (x^2 + y^2) + 2z^2 - q'(x^2 + y^2 + z^2)$$

$$+ \frac{1}{4}(x^4 + y^4) + \frac{1}{6}z^4 + z^2(x^2 + y^2) + \frac{1}{2}x^2y^2 - \sqrt{2}yz\left(x^2 - \frac{1}{3}y^2\right).$$

This Hamiltonian is non-integrable. The soliton states of (9.7) and their bifurcations can be analysed using methods analogous to those for birefringent fibers or couplers (Chapter 8). The main difference is that one more variable is involved in this problem, and this makes the calculation of the soliton states more complicated.

Soliton states are the solutions of the set (9.7) which correspond to the separatrix trajectories of the Hamiltonian. The separatrix must include at least one unstable hyperbolic point. For solitons, the trajectories must start and finish at the hyperbolic point at the origin $(x = y = z = 0)$, as we require $\dot{x}, \dot{y}, \dot{z} \to 0$ and $x, y, z \to 0$ when $t \to \pm\infty$. Hence, we restrict ourselves to the case $H = 0$. This means that the trajectories corresponding to the soliton states must be contained in the volume bounded by the surface

$$-(x^2 + y^2) + 2z^2 - q'(x^2 + y^2 + z^2) + \frac{1}{4}(x^4 + y^4)$$

$$+ \frac{1}{6}z^4 + z^2(x^2 + y^2) + \frac{1}{2}x^2y^2 - \sqrt{2}yz\left(x^2 - \frac{1}{3}y^2\right) = 0.$$

The set of equations (9.7) has two types of simple soliton solutions:

$$x = 0, \qquad y = 0, \qquad z(t, q') = \sqrt{3}\alpha \ \text{sech}(\alpha t), \qquad (9.8)$$

$$x(t, q') = 2\sqrt{q' + 1} \ \text{sech}(\sqrt{2(q' + 1)}t), \qquad y = 0, \qquad z = 0, \qquad (9.9)$$

which correspond to symmetric (9.8) and anti-symmetric (9.9) states in the initial variables f, g and h, respectively. Note that the soliton states of system (9.3), which correspond to solution (9.8), exist for $q/K > 2$, while those which correspond to solution (9.9) exist for $q/K > -1$ (with $K > 0$).

In order to find the bifurcation from solution (9.8), we represent the solution of (9.7) as a solution (9.8) with a small added perturbation:

$$z = \sqrt{3}\alpha \ \text{sech}(\alpha t) + \epsilon^2 F,$$

$$x = \epsilon G_1, \qquad (9.10)$$

$$y = \epsilon G_2,$$

where ϵ is a small parameter. The term in ϵ in the z equation is zero. Substituting (9.10) into (9.7), and linearizing with respect to the small parameter ϵ, we find that the equations for G_1 and G_2 are identical and have the form:

$$G_{tt} - 2(q' + 1)G + \frac{6\alpha^2}{\cosh^2(\alpha t)} G = 0, \qquad (9.11)$$

where $G = G_1 = G_2$.

Equation (9.11) has exactly two solutions which decay at infinity. They are an odd solution,

$$G = \frac{\sinh(\alpha t)}{\cosh^2(\alpha t)} \qquad (\text{at} \quad q' = \infty); \tag{9.12}$$

and an even solution,

$$G = \text{sech}^2(\sqrt{2}t) \qquad (\text{at} \quad q' = 3). \tag{9.13}$$

The odd solution (9.12) does not correspond to any bifurcation because it requires $K/q = 0$, and so x, y and z become decoupled in this limit. The even solution (9.13) transforms the symmetric soliton state into an asymmetric solution at the point when $q/K = 3$. Hence $q/K = 3$ is the point of bifurcation. Moreover, two new branches of soliton states appear at this point. One of them corresponds to positive x, y ($f = g > h$) and the other one to negative x, y ($f = g < h$). Both solutions are partly symmetric soliton states. We shall call these states 'A_1-type' and 'A_2-type' partly symmetric states. When q/K is far from the bifurcation value, the components f and h become noticeably different (as seen in Fig. 9.2).

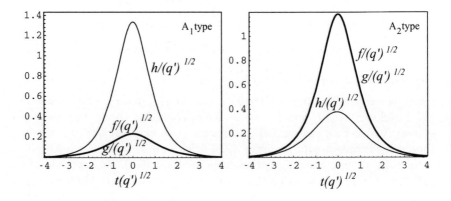

Figure 9.2 *Examples of A_1-type and A_2-type soliton states. The bold line represents the envelopes for $f/\sqrt{q'} = g/\sqrt{q'}$, while the thin line represents the envelope for the function $h/\sqrt{q'}$. The parameter $q/K = 5$.*

Another asymmetric soliton branch bifurcates from the anti-symmetric states at $q/K = 2$. We call these states 'B-type' asymmetric soliton states. This is a more complicated type of bifurcation, and it cannot be analysed using perturbative analysis. The reason is that the value α in (9.7) goes to zero when $q/K \to 2$. This means that the decay of z at $t \to \pm\infty$ can be much slower that the decay of x at y. In other words, the condition $z \ll x$ or $z \ll y$ cannot be fulfilled for all t simultaneously. Note that $q/K = 2$ is the point where symmetric soliton states appear. The appearance of the symmetric state at $q/K = 2$ is the reason for the bifurcation of the B-type state at the point N in Fig. 9.3. One of the consequences is that the shape of the B-type soliton when $t \to \pm\infty$ is the same as the shape of the symmetric state when $t \to \pm\infty$. In the B-type soliton states, all three components are unequal and have complicated profiles (shown in Fig. 9.5).

At points far from the points of bifurcation, the asymmetric soliton states have to be studied numerically. To find the separatrix trajectories, a standard shooting method can be used. As a general rule, these solutions can be classified by constructing their energy-dispersion diagram. The total energy carried by the coupled soliton states is given by:

$$Q = \int_{-\infty}^{\infty} (|U|^2 + |V|^2 + |W|^2) \, d\tau = \sqrt{K} \int_{-\infty}^{\infty} (x^2 + y^2 + z^2) \, dt.$$

The curves $Q(q)$, for the different types of soliton states, are shown in Fig. 9.3. Examples of envelopes of soliton states, corresponding to different branches in Fig. 9.3, are given in Fig. 9.2, 9.4 and 9.5.

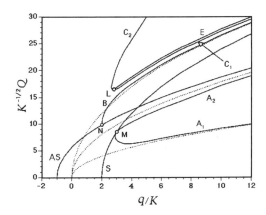

Figure 9.3 *Bifurcation diagram on the (Q, q) plane. Both variables are normalized. Solid curves correspond to different types of soliton states of the set (9.3). Circles M, N and L are the points of bifurcation. Dashed lines correspond to the 'one-soliton', 'two-soliton' and 'three-soliton' solutions of the decoupled set (9.3).*

The S and AS curves in Fig. 9.3 correspond to the symmetric and antisymmetric soliton states, respectively. Two new branches of partly symmetric soliton states (corresponding to types A_1 and A_2) split off from the dispersion curve S at the bifurcation point M, and B-type asymmetric soliton states split off from the dispersion curve AS at the bifurcation point N. M is a point of double bifurcation, as we discussed above. Three types of physically different solutions meet at this point. This is a consequence of the permutational symmetry of the initial equations (9.3), where the coupling parameter K has the same value in each of three equations. In the case of unequal coupling parameters in (9.3), the nature of this bifurcation can change significantly. Loss of system symmetry tends to mean that bifurcations are replaced by gradual transformations, as explained in section 8.14.

The solutions considered above are the simplest ones. They have the lowest values of energy. There are an infinite number of soliton states which can be considered as nonlinear combinations of these simple ones. All of them have energies higher than the energy of the B-type soliton states. Therefore, dispersion curves corresponding to them are located above the curve for the B-type states. Some of them (those with sufficiently low energy, Q) are shown in Fig. 9.3 and are denoted C_1, C_2, and E. The C_2 - branch is a continuation of the C_1-branch. The combined C_1-C_2 branch of soliton states appears at $q/K \approx 2.632$.

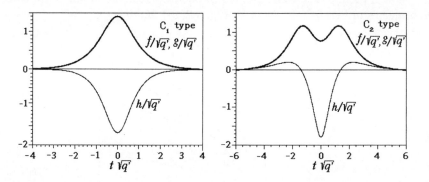

Figure 9.4 *Examples of C_1-type and C_2-type soliton states. These have unequal components. The parameter $q/K = 4$.*

The E-type soliton states bifurcate from the combined C_1-C_2 branch at $q/K \approx 2.747$, (i.e. the bifurcation point L). Almost all the soliton states we have considered are even (symmetric) in time. The soliton states of the E branch are not time-symmetric. Hence, the bifurcation at the point L breaks the symmetry (in time) of the soliton states. For the E-type states, the functions f and g are related by $f(t) = g(-t)$.

In contrast to the curves of the Kerr-law twin-core coupler (Chapter 8), these energy-dispersion curves can intersect. The curves may cross each other, so that the same point in this plot (Fig. 9.3) can correspond to two completely different soliton states. This is partly a consequence of the increased number of degrees of freedom in this problem, but two states can also have the same energy in the twin-core parabolic nonlinearity coupler (as seen in Chapter 8).

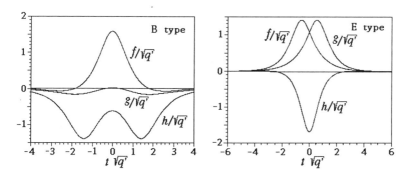

Figure 9.5 *Examples of B-type and E-type soliton states which have all three components unequal. The parameter $q/K = 5$.*

We can also classify all soliton branches in this problem by comparing their energies in the limit $q \to \infty$. For example, the energy of A_1-type solutions converges to the energy of the 'one-soliton' solution of the decoupled $(K = 0)$ set (9.3):

$$U = \sqrt{2q}\,\mathrm{sech}(\sqrt{2q}\tau)\exp(iq\xi), \quad V = W = 0.$$

The energy of A_2 and AS-type solutions converges to the energy of the 'two-soliton' solution of the decoupled set (9.3):

$$U = V = \sqrt{2q}\,\mathrm{sech}(\sqrt{2q}\tau)\exp(iq\xi), \quad W = 0,$$

while the energies of solutions of types S, B, C_1 and E converge to the energy of 'three-soliton' solution of the decoupled system (9.3):

$$U = V = W = \sqrt{2q}\,\mathrm{sech}(\sqrt{2q}\tau)\exp(iq\xi).$$

The energy of more complicated solutions may converge to the energy of the 'multi-soliton' solutions of the decoupled set (9.3). The energy–dispersion curves for 'one-soliton', 'two-soliton' and 'three-soliton' solutions of decoupled NLSE are presented by the dashed lines in Fig. 9.3.

The soliton states which are not presented in Fig. 9.3 have energy-dispersion curves lying above the E branch. Some of these solutions have complicated envelopes which are asymmetric in time. An example of a solution of this type is given in Fig. 9.6. The time symmetry of the problem requires that the solution obtained from this soliton solution using the transformation $\tau \to -\tau$ must also be a solution of (9.3).

Although many stationary states exist, and they can play a certain role in the dynamics when the energy in the input pulse is high enough, not all of them are stable. Using the standard rule, we can conclude that symmetric solitons are stable up to the point of bifurcation (point M) and that A_1-

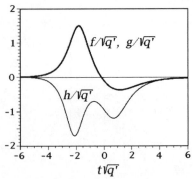

Figure 9.6 *Example of a soliton state having a complicated time-asymmetric shape. The branch for this soliton state is not shown in Fig. 9.3. The bold line represents the functions $f/\sqrt{q'} = g/\sqrt{q'}$, while the thin line represents the function $h/\sqrt{q'}$. For this example, $q/K = 3.333$ and $Q \approx 22.4\sqrt{K}$.*

type solitons are stable to the right of the point of minimum energy in Fig. 9.3. These states play the major role in switching phenomena. To switch the energy to one of the cores, the energy in the soliton must exceed the energy at the minimum point. The higher the value q/K, the more the energy is confined to a single core. In the extreme case $q/K \to \infty$, the energy collapses into a single core. These states have been considered by Aceves et al. (1994). To consider particular switching phenomena in detail, other numerical simulations are needed.

CHAPTER 10

The influence of higher-order dispersion on solitons

Higher-order dispersion effects become important in optical fibers when the carrier frequency is close to the zero dispersion point. For solitons with narrow spectra (long pulses, i.e. tens of picoseconds) the effect can be ignored. However, if the spectrum extends beyond the zero dispersion point, we have to take radiation effects into account. Usually, it is sufficient to take into account only third-order dispersion effects (Wai et al., 1987). In some cases, for specially designed fibers (Cohen and Mammel, 1982), fourth-order effects also become important (Cavalcanti et al., 1991).

In this Chapter, we consider both third- and the fourth-order dispersions, and we concentrate on the effects caused by them. In the case of third-order dispersion, the only effect is reshaping of the soliton and radiation, independent of the sign of the dispersion. Simple perturbation theory shows that, to first order, solitons are 'robust' in the presence of these dispersions, i.e. the soliton amplitude and frequency do not change. In the case of pure fourth-order dispersion, when the spectrum of the soliton is centred on the point of maximum dispersion, the soliton can radiate, and may have radiationless oscillating tails, depending on the sign of the dispersion. In the first few sections of this chapter, we consider solitons emitting radiation which can be called 'Cherenkov' radiation. Later, we will consider solitons with oscillating tails and also bound states of solitons.

The governing equation for 'bright' optical solitons in fibers is the modified NLSE (Chapter 1):

$$i\psi_\xi + \frac{1}{2}\psi_{\tau\tau} + |\psi|^2\psi = \beta_j\,\hat{P}(\psi)\,, \tag{10.1}$$

where we have included a perturbative operator $\hat{P}$ on the right-hand side. The parameter β_j ($j = 3, 4$) is assumed to be sufficiently small for solitons to exist.

In this chapter we will consider $\hat{P} = i\,\partial^3/\partial\tau^3$ and $\hat{P} = -\partial^4/\partial\tau^4$, corresponding to third-order dispersion (3OD) and fourth-order dispersion (4OD), respectively. These two effects usually must be considered simultaneously. However, for a special choice of the carrier frequency, the third-order term can be cancelled. Then we can consider their influence separately. This is important for a qualitative understanding of the phenomena.

Combining the two into one perturbation is then a straightforward task. Third-order dispersion influences both the soliton shape and the emission of radiation. First let us consider the former.

10.1 Soliton renormalization

The unperturbed ($\hat{P} = 0$) soliton solution of (10.1) is

$$\psi_{sol}^0(\xi, \tau) = A \operatorname{sech}(A\tau) \exp(ik_{sol}\xi), \tag{10.2}$$

where $k_{sol} = A^2/2$ is the soliton wavenumber. In the particular case of 3OD, the corrected form of the soliton, to first order in β_3, is

$$\psi_{sol}^{3OD}(\xi, \tau) = A \operatorname{sech}(A\,y) \exp\{ik_{sol}\xi + i\beta_3[2A^2\tau - 3\tanh(Ay)]\} \tag{10.3}$$

where $y = \tau - \beta_3 A^2 \xi$. Thus, the first-order corrections to the soliton (10.2) only affect the phase and the velocity of the soliton, leaving the amplitude, width and shape unperturbed. Such a soliton can exist if a sufficiently large part of the soliton spectrum lies in the anomalous group velocity dispersion (GVD) regime. (This range is marked in Fig. 10.4.)

The effect of 3OD on solitons was first considered by Wai *et al.* (1986). It manifests itself as radiation at the specific frequency $\omega_0 \approx -1/2\beta_3$. This lies in the normal GVD regime, and is separated from (but still overlaps) the soliton spectrum. It is this overlap that is the main cause of the radiation. In the case of 4OD, it was shown by Höök and Karlsson (1993) that the soliton radiates at the two frequencies, $\pm 1/\sqrt{2\beta_4}$.

Analytically, the problem of soliton radiation due to higher-order dispersion has been considered both for NLSE solitons (Kuehl and Zhang, 1990; Wai *et al.*, 1990; Elgin, 1993; Kodama *et al.*, 1994; Grimshaw, 1994; Karpman, 1993a,b) and for solitons governed by the Korteweg–de Vries (KdV) equation (Karpman, 1993a,b; Benilov *et al.*, 1993; Pomeau *et al.*, 1988). The primary interest is in two properties of the radiation, viz. its frequency and its intensity. In the case of NLSE solitons, the frequency has been estimated, to first order in β_j, in the early numerical work (Wai *et al.*, 1986), but more accurate expressions have been obtained using perturbative inverse scattering methods (Kodama, 1994; Grimshaw, 1994; Karpman, 1993a,b). We will point out below that the same result follows from simple physical arguments.

The calculation of the radiation intensity has been carried out by Kuehl and Zhang (1990), Wai *et al.*, (1990) and Grimshaw (1994) using combined numerical and analytical methods. These approaches are rather involved, and the physics may, in some cases, have been obscured by cumbersome mathematics. In the case of KdV solitons, similar methods have been applied (Benilov *et al.*, 1993; Pomeau *et al.*, 1988), although another somewhat simpler method has been proposed (Karpman, 1993a,b). As this radiation is emitted from a wavepacket (the soliton) with a phase velocity exceeding the linear phase velocity of the medium, Akhmediev and Karlsson (1995) pointed out the formal equivalence between this radiation and the well-known Cherenkov radiation processes in nonlinear optics (Bolotovskii and Ginzburg, 1972; Askaryan, 1962; Tien *et al.*, 1970). Here, following Akhmediev and Karlsson (1995), we consider this effect using a simple approach and some rough approximations.

10.2 Radiation frequency

We have seen, in Chapter 2, that NLSE solitons do not interact with radiation because the wavenumbers of the solitons lie in a range which is forbidden for linear dispersive waves. Therefore, linear waves cannot be in resonance with the soliton, and energy cannot be transferred from the soliton to the linear waves. When 3OD is taken into account, the spectrum of linear waves extends into the region allowed for solitons. As a result, solitons can interact with radiation and emit small-amplitude linear waves. More specifically, the soliton of (10.2) has a wavenumber $k_{sol} = A^2/2$, while the linear dispersion relations corresponding to (10.1) are

$$k_{lin}(\omega) = -\frac{1}{2}\omega^2 - \beta_3\,\omega^3, \tag{10.4}$$

$$k_{lin}(\omega) = -\frac{1}{2}\omega^2 + \beta_4\,\omega^4. \tag{10.5}$$

Our convention for these and all subsequent pairs of equations in these sections of this chapter is that the first refers to the 3OD case while the second refers to the 4OD case. In the absence of the perturbation ($\hat{P} = 0$), $k_{lin} < 0$, and, since $k_{sol} > 0$, the soliton is stable. In the presence of the perturbation, however, the soliton is in resonance with the dispersive waves when $k_{sol} = k_{lin}$, i.e. at the frequency ω_0 determined by

$$A^2/2 = -\frac{1}{2}\omega_0^2 - \beta_3\,\omega_0^3, \tag{10.6}$$

$$A^2/2 = -\frac{1}{2}\omega_0^2 + \beta_4\,\omega_0^4, \tag{10.7}$$

which are the phase-matching conditions for the radiation. To lowest orders in β_j, this frequency is given by

$$\omega_0 = -\frac{1}{\beta_3}\left[\frac{1}{2} + 2(\beta_3 A)^2 + O(\beta_3 A)^4\right], \tag{10.8}$$

$$\omega_0 = \frac{1}{\sqrt{2}\beta_4}\left[1 + (A\sqrt{\beta_4})^2 + O((A\sqrt{\beta_4})^4)\right]. \tag{10.9}$$

In the case of 3OD, there is only one unstable frequency (Fig. 10.1a), but for 4OD, the problem is spectrally symmetric and the unstable frequencies are $+\omega_0$ and $-\omega_0$ (Fig. 10.1b). The inverse group velocities at the unstable

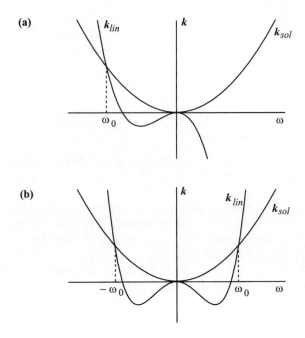

Figure 10.1 *Resonances for (a) 3OD and (b) 4OD.*

frequencies are easily calculated to be

$$V_g^{-1} = \left(\frac{\partial k_{lin}}{\partial \omega}\right)_{\omega=\omega_0} \approx -\frac{1}{4\beta_3}, \qquad (10.10)$$

$$V_g^{-1} = \left(\frac{\partial k_{lin}}{\partial \omega}\right)_{\omega=\pm\omega_0} \approx \pm\frac{1}{\sqrt{2\beta_4}}. \qquad (10.11)$$

Since we are in the frame of reference in which the soliton is stationary, these give the velocities at which the linear radiation leaves the soliton. It is evident that 3OD causes emission of radiation behind (in front of) the soliton when $\beta_3 > 0$ ($\beta_3 < 0$). The radiation from 4OD will be in both temporal directions, due to the symmetry of (10.1). We also see that the resonant frequency ω_0 depends on the soliton amplitude A, as pointed out by Elgin (1993). However, this dependence is rather weak, due to the condition that $A\beta_3$ needs to be sufficiently small for soliton creation. Nevertheless, the dependence must be taken into account when integrating the total radiated energy.

10.3 Transition radiation

Before proceeding with calculations, we note that, as in any initial value problem, we have to separate two distinctly different processes. If an initial condition is not a 'stationary solution' (approximated by (10.3)) or a pulse close to it, the pulse will initially radiate some amount of energy in the form of a wavepacket, in order to adjust itself to that stationary profile. This radiation have the same direction of propagation as Cherenkov radiation which is generated continuously. We call this initial stage the 'transition radiation'. This radiation is clearly seen in Fig. 10.2. By analogy with Cherenkov radiation emitted by particles in a medium, the radiation of light when a particle enters a medium is different from true Cherenkov radiation, where a particle moves with high speed in the medium. After this transition radiation ends, the soliton will emit true Cherenkov radiation. In all theoretical work so far, this latter case has been considered. However, numerical results presented in many papers seem to relate to the above-mentioned transition radiation, rather than to Cherenkov radiation. In what follows, we concentrate on Cherenkov radiation, which does not depend on the initial conditions, and will not consider transition phenomena, as the dependence on initial conditions is then crucial. Depending on initial conditions, the transition radiation may actually remove a greater amount of the pulse energy than Cherenkov radiation.

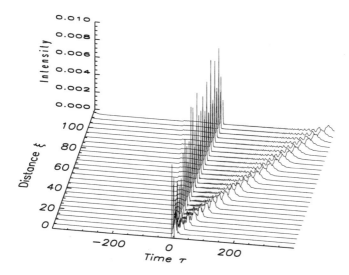

Figure 10.2 *Low-amplitude transition radiation emitted to the right from the pulse with initial condition $\psi = A\,\mathrm{sech}(A\tau)$. [We thank V. Afanasjev for generating this illustration.]*

10.4 Relation to Cherenkov radiation

Classical Cherenkov radiation appears when a small object (particle) moves in a medium with a velocity exceeding the phase velocity of the waves in the given medium (Bolotovskii and Ginzburg, 1972; Askaryan, 1962; Tien *et al.*, 1970). It is assumed that the source has dimensions much smaller than the wavelength. The source of radiation does not necessarily have to be a real particle, but can be, for example, waves of polarization induced in a nonlinear medium by external fields. If the size of the source is finite and comparable with the wavelength in at least one direction, then the radiation is defined in this direction by the 'Cherenkov conditions' rather than the full phase-matching conditions. Usually in nonlinear optics, the range of the induced polarization is large in comparison with the wavelength, so that satisfying the full phase-matching conditions is necessary for the emission of radiation. In the latter case, the direction of the radiation is defined by these phase-matching conditions.

In nonlinear optics, the concept of Cherenkov radiation was introduced by Tien *et al.* (1970). Their experiment shows that a thin-film pump beam can generate a second-harmonic (SH) beam in the substrate below the film. The phase velocity of the induced nonlinear polarization is higher than the phase velocity of the waves in the substrate, or, equivalently, the effective wavenumber of the polarization in the film must be shorter than the wavenumber of the waves in the substrate at the SH frequency. The emerging SH beam is therefore tilted with respect to the longitudinal direction of the pump, due to the longitudinal phase matching of the wavevectors. This is, therefore, an example of Cherenkov radiation with longitudinal phase matching, but without transverse phase matching.

The same considerations apply to the radiation emitted by solitons in fibers. In this case the radiation can be considered as phase-matched along the ξ direction but not phase-matched along the 'transverse' τ axis (Fig. 10.3). Moreover, the phase velocity of the soliton in the ξ direction is higher than the phase velocities of linear waves outside the soliton. The physical consequence of this fact is that radiation is emitted in a 'direction' which is inclined to the ξ direction at an angle ϕ, defined by

$$\sin \phi = \omega_0 / \sqrt{k_{sol}^2 + \omega_0^2}.$$

The intensity of this radiation is proportional to the square of the spectral component of the source. It decreases when the width of the source increases, but becomes higher for small soliton widths, as would be expected for Cherenkov radiation without phase matching. In this sense, all perturbations that phase-match linear waves to solitons in at least one direction can be considered as examples of Cherenkov radiation.

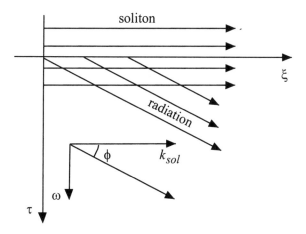

Figure 10.3 *Interpreting the retarded time τ as a transverse coordinate, we can view the radiation process as Cherenkov radiation under the longitudinal phase-matching condition $k_{sol} = k_{lin}$. The radiation is emitted at the Cherenkov angle ϕ defined by $\sin\phi = \omega_0/\sqrt{k_{sol}^2 + \omega_0^2}$.*

10.5 Radiation intensity

An accurate estimate of radiation intensity can be found by using the perturbation method. The solution of (10.1) is given by

$$\psi(\xi, \tau) = \psi_{sol}(\xi, \tau) + f(\xi, \tau),$$

where $|f(\xi, \tau)| \ll |\psi_{sol}(\xi, \tau)|$, and $\psi_{sol}(\xi, \tau)$ is given by (10.2). Linearizing (10.1) in f, we obtain the equation

$$i\frac{\partial f}{\partial \xi} + \frac{1}{2}\frac{\partial^2 f}{\partial \tau^2} - \beta_j \hat{P}(f) + 2|\psi_{sol}|^2 f + (\psi_{sol})^2 f^* = \beta_j \, \hat{P}(\psi_{sol}), \quad (10.12)$$

where ψ_{sol} can be given by either (10.2) or (10.3), depending on the accuracy we desire, and $j = 3, 4$.

Away from the soliton, $k_{sol} \to 0$ and (10.12) becomes homogeneous, and thus has solutions in terms of linear dispersive waves. We can interpret the term on the right-hand side as a source term for these waves. The terms $2|\psi_{sol}|^2 f$ and $\psi_{sol}^2 f^*$ are responsible for frequency variations along τ inside the range where $\psi_{sol} \neq 0$, i.e. inside the soliton. The radiated energy is mainly governed by the source term. Hence, in principle, in order to estimate the radiated energy, we can omit these two terms in favour of the source term. Obviously, in our approximation, the linear waves must have the same dependence on ξ as the source term, viz. $f \propto \exp(ik_{sol}\xi)$. After cancelling the common $\exp(ik_{sol}\xi)$ factors, we rewrite (10.12) in terms of its Fourier components:

$$[k_{lin}(\omega) - k_{sol}]F(\omega) + \frac{1}{\pi} \int\limits_{-\infty}^{\infty} U(\omega - \omega')F(\omega')d\omega'$$

$$+ \frac{1}{2\pi} \int\limits_{-\infty}^{\infty} U(\omega - \omega')\hat{F}(\omega')d\omega' = \beta_j\, P(\omega) \tag{10.13}$$

where

$$F(\omega) = \int\limits_{-\infty}^{\infty} f(\tau)\, \exp(-i\omega\tau)\, d\tau, \tag{10.14}$$

$$\hat{F}(\omega') = \int\limits_{-\infty}^{\infty} f^*(\tau)\, \exp(-i\omega\tau)\, d\tau, \tag{10.15}$$

$$U(\omega) = \int\limits_{-\infty}^{\infty} \psi_{sol}^2(\tau)\, \exp(-i\omega\tau)\, d\tau, \tag{10.16}$$

and

$$P(\omega) = \int\limits_{-\infty}^{\infty} \hat{P}[\psi_{sol}(\tau)]\, \exp(-i\omega\tau)\, d\tau. \tag{10.17}$$

According to the discussion above, we can omit the convolution integrals, and the function F may then be written as

$$F(\omega) = \frac{\beta_j\, P(\omega)}{k_{lin}(\omega) - k_{sol}}. \tag{10.18}$$

The function $F(\omega)$ can be transformed back to the time domain by contour integration in the complex plane. The contour consists of the real axis, a small semi-circle round the pole at ω_0, and a large semi-circle in the upper (lower) half-plane for $\tau > 0$ ($\tau < 0$). The main contribution to the integral comes from the pole at $\omega = \omega_0$ ($\omega = \pm\omega_0$ for 4OD). All other poles in the upper half-plane correspond to evanescent waves that vanish for large τ. We therefore neglect their contribution, and obtain:

$$f(\tau) = a\, \exp(i\omega\tau) = \frac{i}{2}\beta_3\, \frac{P(\omega_0)}{V_g^{-1}}H(\tau\beta_3)\, \exp(i\omega\tau), \tag{10.19}$$

$$f(\tau) = a_+ \exp(i\omega\,\tau) + a_- \exp(-i\omega\tau)$$

$$= \frac{i\beta_4}{2}\left[\frac{P(\omega_0)}{V_g^{-1}(\omega_0)}H(-\tau)\, \exp(i\omega\tau)\right.$$

$$+\frac{P(-\omega_0)}{V_g^{-1}(-\omega_0)}H(\tau)\exp(-i\omega\tau)\bigg],\qquad(10.20)$$

where $H(x)$ is the Heaviside step function. In the 3OD case, its appearance shows that the radiation only exists on one side of the soliton.

The amplitude factor in front of the exponential term $\exp(i\omega\tau)$ is what we need for further calculations. Taking into account additional poles and the two terms omitted from (10.12) and (10.13) will change the step function into a smooth transition function, and it will also alter ω_0 inside this transition region. We now treat 3OD and 4OD separately. Equation (10.18) shows that the radiation amplitude is proportional to $\exp(-1/\beta_j)$. This implies, in the case of 3OD, that we can use the first-order expression (10.18) to calculate $P(\omega_0)$. In the 4OD case we use the unperturbed soliton (10.2). We find

$$|P(\omega_0)| = \frac{\pi}{4\beta_3^3}\left[\frac{5}{2} - \pi A\beta_3 + O(A\beta_3)^2)\right]\exp\left(-\frac{\pi}{4A\beta_3}\right),\qquad(10.21)$$

$$|P(\omega_0)| = \frac{\pi}{2\beta_4^2}\left[1 + 4A^2\beta_4 + O(A^4\beta_4^2)\right]\exp\left(-\frac{\pi\omega_0}{2A}\right).\qquad(10.22)$$

In deriving (10.21), we have used the first-order expressions for ω_0 from (10.8) and (10.9). In the exponential in (10.22), however, we retain the full dependence of A in ω_0. This will be important in section 10.6 when integrating to obtain the total energy. Thus, the moduli of the Fourier amplitudes of the radiation become:

$$|a| = \frac{5\pi}{4\beta_3}\left(1 - \frac{2\pi}{5}A\beta_3\right)H(\tau)\exp\left(-\frac{\pi}{4A\beta_3}\right),\qquad(10.23)$$

$$|a_\pm| = \frac{\pi}{2\sqrt{2\beta_4}}H(\mp\tau)\exp\left(-\frac{\pi\omega_0}{2A}\right).\qquad(10.24)$$

We can see that the dominant factor in these expressions is the spectral value of the source term at ω_0. This is satisfying from a physical point of view, as the radiation amplitude is proportional to the spectral amplitude of the soliton at the radiation frequency. It is thus the spectral tail of the soliton in the normal GVD regime that boosts the radiation (Fig. 10.4).

More elaborate calculations can be used to given an improved expression to replace the function $5\pi(1 - \frac{2\pi}{5}A\beta_3)/4\beta_3$ in front of the exponential in (10.23).

10.6 Radiated energy

Using the formulae for the radiation given above, we may now calculate the energy loss of the solitons. The first invariant of (10.1) can be expressed as

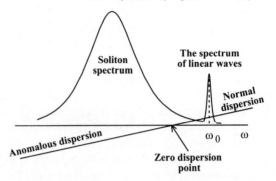

Figure 10.4 *Spectra of the soliton and linear waves and their location relative to the zero dispersion point.*

a continuity equation in integral form (2.20):

$$\frac{\partial}{\partial \xi} \int_{-\tau_0}^{\tau_0} |\psi|^2 \, d\tau = - [j_{rad}]_{\tau=-\tau_0}^{\tau=\tau_0}, \qquad (10.25)$$

where

$$j_{rad} = \frac{i}{2}(\psi \, \psi_\tau^* - \psi_\tau \, \psi^*) - \beta_3(\psi \, \psi_{\tau\tau}^* + \psi_{\tau\tau}\psi^* - |\psi_\tau|^2), \qquad (10.26)$$

$$j_{rad} = \frac{i}{2}(\psi \, \psi_\tau^* - \psi_\tau \, \psi^*) \qquad (10.27)$$
$$- i \, \beta_4(\psi^* \, \psi_{\tau\tau\tau} - \psi_{\tau\tau\tau}^* \psi + \psi_{\tau\tau}^* \, \psi_\tau - \psi_{\tau\tau}\psi_\tau^*),$$

defines the energy flux at any τ, and $\pm \tau_0$ are arbitrary boundaries of integration. Results (10.26) and (10.27) are the extensions of (2.17) to 3OD and 4OD, respectively. Equation (10.25) tells us that the decrease of energy in the range $[-\tau_0, \tau_0]$ is equal to the energy flow out of this region. It is reasonable to set $\pm \tau_0$ beyond the main body of the soliton, where only small-amplitude radiation exists. Then the integral in (10.25) is, to a good approximation, the soliton energy, $2A$, and the energy flow is determined by the dispersive waves, i.e.

$$[j_{rad}]_{\tau=-\tau_0}^{\tau=\tau_0} = |a|^2 \, \omega_0(1 + 3\beta_3 \, \omega_0)$$

$$= \frac{25\pi^2}{64\beta_3^3}\left(1 - \frac{4\pi}{5} \, A\beta_3\right) \exp\left(- \frac{\pi}{2\,A\,\beta_3}\right), \qquad (10.28)$$

$$[j_{rad}]_{\tau=-\tau_0}^{\tau=\tau_0} = 2|a_+|^2 \, \omega_0 \, (4\beta_4 \, \omega_0^2 - 1) = \frac{\pi^2}{8\beta_4 \, \sqrt{2\beta_4}} \exp\left(- \frac{\pi\omega_0}{A}\right). \qquad (10.29)$$

This means that the energy which radiates away is proportional to the inverse group velocity, i.e. $j_{rad} = |a|^2 \, V_g^{-1}$. It is now a straightforward task to find $A(\xi)$ numerically from (10.25) for every case of interest. This works well in the case of 4OD, but for 3OD complications arise due to the temporal (and spectral) asymmetry. These complications, due to the spectral shift, are briefly outlined below.

10.7 The spectral shift of the soliton

The second invariant of (10.1) reflects the conservation of momentum (2.8), i.e.

$$\frac{\partial}{\partial \xi} \int_{-\infty}^{\infty} (\psi^* \psi_\tau - \psi_\tau^* \psi) \, d\tau = 0. \tag{10.30}$$

Physically, this means that the spectral centre of mass is invariant. Thus, if a soliton loses energy by emitting linear waves in the normal dispersion regime, it will be shifted into the anomalous regime of the spectrum. This is the physical reason for the frequency shift (or temporal velocity shift) of order β_3 that was given in (10.3). The soliton radiates at w_0 and shifts further into the anomalous regime. Since this effect pushes the soliton spectrum away from the frequency of radiation w_0, the amplitude of the radiation decreases. Finally, a quasi-stationary state is reached, where the radiation rate is so small that the spectral recoil of the soliton is negligible. This occurs when $A\beta_3 \cong 0.04$. The soliton therefore stabilizes itself through radiative losses.

Using $A = 1$ and $\beta_3 = 0.04$ in equation (10.28), we find $j_{rad} \approx 10^{-12}$. This effect is negligible in comparison with the effects of fiber loss. The important conclusion is that radiative losses due to 3OD are negligible in communication systems utilizing stationary solitons. It should be noted that the spectral recoil effect precludes integration of the energy loss of the soliton using (10.26) and (10.27), since the above analysis neglects the spectral shift of the soliton.

In the case of 4OD, the radiation is spectrally symmetric, and the spectral recoils from the side-bands cancel out. Therefore, the perturbative character of $A\beta_4$ is less important than the corresponding term in the 3OD case, and we may use (10.26) and (10.27) to calculate the radiative losses. The most significant contribution to j_{rad} comes from the exponential, and it is therefore important to retain the full functional dependence of A in w_0:

$$w_0 = \sqrt{\frac{1 + \sqrt{1 + 8\beta_4 A^2}}{4\beta_4}}. \tag{10.31}$$

In Fig. 10.5, the peak intensity $A^2(\xi)$ is shown for various values of β_4. It is clearly seen in Fig. 10.5 that the asymptotic decay at high ξ is very slow. As the soliton loses energy, its spectral width decreases. The intensity of the radiation, which is proportional to the spectral intensity of the soliton

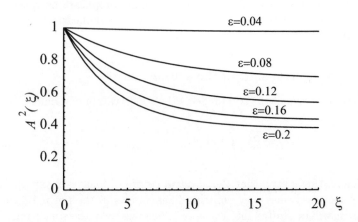

Figure 10.5 *Decaying intensity,* $A^2(\xi)$, *of a soliton perturbed by fourth-order dispersion, for* $\beta_4 = 0.04, 0.08, 0.12, 0.16, 0.2$, *with* $A(\xi = 0) = 1$.

at the unstable frequency ω_0, will therefore decrease. Thus the soliton will radiate less as its total energy loss becomes greater, by analogy with the 3OD scenario outlined above. Similarly, a quasi-stationary state will be reached with negligible radiation loss. Due to the absence of spectral recoil, however, a longer propagation distance is required to reach this state than that for the 3OD soliton.

10.8 Negative fourth-order dispersion and zero third-order dispersion

It is possible to choose the carrier frequency of the soliton at the minimum (or maximum) of the GVD, where 3OD vanishes. The next higher-order dispersive contribution then comes from 4OD (Golovchenko and Pilipetskii, 1994; Blow and Wood, 1989). In the regime of positive 4OD and anomalous GVD, soliton-like solutions radiate. However, soliton propagation influenced by a negative 4OD term is radiationless (Karlsson and Höök, 1994). The reason for this is the fact that the entire soliton spectrum lies in the anomalous GVD regime. In this section, we consider stationary soliton-like solutions of the generalized NLSE, where a negative 4OD term is the only additional term.

The NLSE with an additional 4OD term has the following form:

$$i\frac{\partial\psi}{\partial\xi} + \frac{1}{2}\frac{\partial^2\psi}{\partial\tau^2} + |\psi|^2\psi = |\beta_4|\frac{\partial^4\psi}{\partial\tau^4}. \tag{10.32}$$

Here we have normalized out the factor $|k''(\omega_0)|/2\tau_p$ (> 0) in front of the second-derivative term, but retained $|\beta_4| = |k''''(\omega_0)|/24\tau_p$ (> 0), where $k''(\omega_0)$ and $k''''(\omega_0)$ are derivatives of the fiber material dispersion relation taken at the carrier frequency $\omega = \omega_0$, and τ_p is the pulse duration. In (10.32) we assume $k'''(\omega_0) = 0$.

Höök and Karlsson (1993) demonstrated the existence of a family of stationary soliton-like solutions of (10.32), including an exact solution:

$$\psi(\tau) = \sqrt{\frac{3}{40|\beta_4|}} \, \text{sech}^2 \left(\frac{\tau}{\sqrt{40|\beta_4|}} \right) \exp \left(i \frac{1}{25|\beta_4|} \xi \right). \tag{10.33}$$

Other approximate solutions of this family have been found by using a variational approach, assuming that the solutions have a form close to the sech^2 shape. We now present an exact analysis of stationary localized solutions of (10.32).

The parameter $|\beta_4|$ in (10.32) can be removed using the rescaling transformations $t = \tau/\sqrt{2|\beta_4|}$, $x = \xi/4|\beta_4|$ and $U = 2\psi\sqrt{|\beta_4|}$. They transform (10.32) into

$$i\frac{\partial U}{\partial x} + \frac{\partial^2 U}{\partial t^2} + |U|^2 U = \frac{\partial^4 U}{\partial t^4}. \tag{10.34}$$

This equation has three integrals of motion: the energy of the pulse

$$Q = \int_{-\infty}^{\infty} |U|^2 dt, \tag{10.35}$$

the momentum

$$M = i \int_{-\infty}^{\infty} (U_t U^* - U_t^* U) dt, \tag{10.36}$$

and the Hamiltonian

$$H = \int_{-\infty}^{\infty} (|U_t|^2 + |U_{tt}|^2 - \frac{1}{2}|U|^4) dt. \tag{10.37}$$

Stationary pulse-like solutions of (10.34) have the form

$$U(x,t) = u(q,t) \exp(iqx), \tag{10.38}$$

where q is the soliton propagation constant (nonlinear shift of the wavenumber) and $u(q,t)$ is a real function of its parameters. Substitution of the ansatz (10.38) into (10.34) gives

$$\frac{\partial^2 u}{\partial t^2} + u^3 - qu = \frac{\partial^4 u}{\partial t^4}, \tag{10.39}$$

where q is now the only parameter in the problem. This is a nonlinear ordinary differential equation of fourth order, and its localized solutions give the stationary soliton-like solutions of (10.34).

10.9 Solitons with oscillating tails

The full family of exact solutions to (10.39) is not known. We are interested only in soliton-like solutions. Let us first find the decaying asymptotics of the localized solitons of (10.39) in the limit $t \to \infty$. The limit $t \to -\infty$ can be analysed in the same way and gives identical results. In this limit, the nonlinear term is negligible, so we can analyse the asymptotics of the exponentially decaying solutions of the following linear equation:

$$\frac{\partial^2 u}{\partial t^2} - qu = \frac{\partial^4 u}{\partial t^4}. \tag{10.40}$$

The general form of solutions to (10.40) is $u = A\,e^{\lambda t}$, where λ is the root of the corresponding fourth-order algebraic equation. Decaying solutions, however, require $\mathrm{Re}(\lambda) < 0$, which leaves us with only two possible values of $\lambda(q)$:

$$\lambda = -\left(\frac{1}{2} \pm \frac{1}{2}\sqrt{1 - 4q}\right)^{1/2}. \tag{10.41}$$

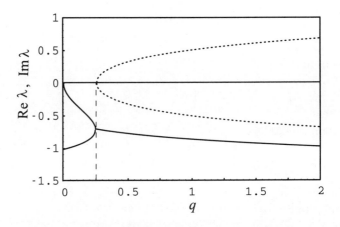

Figure 10.6 *The dependence of λ on q for the decaying solutions of (10.40). The solid curves are $\mathrm{Re}(\lambda)$, and the dashed curves are $\mathrm{Im}(\lambda)$.*

The $\lambda(q)$ dependencies of equation (10.41), with negative real part, are shown in Fig. 10.6. Every localized solution of (10.39) must have one of the asymptotics given by (10.41). However, the existence of a decaying solution of the linear (10.40) is not a sufficient criterion for the existence of the corresponding localized (soliton-like) solution of the nonlinear equation

(10.39). Nevertheless, one may expect that the general form of the $\lambda(q)$ dependencies in Fig. 10.6 will define the structure of the soliton-like solutions of the corresponding nonlinear problem. In particular, there is a critical point at $q = 0.25$ in the $\lambda(q)$ dependence in Fig. 10.6. For $q > 0.25$ the parameter λ has a non-zero imaginary part. Thus, if a localized solution of (10.39) exists in this range of the parameter q, then the corresponding solution must have oscillating tails.

We reformulate (10.39) in a way which simplifies the use of numerical methods. After introducing a new variable, $v = \partial^2 u/\partial t^2$, (10.39) can be presented as a system of two ordinary differential equations of second order:

$$\frac{\partial^2 u}{\partial t^2} = v, \qquad \frac{\partial^2 v}{\partial t^2} = v - qu + u^3. \tag{10.42}$$

The substitution $u = w + v$ converts the system (10.42) into the following form:

$$\frac{\partial^2 w}{\partial t^2} = q(w + v) - (w + v)^3,$$

$$\frac{\partial^2 v}{\partial t^2} = v - q(w + v) + (w + v)^3. \tag{10.43}$$

Equations (10.43) govern the evolution of a Hamiltonian system, with Hamiltonian:

$$H = \frac{\dot{w}^2}{2} - \frac{\dot{v}^2}{2} + \frac{v^2}{2} - q\frac{(w + v)^2}{2} + \frac{(w + v)^4}{2}, \tag{10.44}$$

where dots denote time derivatives. A further simplification of the Hamiltonian (10.44) (and thus system (10.43)) is also possible. For example, the substitution:

$$w = \left(\frac{1 - 2q}{2\sqrt{1 - 4q}} + \frac{1}{2}\right)^{1/2} r + \left(\frac{1 - 2q}{2\sqrt{1 - 4q}} - \frac{1}{2}\right)^{1/2} y,$$

$$v = \left(\frac{1 - 2q}{2\sqrt{1 - 4q}} - \frac{1}{2}\right)^{1/2} r + \left(\frac{1 - 2q}{2\sqrt{1 - 4q}} + \frac{1}{2}\right)^{1/2} y, \tag{10.45}$$

which keeps the variables real for $q < 0.25$, transforms the Hamiltonian (10.44) into

$$H = \frac{\dot{r}^2}{2} - \frac{\dot{y}^2}{2} - a\frac{r^2}{2} + b\frac{y^2}{2} + \frac{1}{1 - 4q}\frac{(r + y)^4}{2}, \tag{10.46}$$

where $a = \frac{1}{2} - \frac{1}{2}\sqrt{1 - 4q}$ and $b = \frac{1}{2} + \frac{1}{2}\sqrt{1 - 4q}$.

To find soliton-like solutions, one has to solve (10.43) numerically. An example is given in Fig. 10.7 (Akhmediev *et al.*, 1994a). As we have seen in previous chapters, a convenient way to analyse stationary solutions is

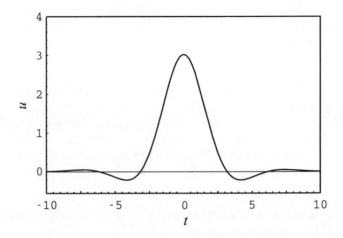

Figure 10.7 *An example of a soliton with oscillating tails.*

to construct the energy-dispersion diagram. This diagram is shown in Fig. 10.8. The energy, Q, of a stationary solution is defined by (10.35).

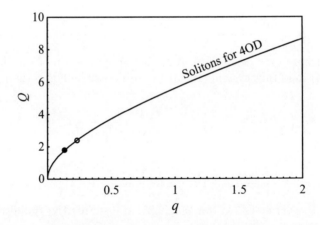

Figure 10.8 *Energy-dispersion diagram $Q(q)$ for a family of solitons of the 4OD case. The bold point corresponds to the exact analytic solution (10.33). Solitons which have oscillating-tails are on the right of the point marked by an open circle, while those to the left of it decrease monotonically for large $|t|$.*

There is only one family of stationary soliton-like solutions of (10.34). This family exists for $q > 0$. The existence of two linear asymptotics in the range $0 < q < 0.25$ does not lead to any bifurcation or to the appearance of a second soliton family. In other words, the lower branch of $\lambda(q)$ in Fig. 10.6, which exists for $q < 0.25$, does not correspond to any finite localized solutions. For $q > 0.25$, soliton-like solutions exist and have the asymptotic forms of the corresponding linear problem (10.40). Thus, solitons in this parameter range have an unusual asymptotic oscillating-tail behaviour in the limits $t \to \pm\infty$ (Fig. 10.7).

Single-soliton solutions are stable for all values of the parameter $q > 0$. All single solitons, with or without oscillating tails, are stable. Simulations show (Akhmediev *et al.*, 1994a) that the oscillating-tail solitons are attractors (stable fixed points) for a wide range of initial conditions.

10.10   Bound states of solitons with oscillating tails

In this section we consider stationary two-soliton bound states (BSs) of bright solitons of the generalized NLSE with a fourth-order derivative. BSs of solitons may exist when single-soliton solutions have non-monotonic asymptotics (radiationless oscillating tails) (Gorshkov and Ostrovsky, 1981). These tails produce local extrema in an effective interaction potential of weakly overlapping solitons, and therefore these solitons can trap each other at certain distances. However, these extrema are quite shallow, so we may anticipate that the soliton will only be weakly bound, and thus not very stable to perturbations. Questions involving the interaction of two (or more) solitary waves and the condition for the existence of soliton BSs in various dynamical systems have been addressed by Kaup (1976) and Karpman and Maslov (1977), and are still topics of active discussion (e.g. Klauder *et al.*, 1993; Cai *et al.*, 1994).

Single solitons can be bound into multi-soliton states. However, the stability of the whole train and, specifically, the stability of two-soliton bound states, are important questions which we need to address. We again use the fact that (10.34) can be written in the canonical Hamiltionian form:

$$iU_x = \frac{\delta H}{\delta U^*}, \quad iU_x^* = -\frac{\delta H}{\delta U}. \tag{10.47}$$

Equations (10.37) and (10.47) define a Hamiltonian dynamical system on an infinite-dimensional phase space of complex functions (U, U^*), whose absolute values decrease to zero at infinity. The equation for finding stationary solutions, in the variational formulation, can be written in the form:

$$\delta(H - qQ) = 0. \tag{10.48}$$

This variational formulation of the problem also defines the stability of stationary states: for any fixed Q, the stationary state is stable if the corresponding H has a local minimum, with q being a Lagrangian multiplier.

To estimate the interaction between the solitons, one can use an approximate solution which has been found using a variational approach (Karlsson and Höök, 1994):

$$u(q,t) = \frac{a(q)}{\cosh^2{[k(q)(t-t_0)]}}, \qquad (10.49)$$

where $a(q)$ and $k(q)$ are defined by:

$$k(q) = \sqrt{\frac{5q}{6 + 2\sqrt{9 + 100q}}}, \qquad (10.50)$$

$$a(q) = \pm\frac{k(q)}{\sqrt{3}}\sqrt{14 + 80\,k^2(q)}.$$

Solution (10.49) is a good approximation of an actual one-soliton solution of (10.34) in its central part (i.e. for $k(q)(t-t_0) \sim 1$). However, it has the wrong asymptotic behaviour. The asymptotics becomes qualitatively wrong when $q > 0.25$. For this range of q, the asymptotics can be obtained from the analysis of the linearized equation (10.40):

$$\lim_{t\to\pm\infty} u(q,t) = 4a(q)e^{-\lambda(q)|t-t_0|}\cos{[\omega(q)|t-t_0| + \psi_0(q)]}, \qquad (10.51)$$

where $\lambda(q)$ and $\omega(q)$ are defined by

$$\lambda(q) = q^{1/4}\cos\left(\frac{\arctan\sqrt{4q-1}}{2}\right),$$

$$\omega(q) = q^{1/4}\sin\left(\frac{\arctan\sqrt{4q-1}}{2}\right), \qquad (10.52)$$

and $\psi_0(q) \approx \pi/2$.

The Hamiltonian of any two interacting solitons, located far enough from each other, can be written in the approximate form:

$$H = H_1 + H_2 + H_{int}, \qquad (10.53)$$

where H_1 and H_2 are the Hamiltonians of the individual solitons, and the small interaction term, H_{int}, is determined from the nonlinear part of the Hamiltonian (10.37) (Gorshkov and Ostrovsky, 1981):

$$H_{NL} = -\int_{-\infty}^{\infty} |U(t,x)|^4 dt. \qquad (10.54)$$

Now, substituting $U = U_1 + U_2$, where U_1 and U_2 stand for unperturbed separate solitons, into H_{NL}, and linearizing relative to U_2, we obtain:

$$H_{int}(\Delta t, \Delta\varphi) = -2\int_{-\infty}^{\infty} |U_1(t,x)|^2 \text{Re}[U_1(t,x)U_2^*(t,x)]dt + (1 \leftrightarrow 2), \qquad (10.55)$$

where the expression responsible for the interaction of the first soliton with the tail of the second one is written down explicitly. Because of the symmetry, the corresponding expression with subscripts swapped, $(1 \leftrightarrow 2)$, has to be added for the interaction of the second soliton with the tail of the first. H_{int} in (10.55) depends on the relative distance, Δt, between the centres of two solitons and their relative phase difference, $\Delta \varphi$.

Assuming that the two interacting solitons are identical, using (10.49) and (10.50) (with $t_0 = 0$ and $\varphi_0 = 0$) for $u_1(t)$ and (10.51) and (10.52) (with $t_0 = \Delta t$ and $\varphi_0 = \Delta \varphi$) for $u_2(t)$ in (10.55), and recalling the form (10.38), we obtain:

$$H_{int}(\Delta t, \Delta \varphi) = -A \cos(\Delta \varphi) e^{-\lambda(q)\Delta t} \cos[\omega(q)\Delta t + \psi_1(q)], \qquad (10.56)$$

where $A \sim a^4/\lambda$ and $\psi_1(q) \approx \psi_0(q)$. We only consider interactions and BSs of two solitons. In principle, this approach can be applied to any number of solitons. Due to the exponential factor, $e^{-\lambda(q)\Delta t}$, in H_{int}, only pair interactions between nearest-neighbour solitons are important.

BSs of two solitons can exist if the interaction part of Hamiltonian (10.56) has local extrema, and these are determined by the equations:

$$\frac{\partial H_{int}}{\partial \Delta \varphi} = 0, \quad \frac{\partial H_{int}}{\partial \Delta t} = 0. \qquad (10.57)$$

For every value of $q > 0.25$, there are two infinite sets of solutions of (10.57) (i.e. there are two sets of families of BSs), viz. a symmetric set with the two solitons in phase, and an anti-symmetric set with π phase difference between the solitons:

$$\Delta \varphi = 0 \quad or \quad \pi$$
$$\Delta t_n = \Delta t_1 + \frac{\pi}{\omega}(n - 1), \qquad (10.58)$$

where $n = 1, 2, 3, \ldots$ and $\Delta t_1 = [\pi - \psi_1(q) + \arccos(\lambda/\sqrt{\lambda^2 + \omega^2})]/\omega$. For the symmetric set, Δt_n with odd (even) subscripts correspond to local maxima (minima) of H_{int}. For the anti-symmetric set of BSs, the situation is reversed. A BS having a relative distance of Δt_n between the component solitons is called a BS of nth order. To a first approximation, the bound states constructed from single-soliton solutions have the same value of q as the individual solitons.

10.11 Hamiltonian versus energy diagram

In order to find stationary soliton solutions of (10.34), one should find localized solutions of (10.39). Champneys and Toland (1993) proved a general theorem showing the existence of multi-hump localized solutions in autonomous Hamiltonian systems of fourth order, based on ordinary differential equations. The results of numerical analysis by Buryak and Akhmediev (1995) are presented in Fig. 10.9 in the form of a Hamiltonian versus energy diagram. In this diagram, Q and H are defined by expressions (10.35) and (10.37), respectively.

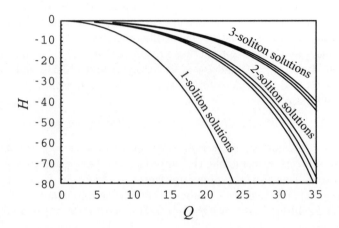

Figure 10.9 *Hamiltonian versus energy diagram for various families of soliton states.*

Each curve in Fig. 10.9 corresponds to a family of soliton solutions of (10.39). The family of single-soliton solutions exists for every $q > 0$. In the interval $0 < q < 0.25$, one-soliton solutions do not have oscillating tails. On the other hand, for $q > 0.25$, the asymptotics of one-soliton solutions are nonmonotonic (oscillating tails). Two- and multi-soliton BSs can exist in the region of parameters where single solitons have oscillating tails. Hence the curves for them start at certain non-zero energies. In the range $q > 0.25$, there are many families of two- (or multi-) soliton BSs. Among the two-soliton BS families, there are both symmetric and anti-symmetric ones. In Fig. 10.9, only the curves corresponding to families of two- and three-soliton BSs are shown, but multi-soliton BSs, with any number N of partial solitons, exist as well. For every number N, there are infinitely many families of BSs, but some of their associated curves are located very close to each other and thus cannot be distinguished at the scale of Fig. 10.9.

10.12 Stability criterion for bound states

For two identical well-separated solitons, local extrema in H_{int} in (10.56) result in extrema in the total Hamiltonian (10.37) at the same values of Δt_n and $\Delta \varphi$ and at constant q. The Hamiltonian (10.56) also has periodically located extrema at any constant energy Q. The only difference is that the whole set of these extrema can be slightly shifted relative to those found in the previous section. If the Hamiltonian has a local maximum at a given energy, the corresponding stationary state is unstable relative to a transverse shift of the two solitons. If the Hamiltonian has a local minimum, then the corresponding stationary solution is stable relative to this type of perturbation. However, the whole analysis was carried out for two solitons with equal energies. Thus, even if the interaction part of the Hamiltonian has a local minimum, other types of perturbation (including those which change the first two terms of Hamiltonian (10.53)) must be considered before drawing a general conclusion about stability.

Suppose that the Hamiltonian and the energy for a family of single-soliton solutions are related by

$$H = f(Q), \tag{10.59}$$

where $f(Q)$ is some functional dependence which can be found numerically. The Hamiltonian for the combined state of two identical solitons is given by

$$H = 2f(Q) + H_{int}(Q). \tag{10.60}$$

Let us consider perturbations which cause energy exchange between the two solitons. Due to such a perturbation, the energy Q_0 of one soliton is increased by small amount ΔQ:

$$Q_1 = Q_0 + \Delta Q. \tag{10.61}$$

Since we are interested in perturbations which conserve total energy, the energy of the second soliton has to decrease by the same amount:

$$Q_2 = Q_0 - \Delta Q, \tag{10.62}$$

so that the change of the Hamiltonian is approximately given by

$$\Delta H = f''(Q_0)(\Delta Q)^2, \tag{10.63}$$

where $f''(Q) = \frac{\partial^2 f}{\partial Q^2}$. The value of $H_{int}(Q)$ also depends on Q, but this dependence is relatively weak because $H_{int}(Q)$ itself is exponentially small. Without loss of generality, we can consider it as a constant in the vicinity of any particular $Q = Q_0$. Hence, the sign of $f''(Q)$ at the point $Q = Q_0$ determines the stability of BSs. If $f''(Q)$ is negative, the Hamiltonian has no local minimum at fixed $Q = Q_0$, and the corresponding BS is unstable. For the solitons of (10.34), $f''(Q)$ is negative for all possible values of Q, and thus we expect two- (and multi-) soliton BSs to be *unstable*.

The stability criterion derived above for our particular problem can be generalized: for conservative non-integrable systems having a family of single soliton solutions with oscillating tails, the Hamiltonian versus energy curve for the family defines the stability of two-soliton bound states. The BS is stable if the second derivative of this curve at the point of interest is positive, and unstable if the second derivative is negative. Hence, BSs are unstable relative to asymmetric perturbations. The exponent for a growing perturbation is usually very small for BSs of sufficiently high order $(n > 3)$. The 'binding energy' of these BSs is also exponentially small $(\sim e^{-\lambda(q)\Delta t_n})$.

As a consequence of the above analysis, all stationary two-soliton BSs are unstable. For all two-soliton bound states, an exponentially growing mode always exists. BSs which correspond to local maxima of H_{int} (the upper curves in each N-soliton solution case in Fig. 10.9), are unstable with respect to perturbations which change the relative distances between the centres of the component solitons. BSs which correspond to local minima of H_{int} (the lower curves in each N-soliton solution case in Fig. 10.9) are also unstable. The perturbation corresponding to this instability keeps the relative distance between the two interacting solitons intact, but it leads to an increase in the amplitude of one soliton and to a decrease in the amplitude of the other. All multi-soliton BSs are unstable, as a direct consequence of the instability of two-soliton bound states.

10.13   Interactions of solitons with oscillating tails

In the previous sections, it has been shown that solitons can form unstable BSs. On the other hand, the oscillatory tails can establish a potential barrier in the interaction between two neighbouring solitons. This potential barrier exists for equal, as well as for slightly unequal, solitons, provided both solitons have oscillatory tails. Thus solitons cannot approach each other more closely than a certain minimum distance, and two adjacent solitons cannot fuse together, even though they are interacting strongly.

The effective potential for the interaction of two identical stationary solitons is given by (10.56). In general, a finite potential barrier between the solitons exists at any value of the initial phase difference, $\Delta\varphi$, except $\Delta\varphi = \pi/2$. Hence, we can expect that, at small velocities, the collision between two solitons should be elastic, for both the interaction of two in-phase and two out-of-phase solitons. Moreover, the solitons in such an elastic collision cannot come closer to each other than the distance corresponding to the smallest potential minimum (or maximum).

This phenomenon has been demonstrated, using numerical simulations, by Buryak and Akhmediev (1994b). Using single moving-soliton solutions, this work investigated pair interactions, starting with two well-separated solitons moving towards each other with the same velocity. Examples of soliton interactions are shown in Fig. 10.10. It can be seen that the two solitons repel each other at a distance where they interact via their tails. The case of zero phase difference between the solitons is shown in Fig. 10.10a. Note the different scales along the t and x axes. The relative velocities are small. The attractive forces between the solitons increase the relative velocities close to the point of interaction, $(x \approx 188)$. Nevertheless, the solitons repel each other after the collision so that the whole picture is

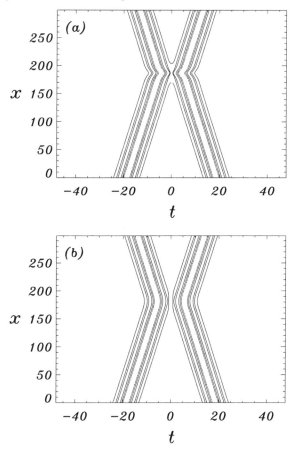

Figure 10.10 *Examples of the interaction of two equal solitons with oscillating tails when $\Delta\varphi$ equals (a) 0, (b) π. Intensity contours are shown.*

symmetric relative to the line $x = 188$. The strong interaction area, which appears in the case of NLSE solitons, is absent.

The case of π phase difference between the solitons is shown in Fig. 10.10b. This interaction is essentially the same as for $\Delta\varphi = 0$, except for a different minimum distance between the solitons. The solitons now repel each other, so that their behaviour in the collision region is slightly different. Note that the interaction in both cases is qualitatively different from the interaction between two NLSE solitons, where the solitons pass through each other without change, except for a phase shift. In the present case, the solitons feel a potential barrier between them.

For $\Delta\varphi = \pi/2$, the effective potential between the solitons vanishes, and they then interact with each other more strongly, leading to the appearance of asymmetry (Fig. 10.11). Because the system is not integrable, the energy and Hamiltonian of each soliton can change during the collision. The two final solitons are very different from each other and from the initial solitons. Their velocities are also different. Radiation of small-amplitude waves from the impact region of collision does occur, but cannot be seen at the scale of Fig. 10.11.

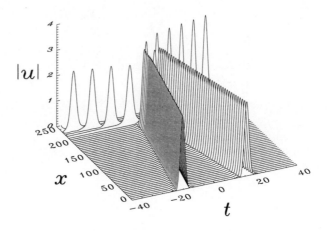

Figure 10.11 *An example of the interaction of two equal solitons with oscillating tails with* $\Delta\varphi = \pi/2$.

For $\Delta\varphi = \pi/2$, a finite potential barrier between solitons can be maintained by using slightly unequal solitons. The reason for this is that for two moving, slightly unequal solitons, the relative phase $\Delta\varphi$ is not fixed but changes slowly. Some examples are given by Buryak and Akhmediev (1995).

CHAPTER 11

Beam dynamics

In this chapter we consider beam propagation in homogeneous media, i.e. cases where no pre-existing waveguide is required, due to the fact that the high-intensity light increases the local index of refraction and thus creates its own guide. As we saw in Chapter 1, the multi-dimensional NLSE (1.18) can be written in the form

$$i\frac{\partial \psi}{\partial \xi} + \frac{\partial^2 \psi}{\partial x^2} + \frac{\partial^2 \psi}{\partial y^2} - \frac{k_0''}{2k_0}\frac{\partial^2 \psi}{\partial \tau^2} + N(|\psi|^2)\psi = 0. \tag{11.1}$$

where we have allowed for a general nonlinearity law, N. This takes into account, in the slowly varying approximation, the evolution of beams in the transverse plane as well as in the longitudinal direction. This equation has stationary solutions in the form of two-dimensional beams, as well as in the form of pulses localized in three dimensions when we take into account the time dependence. The latter can be called three-dimensional envelope solitons (Kaw et al., 1975). Stability problems for beams in Kerr media have been addressed by Vakhitov and Kolokolov (1974) and Kolokolov (1973), and for more general nolinearities by Zakharov and Rubenchik (1973). The beam dynamics in general depends strongly on the nonlinearity function $N(|\psi|^2)$. In this chapter we consider mainly two cases: Kerr nonlinearity and saturable nonlinearity.

11.1 Stationary (in time) solutions of the (2+1)-dimensional problem

Let us consider first those solutions which are stationary in time. Then we can ignore the time derivative in (11.1) and write the equation in the form:

$$i\frac{\partial \psi}{\partial \xi} + \frac{\partial^2 \psi}{\partial x^2} + \frac{\partial^2 \psi}{\partial y^2} + \alpha|\psi|^2\psi = 0, \tag{11.2}$$

where the nonlinearity is of Kerr type. This equation has been extensively studied in the theory of self-focusing and self-trapping (Chiao et al., 1964; Akhmanov et al., 1964). It is not completely integrable, but solutions can be obtained using geometrical optics, symmetry reductions or numerical methods. First, we consider symmetry transformations for which (11.2) is invariant.

Equation (11.2) has several symmetries. Apart from the obvious coordinate translations along the x, y and ξ axes, and a trivial rotation in the complex plane ($\psi \to \psi e^{i\varphi}$), (11.2) is invariant relative to:

(1) rotation in the (x, y) plane,

$$x = x' \cos \varphi + y' \sin \varphi,$$

$$y = -x' \sin \varphi + y' \cos \varphi;$$

(2) dilatation

$$\xi = q^2 \xi', \quad x = qx', \quad y = qy', \quad \psi = \psi'/q,$$

where q is a constant;

(3) Galilean transformations involving both x and y variables

$$x = x' + v\xi', \quad \psi = \psi' \exp\left[i\frac{v}{4}(v\xi' + 2x')\right],$$

$$y = y' + v\xi', \quad \psi = \psi' \exp\left[i\frac{v}{4}(v\xi' + 2y')\right];$$

and

(4) Talanov's (1970) lens transformation

$$\xi = \frac{\xi'}{1 - \lambda\xi'}, \quad x = \frac{x'}{1 - \lambda\xi'}, \quad y = \frac{y'}{1 - \lambda\xi'},$$

$$\psi = \psi'(1 - \lambda\xi') \exp\left(-i\frac{\lambda}{4}\frac{x'^2 + y'^2}{1 - \lambda\xi'}\right).$$

Clearly, the power remains invariant under the latter transformation. For each symmetry transformation, there is an associated invariant quantity.

In particular, (11.2) conserves the power

$$Q = \int\limits_{-\infty}^{\infty} \int\limits_{-\infty}^{\infty} |\psi|^2 dx\, dy,$$

which is a consequence of the symmetry relative to a phase shift. It also conserves the Hamiltonian

$$H = \int\limits_{-\infty}^{\infty} \int\limits_{-\infty}^{\infty} \left(|\psi_x|^2 + |\psi_y|^2 - \frac{\alpha}{2}|\psi|^4\right) dx\, dy,$$

which is a consequence of the translational invariance in ξ. The conservation of momentum

$$M = i \int\limits_{-\infty}^{\infty} \int\limits_{-\infty}^{\infty} (\psi_x\psi^* - \psi_x^*\psi + \psi_y\psi^* - \psi_y^*\psi)\, dx\, dy$$

is related to the invariance relative to the spatial translations in x and y. We also mention conservation of angular momentum, which is related to the rotational symmetry. Galilean transformations generate the equation for the motion of the centre of mass. The conserved quantity related to Talanov's lens transformation has been found by Kuznetsov and Turitsyn (1985).

Each of the above transformations (and combinations of them) can be used to reduce (11.2) to ordinary differential equations (Tajiri, 1983; Gagnon, 1990), and find selected solutions.

11.2 Radially symmetric solutions

Equation (11.2) has the family of radially symmetric stationary solutions

$$\psi(\rho, \xi) = \sqrt{\frac{q}{\alpha}} R(\sqrt{q}\rho) \exp(iq\xi)$$

where $\rho = \sqrt{x^2 + y^2}$, q is the parameter of the family and $R(r)$ is the solution of the equation

$$R_{rr} + \frac{1}{r}R_r - R + R^3 = 0. \tag{11.3}$$

These are self-trapped beams. The lowest-order solution $R_0(r)$ of (11.3) has been found numerically by Chiao et al. (1964). It has no nodes in its radial field profile. The lowest-order stationary soluton of (11.2) can only be cylindrically symmetric (Gidas et al., 1981). The quantity Q_c is the critical power for the self-focusing. Beams at the critial power propagate as stationary beams, while those with smaller power diffract and those having power above Q_c experience self-focusing. Q_c can be found quite accurately using the 'generalized Gaussian' method for nonlinear optical waveguides (Ankiewicz and Peng, 1991b).

It was found by Yankauskas (1966) that, in addition to the fundamental mode, (11.2) has an infinite number of radially symmetric higher-order modes, $\psi_n(\rho)$, where n is the number of nodes in the radial field. The integral H, for radially symmetric beams, takes the form

$$H = 2\pi \int_0^\infty \left(|\psi_\rho|^2 - \frac{\alpha}{2}|\psi|^4 \right) \rho \, d\rho$$

It is equal to zero for all modes, ψ_n.

The value of the power

$$Q = \frac{2\pi}{\alpha} \int_0^\infty R^2(r) \, r \, dr$$

is independent of q for these solutions.

11.3 Stability of the ground state

For the stability of the ground state, there is a criterion which was first derived by Vakhitov and Kolokolov (1974). They proved that

$$\frac{dQ}{dq} > 0 \qquad (11.4)$$

is a necessary condition for the stability of the ground state. The proof of this result is based on linearization of equation (11.2) for solutions of the form

$$\psi(\rho, \xi) = [R_0(\rho) + [u(\rho) + iv(\rho)] \exp(\delta\xi)] \exp(iq\xi).$$

For the ground state (only), the growth rate, δ, of instability can only be real or purely imaginary. Hence, after linearization, the following eigenvalue problem is obtained:

$$\begin{aligned} L_0 v &= -\delta u, \\ L_1 u &= -\delta v, \end{aligned} \qquad (11.5)$$

where the self-adjoint operators L_0 and L_1 are given by

$$L_0 = \frac{\partial^2}{\partial x^2} + \frac{\partial^2}{\partial y^2} - q + \alpha R_0^2,$$

$$L_1 = \frac{\partial^2}{\partial x^2} + \frac{\partial^2}{\partial y^2} - q + 3\alpha R_0^2.$$

From the eigenvalue equations (11.5), we find the relation

$$\delta^2 = -\frac{< u|L_1|u >}{< u|L_0|u >},$$

where $< u|L_j|u > = \int uL_j u \, dxdy$. Using a variational approach, it can be shown that the maximum of this functional is positive if (11.4) is satisfied. The same stability criterion is valid for the three-dimensional self-focused solutions called 'optical bullets' (section 11.9). This can be proved by a simple generalization of the operators L_0 and L_1 to the three-dimensional case.

The stability criterion (11.4) has been generalized, by Jones and Moloney (1986) to include layered media, and by Mitchell and Snyder (1993) for different types of nonlinearity law. The stability of higher-order solutions is a more complicated matter. The instability growth rate for such a solution is not necessarily real or purely imaginary. Generally speaking, all higher-order states are unstable in homogeneous media. Azimuthal instability is considered below.

The value of q may be difficult to measure in practice. On the other hand, if some other parameter of the solution of the family of soliton solutions, say maximum intensity, I_{max}, depends monotonically on q, then the stability criterion can be reformulated in terms of this parameter, and we can write $dQ/dI_{max} > 0$ instead of (11.4) (Snyder et al., 1996).

11.4 Examples of exact solutions

Although the problem is not integrable, exact solutions for the (2+1)-dimensional Kerr-law problem can be found in special cases (Gagnon, 1990) by using symmetry reductions. An example of a radially symmetric stationary solution is

$$\psi = \frac{2p}{3\sqrt{\alpha}} \rho^{-1/3} \operatorname{cn}(p\rho^{2/3} - \rho_0, \sqrt{1/2}) \exp\left(i\frac{\varphi}{3}\right),$$

where p and ρ_0 are arbitrary constants, cn is one of the elliptic Jacobi functions and its modulus is $k = \sqrt{1/2}$. This solution is singular at $\rho = 0$ unless $\rho_0 = (2n - 1)K(\sqrt{1/2})$, where K is the complete elliptic integral of the first kind and $n = 1, 2, \dots$. In this case the solution is zero at $\rho = 0$ and oscillates in ρ. Unfortunately, the solution does not describe a real beam because it does not have the correct periodicity in the angular variable φ. The solution has a branching point at $\rho = 0$, and it is multi-valued at every point in real space.

Another exact solution for the self-focusing case ($\alpha > 0$) is

$$\psi = \frac{\sqrt{2}}{\rho\sqrt{\alpha}} \frac{k}{\sqrt{1 - 2k^2}} \operatorname{cn}\left[\frac{\varphi}{\sqrt{1 - 2k^2}}, k\right], \qquad 0 < k^2 < 1/2,$$

and for the self-defocusing case ($\alpha < 0$) is

$$\psi = \frac{\sqrt{2}}{\rho\sqrt{-\alpha}} \frac{k}{\sqrt{1 + k^2}} \operatorname{sn}\left[\frac{\varphi}{\sqrt{1 + k^2}}, k\right], \qquad 0 < k^2 < 1.$$

These solutions are also multi-valued at each point of real space. Examples of non-stationary singular solutions can be found in the work by Tajiri (1983).

Exact solutions can be found also for special nonlinearity laws. One of them is the threshold nonlinearity, $N(I) = H(I - I_0)$, where H is the Heaviside step function and I_0 is a constant. Solutions for this law can be found in terms of Bessel functions (Snyder et al., 1991a).

11.5 Collapse of optical beams

Collapse (or blow-up, or self-focusing singularity) of the beams is the phenomenon where the field amplitude increases to infinity and the width of the beam decreases to zero after a finite propagation distance. This phenomenon has been studied extensively over the years, with the first work being done by Vlasov et al. (1971). A review of these studies can be found in the article by Rasmussen and Ripdal (1986). Collapse occurs if the exponent σ in the nonlinearity law $N(|\psi|^2) = |\psi|^{2\sigma}$ is larger than or equal to $2/D$, where D is the dimensionality of the beam. The case $\sigma = 2/D$ is the critical one. The example we are dealing with is critical because $\sigma = 1$ and $D = 2$.

For a two-dimensional beam and Kerr-law ($\sigma = 1$) medium, it follows, from (11.2), that

$$\frac{\partial^2}{\partial \xi^2} \int_0^\infty \rho^2 |\psi|^2 d\rho = 4H.$$

This means that a beam with $Q > Q_0$ in a Kerr medium will collapse if $H < 0$. This occurs at some distance, ξ_0, from the point where the beam enters the nonlinear medium. The value ξ_0 depends on the initial profile and is very sensitive to the detailed structure of it. The behaviour of the beam close to the point of collapse is a controversial issue (Ripdal and Rasmussen, 1986). Several different asymptotic formulae have been found for the amplitude of the field on the beam axis. An example is the asymptotic formula

$$\psi(0, \xi) \rightarrow C(\xi_0 - \xi)^{-2/3}$$

for the field at the centre of the beam. It was found by Zakharov and Synakh (1975). We stress here, however, that when the field amplitude goes to infinity quickly, the approximation of slowly changing amplitude loses its validity in any case. Hence, the asymptotic behaviour is interesting only from a mathematical point of view. As is true more generally in theoretical physics, the fact that a term representing a physical quantity becomes infinite indicates that one or more terms have been omitted from the model. The singularity disappears when we use Maxwell's equations or the wave equation rather than its paraxial approximation (Feit and Fleck, 1988; Manassah and Gross, 1992), or even if we step beyond the approximation of the slowly varying envelope (Akhmediev *et al.*, 1993). (This will be discussed in the next section.) If the nonlinearity law being studied gives a finite change in the refractive index when the intensity in increased to infinity (i.e. if the index saturates), then the collapse also disappears.

11.6 Beyond the paraxial approximation

The paraxial approximation has limitations which are related to the approximation of slowly varying amplitude. For example, the collapse phenomenon is related to this approximation, and this turns out to be in contradiction to the general theory of wave propagation, because the transverse size of any beam cannot be smaller than the light wavelength. The mathematical model needs to be improved, as the collapse is unphysical. We shall do this in this section.

Let us take one step back and start from the wave equation. For simplicity, we start from the scalar wave equation for a three-dimensional field $E(x, y, \xi)$ in a medium:

$$E_{xx} + E_{yy} + E_{\xi\xi} - \frac{\epsilon(|E|)}{c^2} \frac{\partial^2 E}{\partial t^2} = 0, \qquad (11.6)$$

where E is a scalar optical field, and $\epsilon(x, |E|)$ is the intensity-dependent dielectric permittivity:

$$\epsilon(|E|) = \epsilon^{(L)} + \alpha |E|^2. \qquad (11.7)$$

Here $\epsilon^{(L)}$ is the linear part of the dielectric permittivity of the layered medium, and α is the nonlinear susceptibility. In the case of a homogeneous medium, they are constants. The averaged dielectric permittivity does not depend on time. The field is assumed to be stationary (in time) and monochromatic:

$$E = \psi(x, y, \xi) \exp[i\phi(\xi) - i\omega t]. \qquad (11.8)$$

We assume that the amplitude function $\psi(x, y, \xi)$ is slowly varying, and that all fast oscillations are included via the phase function $\phi(\xi)$. Usually, in the approximation of slowly varying amplitude, we set $\phi(\xi) = \beta\xi + \phi_0$, with β constant. In this case, fast oscillations are still implicitly included in the function $\psi(x, y, \xi)$. However, the second derivative of this function is then not small, and cannot be dropped completely.

There can be some degree of arbitrariness in separating the fast oscillatory part from the function $\psi(x, y, \xi)$ if the beam divides into two or more during propagation. In this instance, it is more convenient in numerical simulations to keep β constant, as is usually done. The correction to the real value of β (separate for each beam) can then be extracted from the results by calculating the period of the longitudinal oscillations of $\psi(x, y, \xi)$ at the centre of each beam. On the other hand, the correct initial separation produces physically important consequences, as well as crucial differences in numerical results.

We suppose that we have only one beam as a solution of (11.6), so the function $|\psi(x, y, \xi)|$ has only one maximum at any fixed ξ, and that it approaches zero as $\rho \to \infty$.

Substituting (11.8) into (11.6), we obtain:

$$\psi_{xx} + \psi_{yy} + \psi_{\xi\xi} + 2i\phi_\xi\,\psi_\xi + i\phi_{\xi\xi}\,\psi - \phi_\xi^2\,\psi + \epsilon^{(L)}\psi + \alpha|\psi|^2\,\psi = 0. \quad (11.9)$$

For convenience, we have normalized the coordinates x, y and ξ using the free-space wavenumber $k = \omega/c$. The term $\psi_{\xi\xi}$ can be omitted from (11.9), as is usual in the slowly varying amplitude approximation. However, this can only be done if $\psi(x, y, \xi)$ does not include any fast field changes in the ξ direction. With our definition, the rapidly oscillating part of the solution is included via the function $\phi(\xi)$ in such a way that the function $\psi(x, y, \xi)$ maintains constant phase at the centre of the beam where $|\psi(x, y, \xi)|$ has its maximum. If the beam centre is located at $(x, y) = (0, 0)$, then we can define the function $\phi(\xi)$ in such a way that

$$\arg[\psi(x = 0, y = 0, \xi)] = \text{constant}, \qquad (11.10)$$

where $\psi(0, 0, \xi)$ has been written as $|\psi(0, 0, \xi)| \exp[i \arg \psi(0, 0, \xi)]$. This, in turn, means that the ratio of the imaginary to the real part of $\psi(0,0,\xi)$, i.e. $\mathrm{Im}(\psi(0,0,\xi))/\mathrm{Re}(\psi(0,0,\xi))$, remains constant. Equation (11.10) defines the functions $\phi(\xi)$ and $\psi(x,y,\xi)$ uniquely.

Let us now define the function $\tilde{\beta}$ as:

$$\tilde{\beta}(\xi) = \frac{d\phi}{d\xi}. \tag{11.11}$$

This is the instantaneous propagation constant at a given cross-section ξ. This function is also defined in a unique way. Now we can write (11.9) in the form:

$$2i\tilde{\beta}\,\psi_\xi + i\tilde{\beta}_\xi\,\psi + \psi_{xx} + \psi_{yy} - \tilde{\beta}^2\,\psi + \epsilon^{(L)}\psi + \alpha|\psi|^2\,\psi = 0. \tag{11.12}$$

In the case of constant $\tilde{\beta}$, this equation (after normalizations) coincides with equation (11.2). The second term in (11.12), which can be of the same order as the first one, makes it different from the usual parabolic equation.

Now let us turn to some physical differences which appear when we take this term into account. Consider, for example, the invariants of (11.12). By multiplying (11.12) by ψ^*, taking the complex conjugate of this expression, and subtracting and integrating, we obtain:

$$\frac{d}{d\xi}(\tilde{\beta}Q) = 0, \tag{11.13}$$

where Q is the energy invariant for the standard parabolic equation (i.e. (11.12) without the second term):

$$Q = \int\limits_{-\infty}^{\infty} \int\limits_{-\infty}^{\infty} |\psi(x,y,\xi)|^2\,dx\,dy. \tag{11.14}$$

We can see now that the product $\tilde{\beta}Q$, rather than just Q, is the conserved quantity during propagation:

$$\tilde{\beta}(\xi)\,Q(\xi) = \text{constant}. \tag{11.15}$$

This product, $S_\xi = \tilde{\beta}Q$, is proportional to the integrated energy flow in the ξ direction, i.e. to the ξ component of the Poynting vector $\mathbf{S} = \mathbf{E} \times \mathbf{H}$ integrated over the cross-section. This is the result which we would expect physically, because the energy flow is the quantity which has to be conserved in media without gain or loss. In the standard approximation using the parabolic equation, this conservation law is incomplete, and only the 'energy integral' Q is conserved, as rapid oscillations are neglected when the term $\psi_{\xi\xi}$ is dropped. Of course, in the case of constant β, the energy invariant Q itself is conserved.

Let us now consider the second quantity conserved for the parabolic equation, namely the Hamiltonian. For simplicity, we restrict ourselves to a two dimensional field $\psi(x, \xi)$. The Hamiltonian is now proportional to the y component of the electric field. By multiplying (11.12) by $d\psi^*/d\xi$, taking the complex conjugate of this expression, and adding and integrating over x, we obtain:

$$i\,\tilde{\beta}_\xi \int_{-\infty}^{\infty} dx \, (\psi_\xi\,\psi^* - \psi_\xi^*\,\psi) - 2\,\tilde{\beta}\,\tilde{\beta}_\xi \int_{-\infty}^{\infty} dx \, |\psi|^2$$

$$+ \frac{d}{d\xi} \int_{-\infty}^{\infty} dx \left[|\psi_x|^2 + (\tilde{\beta}^2 - \epsilon_1)|\psi|^2 - \frac{\alpha}{2}|\psi|^4 \right] = 0. \qquad (11.16)$$

The sum of the integrals in (11.16) is proportional to the energy density of the optical field integrated over x, per unit ξ. The first two integrals are the part of the energy density associated with the x component of the magnetic field (H_x). These two integrals become zero when β is a constant in the standard parabolic approximation. Then the third integral is conserved along ξ. It can still be considered as the energy density per unit ξ. So, in the absence of currents, the integrated energy density is conserved along ξ. If $\tilde{\beta}_\xi$ is non-zero, then the whole expression (11.16) cannot be represented in the form of a conservation law. The consequence is that the Hamiltonian (the third integral in (11.16)) is no longer conserved.

The new equation gives results which are different from the standard NLSE, but are consistent with the solutions of the wave equation. An example of a numerical simulation to compare the two approaches is shown in Fig. 11.1. A modification of the Crank–Nicholson scheme which maintains constant phase of $\psi(x, y, \xi)$ at $(x, y) = (0, 0)$ to simulate the solutions of (11.12) is described in Soto-Crespo and Akhmediev (1993). The scheme conserves the invariant given by (11.15). Figure 11.1 shows the on-axis intensity of the cylindrically-symmetric beam which initially ($\xi = 0$) has a Gaussian shape:

$$\psi(x, y, \xi = 0) = \exp\left(\frac{x^2 + y^2}{8}\right). \qquad (11.17)$$

This initial condition has a power flow which is above the critical value, and thus leads to collapse when the standard NLSE is used. The parameters of the simulation are given in the figure caption. It is seen from this figure that the standard NLSE leads to collapse and that the on-axis field goes to infinity at finite ξ (≈ 62.5). The solution of (11.12) deviates from it for on-axis intensities, $|\psi(0, 0, \xi)|^2$, between approximately 5 and 10, and then returns to a smoother beam propagation beyond the point of maximum intensity. This result is in qualitative agreement with the numerical results of Feit and Fleck (1988) and Manassah and Gross (1992), where the authors use non-paraxial algorithms for simulation of the solution of the wave equation to describe beam self-focusing. Another mathematical

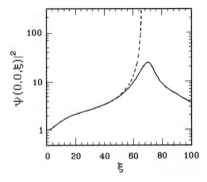

Figure 11.1 *The on-axis intensity versus normalized axial distance ξ for the Gaussian initial condition $(\xi = 0)$ of (11.17) found by using the standard NLSE (dashed curve) and by using (11.12) (solid curve). Parameters chosen for these simulations are: $\epsilon_l = 24.50015104$, $\beta(0) = 5.0$, $\alpha = 1.0$.*

treatment of the problem which shows that small beam nonparaxiality arrests self-focusing has been given by Fibich (1996). Our approach does not take into account longitudinal field components which become important in the neighbourhood of a self-focus point. A generalization of the above approach to vector fields has been developed by Chi and Guo (1995).

11.7 Radially symmetric solutions for the case of saturable nonlinearity

An infinite number of radially symmetric modes exist for both Kerr-type and saturable nonlinearities. To avoid the collapse problem, in this section we consider a nonlinearity with saturation. Examples of the three lowest-order radially symmetric solutions of the equation

$$i\frac{\partial\psi}{\partial\xi} + \frac{\partial^2\psi}{\partial\rho^2} + \frac{1}{\rho}\frac{\partial\psi}{\partial\rho} + \frac{1}{\rho^2}\frac{\partial^2\psi}{\partial\theta^2} - q^2 + \frac{|\psi|^2\psi}{1+|\psi|^2} = 0, \qquad (11.18)$$

which includes a saturable nonlinearity of the form (4.16), are shown in Fig. 11.2. The mode number, n, corresponds to the number of zeros in the field profile, ψ_n.

Figure 11.3 shows the energy flux, given by

$$Q_n = 2\pi \int\limits_0^\infty \psi_n^2(\rho)\, \rho\, d\rho,$$

as a function of q^2 for the first few bound state solutions ψ_n. All these

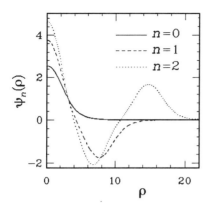

Figure 11.2 *Field profiles for the bound states corresponding to* $q^2 = 0.5$ *and* $n = 0$ *(solid line),* $n = 1$ *(dashed line), and* $n = 2$ *(dotted line).*

curves have positive slopes over the full range of q^2. Thus the condition

$$\frac{\partial Q_0}{\partial q^2} > 0$$

for the stability of the lowest-order mode is fulfilled. In fact, for initial conditions with energies higher than the energy of the ground state, the solution aquires a ring structure (Marburger and Dawes, 1968). This occurs if we ignore the azimuthal variable in the equation. However, the higher-order states are unstable relative to perturbations which break the initial cylindrical symmetry. As a result, the beam splits into several beams of smaller size in a process which is called 'small-scale self-focusing'.

11.8 Loss of cylindrical symmetry

If we perturb the radially symmetric beam to $\psi(\rho, \theta, 0)$, then we expect the evolution to be:

$$\psi(\rho, \theta, \xi) = \psi_n(\rho) + \mu\, f(\rho)\, \cos(m\theta)\, \exp(\delta_{mn}\xi), \tag{11.19}$$

where μ is a small parameter, $f(\rho)$ is a perturbation function, δ_{mn} is the growth rate of the perturbation, and the integer m is the azimuthal index. Inserting (11.19) into (11.18) and linearizing in the small parameter μ, we obtain

$$\nabla^2_{\rho\rho} f - \left[q^2 + \frac{m^2}{\rho^2}\right] + \frac{2\psi_n^2 f + \psi_n^4 f + \psi_n^2 f^*}{(1 + \psi_n^2)^2} = \delta_{mn} f.$$

This equation has many possible types of eigenfunctions and corresponding eigenvalues for each value of m and n. An exception is $n = 0$, because the

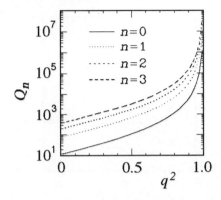

Figure 11.3 *Energy flux versus q^2 for the first four stationary bound states.*

ground state is stable relative to azimuthal perturbations. Solutions for $n \geq 1$ were found numerically by Soto-Crespo and Wright (1991). For the function ψ_1, the amplitude and the growth rate are shown in Figs 11.4 and 11.5, respectively.

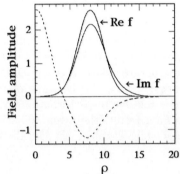

Figure 11.4 *Real and imaginary parts of the perturbation eigenmode with $m = 4$ associated with the stationary solution (dashed line) for $n = 1$ and $q^2 = 0.35$.*

We can see from these figures that the $n = 1$ bound state is unstable for all values of $q^2 < 1$. It can also be seen from Fig. 11.4 that the eigenfunction of the perturbation is mainly localized around the peak of the annular ring of the bound state solution, ψ_1. This shows that instability is related to the ring structure of the solution and that this ring is modulationally unstable. The ring must separate into filaments on propagation. For the chosen parameters of the simulation, the growth rate is maximal for $m = 4$. Hence, it is very likely that the first ring will split into four filaments. However, other forms of evolution are possible, depending on the initial conditions. Because the growth rate for $m = 3$ is close to that for $m = 4$, splitting into three filaments can also occur with a slight variation in the initial conditions. Numerical simulations confirm this fact. For higher-order modes, the distribution of growth rates will be different.

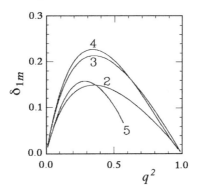

Figure 11.5 *Growth rates δ_{nm} of the predominant perturbation eigenmodes with m-fold symmetry, associated with the stationary solution for $n = 1$, as a function of q^2. Each curve is labelled with its value of m.*

Azimuthal symmetry-breaking of higher-order beams has been experimentally observed by Tikhonenko *et al.* (1995).

11.9 Optical bullets

Let us now consider pulse-beam propagation effects in three dimensions. Pulse beam-like solutions of the wave equation for this case have been studied by Kaw *et al.* (1975), where they were called '3-D envelope' solitons. Later, they were studied by Silberberg (1990), who used the term 'optical bullets'. The stability properties of these solutions have been studied by Akhmediev and Soto-Crespo (1993). In all cases, a saturable nonlinearity has been used to avoid the problem of collapse .

The radially symmetric nonlinear wave equation (11.1), with N given by (4.16) and with suitable normalizations, has the form:

$$i\psi_\xi + \Delta_\rho\psi + \psi_{\tau\tau} + \frac{|\psi|^2\psi}{1 + \gamma|\psi|^2} = 0, \qquad (11.20)$$

where the transverse Laplacian, $\Delta_\rho\psi = \psi_{\rho\rho} + \frac{1}{\rho}\psi_\rho$, is independent of the angular variable, ξ is the normalized longitudinal coordinate, τ is the normalized retarded time, and γ is the saturation parameter. We assume that the group velocity dispersion, k'', is negative. The function ψ and all variables in (11.20) are normalized to avoid most of the parameters of the problem. The only parameter left in the equation is γ.

As in the one- and two-dimensional cases, we explicitly separate the propagation constant q from the solution, setting

$$\psi(\rho, \tau, \xi) = U(r) \exp(iq\xi) \qquad (11.21)$$

where $r = \sqrt{\rho^2 + \tau^2}$. Now (11.20) reduces to:

$$U_{rr} + \frac{2}{r} U_r - qU + \frac{|U|^2 U}{1 + \gamma |U|^2} = 0. \qquad (11.22)$$

The energy invariant, Q, of optical bullets, in spherical coordinates, takes the form:

$$Q = 4\pi \int_0^\infty |U|^2 \, r \, dr. \qquad (11.23)$$

The energy invariant is plotted against propagation constant q for three-dimensional solitons described by (11.22) in Fig. 11.6. For any value of the energy invariant higher than some Q_{min}, there are corresponding values of q. The branch of this curve with positive slope (solid curve) is stable and the branch with negative slope (dashed curve) is unstable. Thus, the plot has the same form as that of a surface-guided wave in a nonlinear half-space.

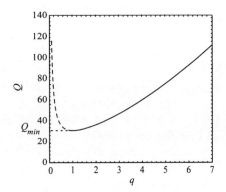

Figure 11.6 *Energy versus propagation constant diagram for three-dimensional optical solitons. The curve has a stable (positive slope) as well as an unstable (negative slope) branch.*

Initial conditions in the form of three-dimensional solitons

$$\psi(\rho, \tau, \xi = 0) = U(r, q), \qquad (11.24)$$

allow propagation over long distances without change in q on the branch of the energy-dispersion curve (Fig. 11.6) with positive slope, because optical bullets are stable on this branch of the curve.

11.10 Longitudinal modulation instability of self-trapped beam

Equation (11.20) can have solutions in the form of pulses localized in three dimensions. For any solutions localized in three dimensions, (11.20) has the invariant

$$Q = 2\pi \int_0^\infty \rho \, d\rho \int_{-\infty}^\infty d\tau |\psi(\rho, \tau)|^2, \qquad (11.25)$$

which does not depend on ξ. It represents the total energy of the localized solution.

Equation (11.20) also has stationary solutions (independent of ξ and τ) localized in two transverse dimensions. These stationary solutions, $\psi_0(\rho)$, can be found by solving a modified (11.20), where the terms with derivatives in ξ and τ have been dropped:

$$\Delta_\rho \psi_0 - \psi_0 + \frac{|\psi_0|^2 \psi_0}{1 + \gamma |\psi_0|^2} = 0. \qquad (11.26)$$

We consider only the lowest-order (nodeless) solution of (11.26). For the stationary solution $\psi_0(\rho)$, the quantity

$$Q = 2\pi T \int_0^\infty |\psi_0(\rho, \tau)|^2 \, \rho \, d\rho \qquad (11.27)$$

represents the energy of the beam for a length T along the τ axis. In the case of solutions which are periodic in the τ variable, T is equal to the period.

We are interested in solutions of (11.20) which evolve as

$$\psi(\rho, \tau, \xi) = \psi_0(\rho) + \mu f(\rho) \cos(\Omega \tau) \exp(\delta \xi) \qquad (11.28)$$

for small ξ, due to an initial condition $\psi(\rho, \tau, 0)$. Here μ is a small parameter, $\Omega = 2\pi/T$ is the frequency of a periodic modulation, $\delta = \delta(\Omega)$ is the modulation growth rate, and $f(\rho)$ is the perturbation function. This function is, in general, complex. An example of the function $f(\rho)$ for the case of the maximum growth rate for $T = 4.4$ and $\Omega = 0.05$ is given in Fig. 11.7.

Figure 11.8 gives an example of the evolution (in ξ) of the field intensity $|\psi|^2$, with an initial condition given by (11.28) with $\xi = 0$, at its maximum amplitude ($\rho = 0, \tau = 0$). The field retains the shape of the stationary solution for a certain distance ξ ($\xi \sim 5$ in this case), until the exponential factor in (11.28) becomes large. Then the field intensity increases rapidly, and the field becomes oscillatory, with its amplitude decaying smoothly after reaching a maximum amplitude at some point (point A in Fig. 11.8). The first oscillation resembles the behaviour of modulation instability of the basic NLSE. At point A, the field in each period is highly compressed in three dimensions and the whole solution looks like a train of three-dimensionally self-compressed pulses. Starting from point A, the degree of

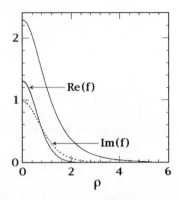

Figure 11.7 *Real and imaginary parts of the perturbation eigenfunction (as labelled) along with the shape of the initial self-focused beam, ψ_0 (upper curve).*

compression decreases up to the minimum point of the curve in Fig. 11.8, where pulses become more spread out, but are still well separated. There is no recurrence back to the initial beam because an appreciable amount of energy has irreversibly been lost as radiation.

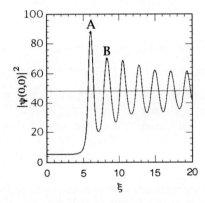

Figure 11.8 *Field intensity $|\psi|^2$ at $\tau = 0, r = 0$ versus propagation distance ξ for initial condition (11.28). The modulation period, T, is 4.4. The length of the initial part of this curve (with exponentially growing perturbation) depends on the initial amplitude of modulation, μ. The horizontal line in this plot corresponds to the maximum intensity of the optical bullet at $q = 3.0$. The saturation parameter $\gamma = 0.05$.*

After the minimum point in Fig. 11.8, the pulses in each period are compressed again up to the point of the next maximum (point B in Fig. 11.8).

The degree of compression is now less than that at point A. The process is repeated again in the next period but with decreasing amplitude. The initially c.w. beam has disintegrated into separate three-dimensional optical solitons which oscillate on propagation.

The horizontal line in Fig. 11.8, around which oscillations occur, corresponds to the peak intensity $|U(r = 0)|^2$ of the optical bullet with propagation constant $q = 3.0$. The oscillations in Fig. 11.8 clearly converge to this value. Field envelope oscillations also occur around a mean shape corresponding to this optical bullet.

As a consequence of the radiation process, the energy in each period in τ decreases with each oscillation. Up to 25% of the energy can be radiated during the transition process from the c.w. beam to the train of optical bullets. The amount of radiation at each oscillation decreases to almost zero even if the oscillations continue with finite amplitude.

Thus, the initially homogeneous self-focused beam breaks down into separate 'clumps' as a result of modulation instability, and each of these 'clumps' converges to a three-dimensional soliton (optical bullet). Because of radiation, the process is apparently irreversible, and this makes it different from the process of modulation instability in one dimension. There is no recurrence to the initial c.w. wave which takes place in the one-dimensional case, or even pseudo-recurrence, as in the case of transverse modulation instability in two dimensions (Akhmediev et al., 1990).

A trajectory starting from a saddle point on one energy level cannot come back to any point on the same energy level. Instead, such a trajectory can come to some special point on a different energy level. The final state can be a stationary solution or an oscillatory one close to a three-dimensional soliton. Therefore, in terms of nonlinear dynamical systems, our system evolves from a saddle-type point at a high energy level, to a focal point, or to a limit cycle, on a lower energy level.

The process of conversion to an optical bullet can be visualized using a simple technique. Figure 11.9 shows the trajectory of the field ψ at ($\tau = 0, r = 0$) on the complex plane. The field, on evolution, accumulates phase, and the resulting field at the end of the process has an additional rapidly oscillating factor, $\exp(i(q-1)\xi)$, relative to the initial field. The final limit cycle is then motionless on the complex plane which rotates relative to the initial plane, where the saddle-type point is motionless.

Physically, this means that, on evolution, the propagation constant changes by a certain amount, $q - 1$. Figure 11.9 is plotted in a rotating frame with the additional phase factor $\exp(i1.92\xi)$. The corresponding three-dimensional solitons can be represented in this plot by any point at a fixed distance from the zero point. (The constant phase is arbitrary.) For a given q, these points are located on the circle designated by the dashed line in Fig. 11.9. The trajectory of the field converges to one of these points. So, the optical bullets are foci in a Hilbert space, in the sense that they

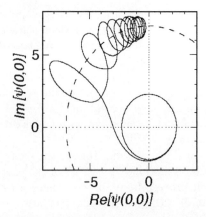

Figure 11.9 *Trajectory of the solution at $\rho = 0$ and $\tau = 0$ on the complex plane.*
The data are presented in the frame with the additional phase factor $\exp(i1.92\xi)$.
Trajectories for different points r have the same qualitative behaviour as for $\rho = 0$.
The dashed line corresponds to the maximum amplitude of 3-D soliton.

can attract trajectories which pass near to them. In Fig. 11.9 we have an
example of motion from an unstable stationary saddle point to a stable
focal point. Three-dimensional solitons are attractors for the solutions of
(11.20). We may expect that an arbitrary three-dimensional initial condition may contain a certain number of three-dimensional solitons plus a
certain amount of radiation. The interactions between three-dimensional
solitons have been studied by McLeod *et al.* (1995), with the intention of
using them in logic gates.

CHAPTER 12

Planar nonlinear guided waves

In this chapter, we consider several examples of planar nonlinear guided waves and their stability. The first examples were presented by Litvak and Mironov (1968) and then by Agranovich et al. (1980), Tomlinson (1980) and Maradudin (1981a,b). Nonlinear surface waves are self-focused beams which are guided by an interface, where the refractive index on the linear side of the interface is higher than the low-power refractive index on the nonlinear side. Although the existence of nonlinear surface waves is easily explained in physical terms, their properties are not trivial. They become even more complicated in the case of waveguide geometry (Akhmediev 1982; Stegeman 1992). The first topic to be addressed is stationary solutions for a given geometry. The second subject is their stability. Stability results for the lowest mode of nonlinear wavegudes have been summarized by Mitchell and Snyder (1993). The third area to be addressed is the dynamics of self-focused beams in layered structures. Results relating to this third topic can be found in the book by Newell and Moloney (1992).

Understanding the stability of nonlinear guided waves is essential for the future exploitation of these waves in optical devices such as power limiters (Seaton et al., 1985; Maradudin, 1981a,b; Ankiewicz and Tran, 1991). Much experimental work on planar nonlinear guides is described in the review article by Mihalache et al. (1989). Needless to say, this problem has not been solved completely, in the sense that we do not have a comprehensive stability criterion which can be applied to any case without detailed calculations (Akhmediev, 1991). The authors of some early papers hoped that Kolokolov's specific criterion (Kolokolov, 1973) for a homogeneous medium could serve as a general one. However, it has turned out that this criterion can be applied only to parts of the dispersion curve of the lowest-order nonlinear wave. Each of the higher-order 'nonlinear modes' has to be considered separately, in order to decide whether it is stable or not. Only a few of the simplest cases have been considered before, either by numerical or analytic means. An analysis, based on generalizing Kolokolov's technique to inhomogeneous media, is given by Jones and Moloney (1986). This technique uses the set of linear operators of the linearized problem, but is strictly limited to the nodeless (viz. TE_0 symmetric and asymmetric ones) solutions, because Jones and Moloney (1986) only considered the case of purely real or imaginary eigenvalues of those operators. For the anti-symmetric TE_1 solution, the instability growth rate can be complex (Tran et al., 1992).

We give fields and powers for simple cases, explain the stability behaviour of stationary solutions and concentrate on the stability of higher-order modes. This is not a simple topic, and each case must be considered separately.

12.1 Nonlinear waves in a layered medium

We suppose that the medium is stratified in the x direction and that waves are propagating in the ξ direction. Exact solutions of Maxwell's equations can be obtained for TE waves, i.e. when the electric field is oriented along the y axis. We seek a stationary solution of this two-dimensional problem in the form

$$E_y(x,\xi) = \psi(x,\xi) \exp(i\beta\xi), \qquad (12.1)$$

where β is the propagation constant, and $\psi(x,\xi)$ is the transverse field profile. The equation for the field ψ, in the approximation of slowly varying amplitudes, is found from (1.29) (with $k \to \beta$):

$$2i\beta\psi_\xi + \psi_{xx} - g^2(x)\psi + h(x)|\psi|^2\psi = 0, \qquad (12.2)$$

where $g(x)$ and $h(x)$ are the transverse linear and nonlinear susceptibility profiles, respectively. The coordinates x, ξ, and propagation constant β in (12.2) have been normalized with the free-space wavenumber $k_0 = \omega/c$, and the field $\psi(x,\xi)$ has been normalized with the nonlinear permittivity, so that they are all dimensionless.

Equation (12.2) has two conserved quantities,

$$Q = \int_{-\infty}^{\infty} |\psi(x,\xi)|^2 \, dx, \qquad (12.3)$$

which is proportional to the power flow at any cross-section ξ, and the Hamiltonian

$$H = -\frac{1}{2\beta} \int_{-\infty}^{\infty} \left[|\psi_x|^2 + g^2(x)|\psi|^2 - \frac{1}{2}h(x)|\psi|^4 \right] dx, \qquad (12.4)$$

which is proportional to the energy density at any cross-section ξ. We note that the power flow is proportional to βQ. Momentum is not conserved because of the presence of interfaces.

As for other Hamiltonian systems considered earlier, (12.2) can be written in the canonical form

$$i\psi_\xi = \frac{\delta H}{\delta\psi^*}. \qquad (12.5)$$

This means that stationary solutions can be found as extrema of the Hamiltonian (12.4), and that they are stable if the Hamiltonian has a local minimum or maximum.

12.2 Waves on a single boundary of a nonlinear medium

A single interface between a linear medium ($n = \sqrt{\epsilon_l}$) and a nonlinear medium can support a nonlinear surface wave (NSW) (Litvak and Mironov, 1968; Tomlinson, 1980; Agranovich *et al.*, 1980; Marududin, 1981a). We set the low-power limit of the nonlinear medium dielectric constant to be $\epsilon_0 (< \epsilon_l)$. Energy near the interface increases the local nonlinear index, thus creating an effective waveguide and hence allowing a solution of Maxwell's equations with a field maximum in the nonlinear medium. The linear and nonlinear susceptibility profiles in this case are:

$$g(x) = \begin{cases} \gamma_l, & x < 0 \\ \gamma_0, & x \geq 0 \end{cases} \qquad h(x) = \begin{cases} 0, & x < 0 \\ 1, & x \geq 0 \end{cases} \tag{12.6}$$

where $\gamma_l = \sqrt{\beta^2 - \epsilon_l}$ and $\gamma_0 = \sqrt{\beta^2 - \epsilon_0}$.

The stationary solution of (12.2) for a single interface is:

$$\psi = \begin{cases} \sqrt{2(\epsilon_l - \epsilon_0)} \exp(\gamma_l x), & x < 0 \\ \sqrt{2}\,\gamma_0 \operatorname{sech}[\gamma_0(x - x_0)], & x \geq 0 \end{cases} \tag{12.7}$$

This solution is represented schematically in Fig. 12.1. Interestingly enough, the field at the interface is independent of β:

$$\psi^2(x = 0) = 2\,(\epsilon_l - \epsilon_0).$$

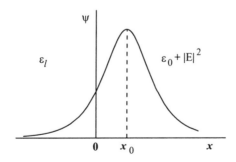

Figure 12.1 *Single interface between two media and the profile of the nonlinear surface wave which it supports.*

Matching boundary conditions, we find the dispersion relation

$$\tanh(\gamma_0 x_0) = \gamma_l/\gamma_0.$$

Hence, the field maximum occurs at

$$x = x_0 = \frac{1}{2\gamma_0} \ln \frac{\gamma_0 + \gamma_l}{\gamma_0 - \gamma_l} = \frac{1}{\gamma_0} \text{arctanh}(t), \qquad (12.8)$$

where, for convenience, we have defined $t = \frac{\gamma_l}{\gamma_0}$ so that $0 < t < 1$. Writing x_0 in terms of t,

$$x_0 = \frac{1}{2} \sqrt{\frac{1 - t^2}{\epsilon_l - \epsilon_0}} \ln \frac{1 + t}{1 - t},$$

shows that x_0 is positive but goes to zero when $\beta \to \sqrt{\epsilon_l}$ (i.e. $t \to 0$) and for high β values (i.e. $t \to 1$).

The integral Q can be shown to have the form

$$Q = \frac{(\epsilon_l - \epsilon_0)}{\gamma_l} + 2(\gamma_l + \gamma_0) = (t + 1)^2 \frac{\gamma_0 [\beta(t)]}{t} \qquad (12.9)$$

and the Hamiltonian is:

$$H = - \left[\frac{2}{3} (\gamma_0^3 + \gamma_l^3) + \gamma_l (\epsilon_l - \epsilon_0) \right] / \beta. \qquad (12.10)$$

The Q–β diagram (Fig. 12.2) exhibits a minimum at

$$\beta = \beta_{cr} = \sqrt{(4\epsilon_l - \epsilon_0)/3}.$$

At this point $\gamma_l = \sqrt{(\epsilon_l - \epsilon_0)/3}$ and $\gamma_0 = \sqrt{4(\epsilon_l - \epsilon_0)/3}$. When viewed as a function of t, Q always has its minimum at $t = \frac{1}{2}$. A certain minimum power, viz. $Q_{min} = 3\sqrt{3(\epsilon_l - \epsilon_0)}$, is required for the surface wave to exist.

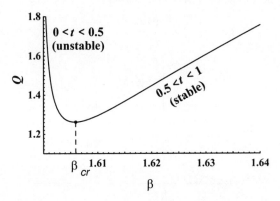

Figure 12.2 *Integral Q versus β for nonlinear surface wave ($\epsilon_0 = 2.5$ and $\epsilon_l = 2.559$).*

The Hamiltonian H is plotted against the power Q for the surface wave in Fig. 12.3. Due to the stability criterion, the NSW with the lower H at given Q in this diagram is stable. Hence, all nonlinear single interface surface waves with $\beta > \beta_{cr}$ (i.e. $1/2 < t < 1$) are stable. The stability criterion is $dQ/d\beta > 0$. This agrees with the stability properties seen in the $H - Q$ diagram (Fig. 12.3).

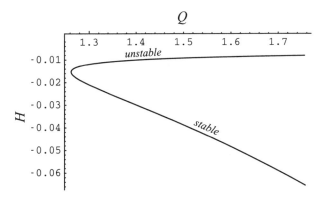

Figure 12.3 *Hamiltonian H versus Q for nonlinear surface wave ($\epsilon_0 = 2.5$ and $\epsilon_l = 2.559$). The leftmost point on the curve corresponds to $t = 1/2$.*

12.3 Nonlinear waves in the three-layer symmetric waveguide

The layered structure which can be used for switching is more complicated. Consider a linear planar core with index $n_l = \sqrt{\epsilon_l}$ and thickness $2\bar{d}$ surrounded by a nonlinear cladding and substrate. To be consistent with section 12.1, we use the normalization $d = \bar{d}k_0$. We call this the 'three-layer' NL-L-NL (i.e. nonlinear-linear-nonlinear) structure. Suppose that the low-power refractive index is $n_0 = \sqrt{\epsilon_0}$ for both cladding and substrate, and that $\gamma_l > 0$ (the region of our analysis). The coordinate origin ($x = 0$) is now the centre of the core. The symmetric (even) solution for this structure is the following:

$$\psi_+ = \left\{ \begin{array}{ll} \sqrt{2}\,\gamma_0\,\mathrm{sech}[\gamma_0(x + d + x_{0s})], & x \leq -d \\[2mm] \sqrt{2}\,\gamma_0\cosh(\gamma_l x)/[\cosh(\gamma_l d)\cosh(\gamma_0 x_{0s})], & |x| < d \\[2mm] \sqrt{2}\,\gamma_0\,\mathrm{sech}[\gamma_0(x - d - x_{0s})], & x \geq d \end{array} \right\}, \quad (12.11)$$

where the normalized distance of the maximum from each interface, x_{0s}, is given by:

$$x_{0s} = \frac{1}{2\gamma_0}\ln\frac{\gamma_0 + \gamma_l\tanh(\gamma_l d)}{\gamma_0 - \gamma_l\tanh(\gamma_l d)} = \frac{1}{\gamma_0}\,\mathrm{arctanh}[t\,\tanh(\gamma_l d)]. \quad (12.12)$$

The power integral for this solution is:

$$Q_+ = [\gamma_0 + \gamma_l \tanh(\gamma_l d)] \left[2 + \frac{\gamma_0 - \gamma_l \tanh(\gamma_l d)}{\cosh^2(\gamma_l d)} \left(\frac{\sinh(2\gamma_l d)}{2\gamma_l} + d\right)\right]. \quad (12.13)$$

The anti-symmetric (odd) solution for the structure is:

$$\psi_- = \begin{cases} -\sqrt{2}\,\gamma_0 \operatorname{sech}[\gamma_0(x + d + x_{0a})], & x \le -d \\[2ex] \sqrt{2}\,\gamma_0 \sinh(\gamma_l x)/[\sinh(\gamma_l d)\cosh(\gamma_0 x_{0a})], & |x| < d \\[2ex] \sqrt{2}\,\gamma_0 \operatorname{sech}[\gamma_0(x - d - x_{0a})], & x \ge d. \end{cases} \quad (12.14)$$

The normalized distance of the maximum from each interface, x_{0a}, in this equation is different from that in the symmetric solution:

$$x_{0a} = \frac{1}{2\gamma_0} \ln \frac{\gamma_0 + \gamma_l \coth(\gamma_l d)}{\gamma_0 - \gamma_l \coth(\gamma_l d)}, \quad (12.15)$$

with power

$$Q_- = [\gamma_0 + \gamma_l \coth(\gamma_l d)] \left[2 + \frac{\gamma_0 - \gamma_l \coth(\gamma_l d)}{\sinh^2(\gamma_l d)} \left(\frac{\sinh(2\gamma_l d)}{2\gamma_l} - d\right)\right]. \quad (12.16)$$

Additionally, a symmetric solution with a single maximum in a nonlinear medium exists for the structure (Akhmediev, 1982).

The expressions in this section can be written, if required, in terms of the standard waveguide parameter $V = d\sqrt{\epsilon_l - \epsilon_0}$ (Snyder and Love, 1983) by noting that $\gamma_0 = \gamma_l/t$ and

$$\gamma_l d = \frac{Vt}{\sqrt{1 - t^2}}. \quad (12.17)$$

12.4 Power–dispersion diagram

For a fixed d, the power–dispersion diagram consists of a plot of Q against propagation constant β for various stationary states. A typical example is shown in Fig. 12.5. Here the curve labelled ψ_+ is plotted for the family of symmetric solutions, that labelled ψ_- represents the family of anti-symmetric solutions and that labelled ψ_0 represents the family of asymmetric solutions. For $Q_\pm = 0$ we have the linear limit – here giving the lowest-order symmetric (ψ_+) and anti-symmetric (ψ_-) linear modes.

These stationary states, which are solutions of the nonlinear wave equation (12.2), may be stable or unstable with respect to propagation along the longitudinal coordinate (ξ axis). The solid lines indicate stability, while dashed lines show instability. We explain the stability behaviour of the three-layer guide (NL-L-NL) at high β in terms of the known behaviour of the single interface. This is feasible because the symmetric and anti-symmetric states of the full waveguide closely resemble the sum and differ-

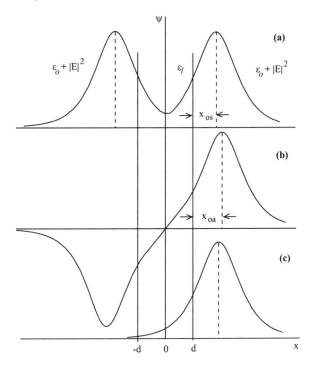

Figure 12.4 *Three-layer symmetric waveguide structure and the profiles of the lowest-order nonlinear guided waves: (a) symmetric solution, (b) anti-symmetric solution and (c) asymmetric solution.*

ence of two mirror-image nonlinear surface waves. The A-type asymmetric state of the three-layer system closely resembles a surface wave when $dQ/d\beta > 0$, because the second interface then has almost no effect on the first.

We explain the stability first by considering a lateral field shift (in section 12.5), and then with a calculation of growth rates (in section 12.8). In doing this, our restriction is that we consider the range $\gamma_l > 0$, where NSWs exist, thus excluding β values lower than $\sqrt{\epsilon_l}$. This is equivalent to $0 < t < 1$. Some analytic expressions and the results of section 12.10 are applicable for high β values ($t \to 1$) only.

12.5 Lateral field shift in nonlinear medium

In the three-layer problem, the symmetric and anti-symmetric solutions in the nonlinear media, at fixed β, are $\sqrt{2}\,\gamma_0\,\mathrm{sech}[\gamma_0(x \pm (d + x_{01}))]$, where x_{01} is x_{0s} or x_{0a}. Thus, in the nonlinear media, the only difference from the

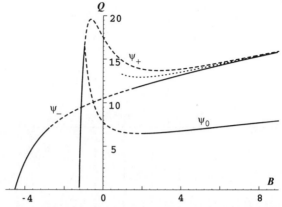

Figure 12.5 *Power–dispersion diagram for planar nonlinear guide for $V = d\,(\epsilon_l - \epsilon_0)^{1/2} = 2.5$. The curves correspond to: symmetric TE_0 solution (curve ψ_+), anti-symmetric TE_1 solution (curve ψ_-), asymmetric solution with single maximum (curve ψ_0). Unstable branches of these curves are indicated by dashed lines, and stable parts by solid lines. The dotted curve indicates double the power level of field ψ_0. Note that it lies between the power levels of ψ_+ and ψ_-. Here B is defined as $d^2(\beta^2 - \epsilon_l)$. Thus high β corresponds to high B (i.e. right-hand part of diagram).*

single-interface problem is that x_0 has been replaced by $x_{0s} = x_0 - \Delta x_+$ or by $x_{0a} = x_0 + \Delta x_-$. In the limit of large $\gamma_l d$ (or high β, where $t \to 1$), the shift in the position of the peaks, $\Delta x_\pm$, is given by

$$\Delta x_\pm \approx \frac{2\gamma_l}{\epsilon_l - \epsilon_0}\exp(-2\gamma_l d). \qquad (12.18)$$

We note that linear coupling between two planar waveguides also decreases exponentially with the separation between them (Marcuse, 1973). In our nonlinear case, the shift may also be expressed in terms of V and t:

$$\Delta x_\pm \approx \frac{2dt}{V\sqrt{1 - t^2}}\exp\left(\frac{-2Vt}{\sqrt{1 - t^2}}\right).$$

If we regard the right-hand interface and nonlinear medium as a perturbation of the single-interface problem, then the effect of a state of the same parity on the left (i.e. the left-hand side of the ψ_+ state) is to increase the index near the right-interface, and thus shift the field peak towards the interface. Thus we can write

$$\psi_+ = \sqrt{2}\gamma_0\,\mathrm{sech}[\gamma_0(x - d - x_0 + \Delta x_+)] \qquad (12.19)$$

in the right-hand region ($x > d$). Correspondingly, a state of opposite parity (i.e. the left-hand side of the ψ_- state) decreases the index on the right near the interface and thus shifts the NSW peak away from the interface. Hence, on the right $(x > d)$,

$$\psi_- = \sqrt{2}\,\gamma_0 \operatorname{sech}[\gamma_0(x - d - x_0 - \Delta x_-)]. \tag{12.20}$$

Now the lateral shift towards the interface means that the energy can 'leak' into the linear region, thus making the NSW less stable. At high values of β the original NSW is stable; we would thus expect the ψ_+ state to be unstable at sufficiently high β. We will verify this by calculating the perturbation mode growth rate.

The ψ_- perturbation causes the NSW to become more like a self-sustaining soliton in the cladding, and thus to become more stable. Since the original NSW was stable, we can deduce that the ψ_- state will also be stable.

This explains the stability physically, and we now turn to some analysis to establish actual growth rates.

12.6 Stability analysis

Let us represent the solution of (12.2) for the three-layer guide as a stationary solution ψ with a small perturbation $f(x, \xi)$. We substitute

$$\psi + f(x, \xi) \tag{12.21}$$

into (12.2) and find that the perturbation equation has the form

$$2i\beta f_\xi + f_{xx} - g^2(x)f + h(x)\psi^2(x)\,(2f + f^*) = 0. \tag{12.22}$$

The perturbation function can be represented in terms of u and v, which are functions only of x, as follows:

$$f(x, \xi) = \frac{1}{2}\left[(u + v)\exp(\delta\xi) + (u^* - v^*)\exp(\delta^*\xi)\right]$$

where δ is the growth rate. A δ with a real part means that the perturbation can grow, and hence that the corresponding state is unstable.

Here u and v satisfy

$$L_0 v = -i\Omega u,$$
$$L_1 u = -i\Omega v, \tag{12.23}$$

where operators are defined by

$$L_0 = d^2/dx^2 - g^2(x) + h(x)\psi^2$$

and

$$L_1 = L_0 + 2h(x)\,\psi^2,$$

where $h(x)$ equals zero in the core and unity elsewhere, and $\Omega = 2\delta\beta$, so that

$$L_0\,L_1 u = -\Omega^2\,u,$$
$$L_1\,L_0 v = -\Omega^2\,v. \tag{12.24}$$

We now discuss the nature of the eigenvalue Ω, as it determines the stability properties of the stationary solutions. Let $< s_1, s_2 >$ be defined as the scalar product:

$$< s_1, s_2 > = \int\limits_{-\infty}^{\infty} s_1^* s_2 \, dx$$

for continuous functions s_1 and s_2 which decrease to zero at infinity. Since L_0 and L_1 are self-adjoint operators, the adjoint of $L_0 L_1$ is $L_1 L_0$, thus allowing us to write

$$(-\Omega^2)^* < u, v > = < L_0 L_1 u, \, v > = < u, L_1 L_0 v > = -\Omega^2 < u, v > \tag{12.25}$$

We thus analyse the problem of stability for the double interface by reducing it to the analogous problem for one interface. We remind the reader that this applies for values of $\beta > \sqrt{\epsilon_l}$. Our aim is to find the growth rates (complex in general) for the symmetric and anti-symmetric states as explicit functions of β and the physical parameters of the waveguide, viz. d and $\epsilon_l - \epsilon_0$.

For TE_0 stationary solutions (including both symmetric (ψ_+) and asymmetric (ψ_0) solutions), which are the fundamental modes of L_0, we note that $< u, v >$ is always non-zero. To prove this, we use the fact that $< L_0 v, v >$ is negative-definite because all eigenvalues of L_0 (including the continuous spectrum) are real and non-positive, and the eigenvalues of L_0 can be chosen to form a complete set. Therefore $< u, v > = -i (-\Omega^{-1})^* < L_0 v, v >$ is a non-zero quantity. In (12.25), this means Ω is either real or purely imaginary.

Let us now consider the TE_1 stationary solutions (ψ_-), for which L_0 is defined by $L_0 = d^2/dx^2 - g^2(x) + h(x) \psi_-^2$ with $L_0 \, \psi_- = 0$. In this case, L_0 has both positive (for the lowest eigenfunction) and negative eigenvalues, including the continuous spectrum. If only odd perturbations are used, i.e. u and v are odd, then $< L_0 v, v >$ is still negative-definite because the only positive eigenvalue of L_0 is not present in the expansion of $L_0 v$ in terms of the eigenfunctions of L_0, and so Ω is real or purely imaginary. The result is that the mode is stable when subject to odd perturbations in the positively sloped part of the power–dispersion curve and unstable elsewhere. This is the same result as that obtained by Tran and Ankiewicz (1992).

A more interesting case occurs when even perturbations are also allowed, as then $< L_0 v, v >$ may vanish, and (12.25) means Ω can thus be complex. This is the crucial factor which provides a complete picture of the stability characteristics, and it is the essence of this section. In such a case, the evolution of an initial perturbation will show both exponential growth and oscillations. We analyse this in section 12.9.

We may deduce some properties of the TE_2 stationary solutions (ψ_2) by employing the argument above. Here, $L_0 = d^2/dx^2 - g^2(x) + h(x)\,\psi_2^2$ with $L_0\psi_2 = 0$, and $< L_0 v, v >$ can be zero for both even and odd perturbation functions. This, in turn, means that the minimum and maximum points in the power–dispersion diagrams of these modes are not necessarily points where the stability characteristics change, for perturbations having any symmetry.

Furthermore, it can be seen from (12.23) and (12.24), that if (Ω, u, v) is a solution, then so are $(-\Omega, u, -v)$ and $(\Omega^*, -u^*, v^*)$. This means that it is sufficient to limit ourselves to the first quadrant of the complex plane when seeking complex eigenvalues.

12.7 Perturbation function

For the single-interface problem, the solution for u and v in the nonlinear medium $(x > 0)$ can be found using (2.40). It is given by Vysotina *et al.* (1987):

$$
\begin{cases}
u = C_1\exp(-px')\left[-i\zeta + 2p\tanh(x') + 2\tanh^2(x')\right] \\[2mm]
\quad + C_2\exp(-\bar{p}\,x')\left[i\zeta + 2\bar{p}\tanh(x') + 2\tanh^2(x')\right], \\[2mm]
v = C_1\exp(-px')\left[2 - i\zeta + 2p\tanh(x')\right] \\[2mm]
\quad - C_2\exp(-\bar{p}\,x')\left[2 + i\zeta + 2\bar{p}\tanh(x')\right],
\end{cases}
\tag{12.26}
$$

where $x' = x - x_0$, $p = \sqrt{1 - i\zeta}$, $\bar{p} = \sqrt{1 + i\zeta}$, with $\zeta = 2\beta\delta/\gamma_0^2$, while C_1, C_2 are constants of integration. In the linear medium $(x < 0)$, we have

$$
\begin{cases}
u = A_1\exp(sx) + A_2\exp(\bar{s}x) \\[2mm]
v = A_1\exp(sx) - A_2\exp(\bar{s}x),
\end{cases}
\tag{12.27}
$$

where $s = \sqrt{t^2 - i\zeta}$, $\bar{s} = \sqrt{t^2 + i\zeta}$, and A_1, A_2 are constants. Now, for a surface wave, ζ is either real or purely imaginary; these cases are considered separately in the following. The more complicated case involving complex roots is considered in section 12.9.

Suppose $\zeta = \zeta_r$ is real (i.e. $\zeta_r^2 > 0$). Then matching u, v and their derivatives at the interface $(x = 0)$ gives a dispersion relation for determining ζ_r:

$$
|p(1+i\zeta_r) - 2i\zeta_r t - 3pt^2 + 2t^3 - s(p-t)^2|^2 - (1-t^2)^2|p - 2t + \bar{s}|^2 = 0, \tag{12.28}
$$

where $t = \tanh(\gamma_0 x_0)$, with x_0 being the position of the peak in the nonlinear medium. This also implies $0 < t < 1$. The dispersion relation for the NSW,

$$
\gamma_l = \gamma_0 \tanh(\gamma_0 x_0),
$$

leads to $t = \gamma_l/\gamma_0$, in agreement with our earlier definition (section 12.2) and so (12.28) is an equation for ζ_r with effectively only one parameter, t. Since we are only interested in small values of ζ_r (i.e. $\zeta_r \ll 1$), equation (12.28) may be solved analytically by expanding each of the two expressions inside the absolute value signs in (12.28) up to fourth order in terms of ζ_r.

Calculating the absolute values of these expressions in (12.28), we obtain the equation for the approximate growth rate, ζ, of the NSW on a single interface:

$$\zeta_r^2[(1+t)^2\zeta_r^2\, y(t) - 16(1-2t)t^4(1-t^2)^2] = 0, \qquad (12.29)$$

where

$$y(t) = 5 - 20t + 33t^2 - 32t^3 + 22t^4 - 4t^5 - 2t^6. \qquad (12.30)$$

In the interval $0 < t < 1$, (12.29) has the root $\zeta_r = 0$ in the whole range. The roots

$$\zeta_r = \pm 4t^2\,(1-t)\,\sqrt{\frac{1-2t}{y(t)}} \qquad (12.31)$$

exist when $\zeta_r^2 > 0$, i.e. for $t < \frac{1}{2}$ (or $\beta < \sqrt{(4\epsilon_l - \epsilon_0)/3}$). Hence the system is unstable for this β range. This is in agreement with previous investigations of stability on one interface (Akhmediev, 1982). The approximate roots given by (12.31) are valid when t is just under $\frac{1}{2}$, in which range we have $\zeta_r^2 \approx 0.533\,(1-2t)$. This suggests that an imaginary root may exist when $t > \frac{1}{2}$. However, to consider possible imaginary roots $\zeta = i\zeta_i$, we need to replace (12.28) and go through the expansions again. We now have

$$a_1\,a_2 - (1-t^2)^2\,b_1\,b_2 = 0, \qquad (12.32)$$

where

$$
\begin{aligned}
a_1 &= p\,(1-\zeta_i) + 2\zeta_i t - 3pt^2 + 2t^3 - s\,(p-t)^2 \\
a_2 &= \bar{p}\,(1+\zeta_i) - 2\zeta_i t - 3\bar{p}t^2 + 2t^3 - \bar{s}\,(\bar{p}-t)^2 \\
b_1 &= p - 2t + \bar{s} \\
b_2 &= \bar{p} - 2t + s.
\end{aligned}
$$

Expanding (12.32) yields an equation for ζ_i which is analogous to (12.29):

$$\zeta_i^2[(1+t)^2\zeta_i^2 y(t) - 16\,(2t-1)\,t^4\,(1-t^2)^2] = 0, \qquad (12.33)$$

where $y(t)$ is the sixth-order polynomial already defined by (12.30). This reaffirms that $\zeta_i = 0$ is a root (for all t). Other roots are

$$\zeta_i = \pm 4t^2\,(1-t)\sqrt{\frac{2t-1}{y(t)}}.$$

Clearly $\zeta_i^2 > 0$ when $t > \frac{1}{2}$, as $y(t) > 0$ for $0 < t < 1$. This indicates stability in the range $t > \frac{1}{2}$. Thus the real root which exists for $t < \frac{1}{2}$ (12.31) becomes imaginary for $t > \frac{1}{2}$. When t is just over $\frac{1}{2}$, we have

$\zeta_r^2 \approx 0.533\,(2t-1)$, showing that the slope of $|\zeta|^2$ is the same on each side of $t = \frac{1}{2}$.

12.8 Three-layer structure

We now consider the three-layer system by regarding it as a perturbed single-interface system, for which the relevant results are presented above. For a double-interface system, the solution in each nonlinear medium at fixed β is the same as for a single interface but with x_0 replaced by $x_0 \mp \Delta x_\pm$, where the upper sign is for the symmetric solution, and the lower sign is for the anti-symmetric one. We follow this convention in all equations henceforth. The solution for the perturbation in the nonlinear medium retains the same form but is also shifted. Strictly speaking, the solution in the linear core is now given by cosh or sinh functions, but this does not influence dispersion relation (12.28). In the range of interest $(0 < t < 1)$, the effect of the perturbation in (12.28) is simply that $t = \tanh[\gamma_0 x_0]$ is replaced by $t_\pm = \tanh[\gamma_0(x_0 \mp \Delta x_\pm)]$. Note that s and $\bar{s}$ are parameters of the linear medium and thus do not change. Expanding $t_\pm$ to first order in $\Delta x_\pm$, we have

$$t_\pm = \tanh[\gamma_0(x_0 \mp \Delta x_\pm)] = \frac{\tanh(\gamma_0 x_0) \mp \tanh(\gamma_0 \Delta x_\pm)}{1 \mp \tanh(\gamma_0 x_0)\tanh(\gamma_0 \Delta x_\pm)}$$

$$\approx [\tanh(\gamma_0 x_0) \mp \gamma_0 \Delta x_\pm]\,[1 \pm \gamma_0 \tanh(\gamma_0 x_0)\,\Delta x_\pm]$$

$$= t \mp \gamma_0(1 + t^2)\Delta x_\pm. \tag{12.34}$$

This expansion is valid when $\gamma_0 \Delta x_\pm \ll 1$. It can be shown that $\gamma_0 \Delta x_+$ is zero at $t = 0$ and $t = 1$, and that it displays a single maximum of about $\frac{0.3}{V}$. So, for V of 1.5 and higher, $\gamma_0 \Delta x_+$ is small in the whole range $0 < t < 1$. On the other hand $\gamma_0 \Delta x_-$ matches $\gamma_0 \Delta x_+$ for t near 1, but increases to values around $\frac{1.1}{V}$ as t approaches to zero. Thus, if V is low, the expansion loses accuracy when t is close to zero. Nevertheless, for V of 2 and higher, it works well even for $t \to 0$. Equation (12.28) can now be expanded, including small linear terms proportional to $\mp \gamma_0(1 - t^2)\,\Delta x_\pm$.

For high β, we note that t is close to 1. In fact,

$$1 - t \approx (\epsilon_l - \epsilon_0)/2\beta^2.$$

Using this fact, we find that, instead of the zero solution for ζ_r and ζ_i in (12.29) and (12.30), we now have

$$\zeta_r^2 = \pm 8(1 - t)^2\,\gamma_0\,\Delta x, \tag{12.35}$$

and

$$\zeta_i^2 = \mp 8(1 - t)^2\,\gamma_0\,\Delta x, \tag{12.36}$$

where Δx here is the common value taken by Δx_+ and Δx_- for high β. In both (12.35) and (12.36), the upper sign is for the symmetric (TE$_0$)

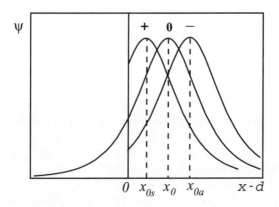

Figure 12.6 *Field shift in nonlinear half-space due to the influence of a distant field. The curve labelled '0' is the original single-interface field, while that labelled '+' shows that a distant field of the same parity shifts the maximum towards the interface. On the other hand, the curve labelled '−' indicates that a distant field of opposite parity shifts the field maximum away from the interface.*

solution, and the lower sign is for the anti-symmetric (TE$_1$) one. Thus ζ_r is real only for the TE$_0$ solution in (12.35), and ζ_i is real only for the TE$_1$ solution in (12.36). These are explicitly given by

$$\zeta_r = \zeta_i \approx 2\left(1 - t\right)\sqrt{2\gamma_0 \Delta x} \approx \frac{\epsilon_l - \epsilon_0}{\beta^2}\sqrt{2\gamma_0 \Delta x}. \tag{12.37}$$

We thus see now that a nonlinear surface wave shifted (due to the interaction) towards the surface is unstable at large β due to the real root $\zeta = \zeta_r$, while a nonlinear surface wave shifted away from the interface will be stable as the root $\zeta = i\zeta_i$ is purely imaginary. This proves the stability of the TE$_1$ solution and the instability of the TE$_0$ one at high β. It agrees with the physical prediction of section 12.5.

We may now use the expression for $\Delta x_\pm$ (12.18) to write the asymptotic growth rates directly. The growth rate for the symmetric (TE$_0$) solution is

$$\zeta_+ = \zeta_r = \frac{2\exp(-\gamma_l d)}{\beta^2}\sqrt{(\epsilon_l - \epsilon_0)\gamma_0 \gamma_l}, \tag{12.38}$$

while that for the anti-symmetric (TE$_1$) solution is

$$\zeta_- = i\zeta_i = i\zeta_+. \tag{12.39}$$

Clearly, $|\zeta_\pm|$ decreases towards zero as β increases. Hence, at very high β, a long propagation length would be required for the symmetric state instability to become apparent. Similarly, the oscillation frequency, $|\zeta_-|$, of a perturbation to the nonlinear TE_1 state would asymptotically decrease to zero as β increases.

12.9   Complex eigenvalues

Let us consider the changes in growth rates caused by shifting the nonlinear single-interface wave peak. We start from the original equations (12.26), (12.27) and match boundary conditions at the interface. We take into account the fact that the growth rates can, in general, be complex (Tran *et al.*, 1992). The results are presented in Fig. 12.7a and Fig. 12.7b for Δx_+ and Δx_-, respectively. Solid lines in each case correspond to one of the roots of the exact solution for a single interface. The root is real roughly in the range $(0 < B < 1.35$, or $0 < t < \frac{1}{2})$ and purely imaginary in the range $(1.35 < B < \infty$, or $\frac{1}{2} < t < 1)$, as we have explained earlier. The second root is equal to zero. Dotted lines in Fig. 12.7a correspond to the perturbed roots which occur when the surface wave is shifted by the value Δx_+ towards the linear medium. This gives one root which is real in the whole range $0 < B < \infty$ and another which is purely imaginary in the same range. On the right-hand side, the upper curve is described by (12.37).

Dotted lines in Fig. 12.7b correspond to the deformed purely real or imaginary roots when the surface wave is shifted by the value Δx_- away from the core. Dashed lines correspond to the complex root which appears in this case in a restricted range around the value $t = \frac{1}{2}$. To the left of this range both roots are real, but to the right of it they are purely imaginary. The curve given by the imaginary root with the smaller absolute value can be described by (12.37) when β is high.

The appearance of complex roots in the case of the TE_1 solution is in agreement with previous results (section 12.6) in the investigation of its stability. It shows that there is a range of β values with complicated instability behaviour showing oscillatory exponential growth of small perturbations. The qualitative behaviour of the roots as functions of β (calculated for small Δx) is also in agreement with the exact results of Tran *et al.* (1992).

We can make some estimates when the condition $|\zeta|/t^2 \ll 1$ holds. The values $\Delta x_+(TE_0)$ and $\Delta x_-(TE_1)$ become different when t is not near 1, and can be defined exactly using (12.8), (12.12), and (12.15):

$$\begin{aligned} \Delta x_+ &= x_0 - x_{0s} \\ \Delta x_- &= x_{0a} - x_0. \end{aligned} \tag{12.40}$$

Using these equations and solving a biquadratic equation which appears on matching boundary conditions, we find that complex roots appear in the interval

$$\frac{1}{2} - 0.484\sqrt{\gamma_0 \Delta x_-} < t < \frac{1}{2} + 0.484\sqrt{\gamma_0 \Delta x_-}. \tag{12.41}$$

Substituting Δx_- in (12.41) and using the V parameter we obtain, for V

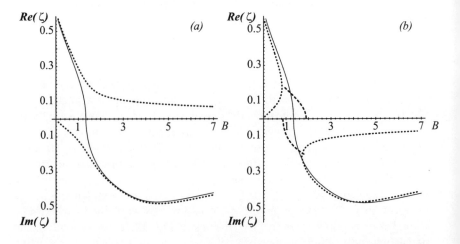

Figure 12.7 *Deformations of the real and imaginary parts of the normalized growth rate ζ introduced by shifting the nonlinear surface wave peak by the value Δx_+ towards to the linear medium (a), and by Δx_- away from the linear medium (b). Here B is defined as $\frac{4(\beta^2 - \epsilon_l)}{(\epsilon_l - \epsilon_0)}$. The value of Δx is the same in both cases and is set equal to $\frac{0.01}{(\epsilon_l - \epsilon_0)}$ to produce this schematic. Note that the two dashed curves in (b) represent the real and imaginary parts of the same root. Dotted curves in (a) and (b) represent two purely real or imaginary roots. Note that $Re(\zeta) = \zeta_r$ is plotted above the B axis and that $Im(\zeta) = \zeta_i$ is plotted below the B axis.*

not too small:

$$\frac{1}{2} - 0.56 \exp\left(-\frac{V}{\sqrt{3}}\right) < t < \frac{1}{2} + 0.56 \exp\left(-\frac{V}{\sqrt{3}}\right).$$

Thus the complex range is centred on $t = \frac{1}{2}$. At lower V, the greater part of the complex range occurs in the interval $t < \frac{1}{2}$.

Thus the qualitative deformation peculiarities shown in Fig. 12.7 are the same regardless of the magnitude of $\Delta x_\pm$. So, the TE$_0$ solution is unstable in the whole region $(\epsilon_l < \beta^2 < \infty)$ which is covered by this analysis. The TE$_1$ solution is unstable in the region of the appearance of complex roots (close to $t = \frac{1}{2}$), unstable to the left of this region, and stable to the right of it.

We have explained the stability ranges of the TE$_0$ and TE$_1$ modes of the three-layer nonlinear guide in terms of the known result for the single-interface problem. In the single-interface case, the non-zero root causes the structure to be unstable when $t < \frac{1}{2}$ and stable when $t > \frac{1}{2}$. The perturbation giving rise to the TE$_0$ solution in the three-layer system mixes the roots and causes the zero root to take real values when $t > \frac{1}{2}$, making the

TE$_0$ solution unstable for the whole interval $0 < t < 1$. The perturbation creating the TE$_1$ solution also mixes the roots, making the zero root take on small imaginary values, forming loops with the (perturbed) original non-zero root, and leaving a small complex (unstable) region around $t = \frac{1}{2}$. Thus the unstable range is only slightly greater than that in the single-interface problem, and in particular, the TE$_1$ is stable for high values of the propagation constant.

12.10 Particle analogy

We now look at the problem from a different viewpoint. In the high β range, the even and odd states can be considered as combinations of two NSWs combined in phase or out of phase. Due to a field interaction term (Ankiewicz, 1988a) the total energy in the states ψ_+ and ψ_-, at some fixed β, is not exactly twice that in the component asymmetric state. The dispersion diagram (Fig. 12.5) clearly shows that $Q_+ > Q_-$ at any fixed high β. It is easy to show, using formulae (12.9), (12.13) and (12.16), that in the region of high β values, the power differences are

$$\Delta Q = Q_+ - 2Q \sim 2Q - Q_-$$

$$\approx 2(\epsilon_l - \epsilon_0)\, \frac{(2d\gamma_l - 1)}{\gamma_l}\, \exp(-2\gamma_l d). \tag{12.42}$$

It is a general principle of physics that when two particles coalesce to form a new bound state, then the new state will be stable if the combined energy is less than that of the two components, and unstable if the combined energy is greater than the sum of the two components (Black and Ankiewicz, 1985). In our case, the single NSW is stable. For the ψ_+ state, the power is greater than the sum of the component surface waves, thus indicating probable instability, while the ψ_- state has a lower power than the sum of the individual surface waves, thus implying stability. So, if we consider the TE$_0$ and TE$_1$ solutions as bound states of two spatial solitons, which in turn can be considered as particles (Aceves *et al.*, 1989), then the stability at high β can be understood in terms of this general principle.

12.11 Analogy with soliton states in couplers

In the NL-L-NL planar guide, anti-symmetric, symmetric and asymmetric solutions exist, as indeed they do in nonlinear couplers (considered in Chapter 8). Thus there is an analogy between the two situations, with the centre linear core replacing linear 'coupling' between two cores. Power and propagation constant in the planar guide play the roles of energy and frequency shift in the coupled-solitons system. Of course, a single function gives the field in the planar guide case, whereas two separate functions are needed to describe the stationary states in the coupler.

In both cases, the asymmetric state is stable when the dispersion curve slope is positive, and the symmetric state is unstable above the bifurcation point. However, the anti-symmetric state is stable at high powers in the planar guide (Fig. 12.5), while it remains unstable in the coupler case. This occurs because the analogy is not exact.

12.12 Amplification of nonlinear guided waves

The amplification of guided waves inside a guide itself is one of the ways which can be used to excite nonlinear guided waves. This can be done in a waveguide where the core incorporates active ions like erbium. External pumping from the lateral interface can then create the gain and thus amplify the linear mode of the guide up to amplitudes where it gradually starts to manifest nonlinear properties.

The standard exponential formulae for mode amplification in linear waveguides cannot be applied in this case, because the nonlinear solutions change their shapes during this process. Moreover, a question arises as to whether or not the waves excited in this manner are stable under adiabatic amplification. We can suppose that, if we start at a stable point, e.g. a linear mode, then the wave amplitude will grow slowly, with its envelope remaining close to the shape of the stable nonlinear mode at low power levels.

12.12.1 Physical interpretation

The difference in the process of amplification in linear and nonlinear cases is that, in the linear case, the transverse field envelope retains its shape, while in the nonlinear case it does not. For the linear guide, the propagation constant remains constant, while for the nonlinear guide it does not, because the propagation constant depends on the power in a nonlinear guide (as seen in Fig. 12.8). As a result, in the linear case, amplification simply produces an exponential increase in amplitude. This is not the case for nonlinear waves. The function ψ can be written as (Akhmediev and Ankiewicz, 1993d):

$$\psi(x,\xi) = \psi_0 + v(x,\gamma)\xi + iu(x) + f(x,\xi) + ..., \qquad (12.43)$$

where the functions $v(x)$, $u(x)$ and $f(x)$ are small. The second term in this equation,

$$v(x)\xi = 2\beta_1 \frac{d\psi_0}{d\beta}\,\xi = \frac{d\psi_0}{d\beta}\Delta\beta(\xi), \qquad (12.44)$$

is responsible for a gradual change in the shape of the nonlinear guided wave along the ξ axis during the amplification process. The shape changes in such a way that the whole function satisfies the quasi-stationary solution to first order:

$$\psi(x,\xi) = \psi_0[x,\beta(\xi)] + iu(x), \qquad (12.45)$$

where β varies with ξ and there is a small imaginary perturbation. The imaginary term, $iu(x)$, is responsible for the phase deformation in the cross-section which necessarily accompanies this change in shape.

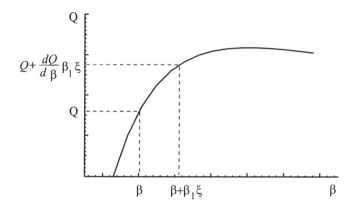

Figure 12.8 *Schematic power–dispersion diagram showing the physical meaning of the growth rate parameter β_1.*

This motion can be seen in Fig. 12.8 on the power–dispersion curve. If at some ξ ($= 0$ in our case), the nonlinear wave has the shape $\psi_0(x, \beta)$, which corresponds to the propagation constant β and power Q, then after propagation for some (infinitesimal) distance ξ, β will be changed to $\beta + \beta_1\xi$ and Q will be changed to $Q + (dQ/d\beta)\beta_1\xi$, corresponding to the new nonlinear function $\psi_0(x, \beta + \beta_1\xi)$. The growth rate parameter β_1, calculated as a function β, for a symmetric nonlinear guide, is shown in Fig. 12.9b, while the power–dispersion curve $Q(\beta)$ is shown in Fig. 12.9a. The parameter β_1 is plotted in Fig. 12.9b for all values of β where it is positive. We remind the reader that it has meaning only on those branches where the nonlinear wave is stable.

If we have gain in a medium, then nonlinear guided waves can be amplified. For adiabatic amplification, β in (12.1) can change with ξ. This means that, instead of (12.2), we have

$$2i\beta(\xi)\psi_\xi + i\beta_\xi(\xi)\psi + \psi_{xx} - g^2(x, \xi)\psi + h(x)|\psi|^2\psi = i\gamma(x)\psi. \quad (12.46)$$

We now consider (12.46) in terms of the energy balance. For constant β and $\gamma = 0$, this equation coincides with (12.2). Consider now a change of power level. By multiplying (12.46) by ψ^*, taking the complex conjugate of that expression, and subtracting and integrating, we obtain:

$$\frac{d}{d\xi}(\beta Q) = \int_{-\infty}^{\infty} \gamma(x)\,|\psi(\xi, x)|^2 dx, \quad (12.47)$$

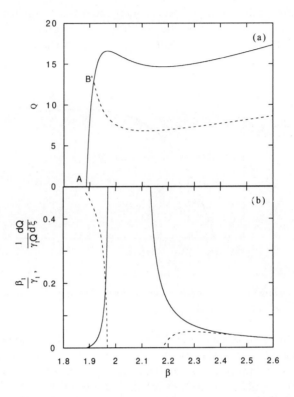

Figure 12.9 *(a) Power Q and (b) normalized growth rate parameter β_1/γ_l, versus propagation constant β for the symmetric wave of a symmetric waveguide. The growth rate for the power, $(1/\gamma_l Q)(dQ/d\xi)$, is shown in (b) by dashed lines. The dashed line in (a) corresponds to the asymmetric solution. Parameters used for creating this diagram are: $\epsilon_l = 4$, $\epsilon_{nl} = 2.5$, $2d = \pi$ (i.e. waveguide parameter $V = d\sqrt{\epsilon_l - \epsilon_{nl}} = 2.078$).*

where Q is the power. Equation (12.47) can be written in the form of an energy balance theorem:

$$\frac{d}{d\xi}\left(S_\xi\right) = W, \qquad (12.48)$$

where $S_\xi = \beta Q$ is the total energy flow in the ξ direction, i.e. the ξ component of the Poynting vector integrated over the cross-section. The right-hand side of (12.47) is the integrated rate of average energy generation by conduction currents at a given cross-section ($\xi = $ constant) per unit time and ξ (or total power generated per unit ξ):

$$W = \int\limits_{-\infty}^{\infty} \gamma(x)\, |\psi(\xi, x)|^2 dx. \qquad (12.49)$$

Equations (12.47) and (12.48) express the energy balance theorem for nonlinear guided waves. Expanding the left-hand side of equation (12.47), and taking into account the fact that

$$\frac{dQ}{d\xi} = \frac{dQ}{d\beta}\frac{d\beta}{d\xi}$$

and $\beta_\xi = 2\beta_1$, we have:

$$\frac{d\beta}{d\xi}\left[Q + \beta\,\frac{dQ}{d\beta}\right] = \int\limits_{-\infty}^{\infty} \gamma(x)\, |\psi(\xi, x)|^2 dx, \qquad (12.50)$$

or

$$\frac{d\beta}{d\xi} = \frac{\int\limits_{-\infty}^{\infty} \gamma(x)\, |\psi(\xi, x)|^2 dx}{Q + \beta\,\frac{dQ}{d\beta}}. \qquad (12.51)$$

The rate of increase of the power invariant Q is given by:

$$\frac{dQ}{d\xi} = \frac{\int\limits_{-\infty}^{\infty} \gamma(x)\, |\psi(\xi, x)|^2 dx}{Q + \beta\,\frac{dQ}{d\beta}}\,\frac{dQ}{d\beta}. \qquad (12.52)$$

We can obtain a general formula for $dQ/d\xi$, allowing for gain in both the linear and nonlinear media:

$$\frac{dQ}{d\xi} = \frac{\gamma_l Q_l + \gamma_{nl} Q_{nl}}{Q + \beta\,\frac{dQ}{d\beta}}\,\frac{dQ}{d\beta}. \qquad (12.53)$$

Here γ_l and γ_{nl} are gain coefficients of the linear and nonlinear media respectively, and Q_l and Q_{nl} are the power components of the linear and nonlinear regions respectively.

CHAPTER 13

Nonlinear pulses in presence of gain, loss and spectral filtering

Optical fiber soliton systems inevitably include losses and so they need gain to compensate for them. A typical example is a soliton communication line or soliton fiber laser. Gain can be applied even in such elements as logic gates (Islam, 1992). Moreover, gain can be useful in planar soliton systems (Khitrova *et al.*, 1993) and with light-guiding-light phenomena for creating tapered two-dimensional beams (Snyder *et al.*, 1991a).

Typically, in such systems, a soliton propagates in the presence of periodic loss and amplification. For example, in communication lines there are linear losses, distributed over the communication fiber, and lumped amplification at the repeaters. Stable soliton propagation is possible when the period of amplification is much shorter than the soliton dispersion length (Mollenauer *et al.*, 1986), although methods to extend the amplification period have also been reported (Kodama and Wabnitz, 1993; Atkinson *et al.*, 1994). If the period of the perturbation related to amplifiers is small, the equations for soliton propagation can be averaged (Hasegawa and Kodama, 1991). Normally, in communication systems, all losses are linear (i.e. they do not depend on the signal intensity) and they do not significantly depend on the frequency within the signal bandwidth. In this case the averaged equation is the NLSE, with neither gain nor loss. Use of filters has been proposed to suppress the random deviation (Gordon–Haus jitter) of the soliton central frequency (Kodama and Hasegawa, 1992b; Mecozzi *et al.*, 1991, 1992; Romagnoli *et al.*, 1994; Atkinson *et al.*, 1994; Matsumoto *et al.*, 1995) and soliton interaction (Afanasjev, 1993). In this case, some excess linear gain should be in the system to compensate for the losses suffered by solitons at filters. Thus, the resulting averaged equation includes linear gain as well as the terms responsible for spectral filtering. The excess linear gain is very small in comparison with the gain which is necessary to overcome the linear losses.

In soliton fiber lasers, the process is basically the same. There is a linear gain element (typically an erbium-doped fiber amplifier), as well as nonlinear and dispersive elements, and losses. However, soliton fiber lasers incorporate some element with an intensity-dependent transmission function (e.g. a fast saturable absorber), which is normally absent in communication systems. The presence of this element creates several important differences between the communication systems and the lasers. To describe this element, the nonlinear gain should be introduced into the averaged equation. Besides, the presence of a mode-locking mechanism means that the linear radiation in the cavity is absorbed. Thus the amplification period may be of the same order as the soliton dispersion length.

However, not all soliton fiber lasers can be described by the distributed equation. For example, the figure-of-eight laser (Duling, 1991; Richardson et al., 1991) is based on intensity discrimination by a nonlinear amplifying loop mirror (NALM). A pulse at the output of a NALM has substantial deformations in its shape, so the distributed model cannot be used. On the other hand, another type of passively mode-locked laser, the polarization mode-locked laser, does not produce any substantial perturbations in the solitons, and the distributed model can be used.

Erbium-doped amplifiers can also be distributed (Kurokawa and Nakazawa, 1992). Such devices are perfectly described by the NLSE with distributed gain and loss, although in real devices the spatial distribution of gain is non-uniform.

So, from this short review, we can see that many soliton systems can be described by the NLSE with gain and loss, where both gain and loss may be frequency- and intensity-dependent. The equation describing these proceses is known in theoretical physics as the complex Ginzburg–Landau equation (CGLE).

Historically, three forms of the CGLE have been analysed. The simplest of them is the cubic CGLE, which includes only cubic terms. This equation has been analysed mainly in the context of plasma physics (Nicholson and Goldman, 1976; Pereira, 1977; Pereira and Stenflo, 1977; Pereira and Chu, 1979; Weiland et al., 1978). It admits an exact solution, which can be found relatively easily. This solution is valid for the whole range of possible values of the parameters δ, β and ϵ. In optics, the solution was introduced by Bélanger et al. (1989). It was soon realized that the pulse-like solutions of this equation are unstable.

Later, the quintic CGLE model was introduced, as it admits stable solutions, i.e. both the pulse and the background are stable (Malomed, 1987; Thual and Fauve, 1988; Brand and Deissler, 1989). At that time, this equation was used mainly as a model for binary fluid convection. The correct modelling of this process requires consideration of a system of two coupled CGLEs (e.g. Brand and Deissler, 1989). The quintic CGLE is more difficult to analyse than the cubic. Although some analytical solutions have been found, most of them do not describe stable pulses.

Another model has been introduced by Haus et al. (1991) to describe passively mode-locked fiber lasers. This proposes using the cubic CGLE, but with a linear gain term which depends on the total pulse energy. This model allows us to use the exact solution of the cubic CGLE, and at the same time it admits stable solutions for some range of the parameters. However, this stability range is very small. As the self-start condition should also be fulfilled, this seriously limits the applicability of this model. Moreover, the model is based on the assumption that there is only one pulse in the cavity. However, real soliton fiber lasers typically support tens of pulses in the cavity. In addition, as the gain is varied, the number of solitons in the cavity also varies, while the parameters of each soliton (pulse width and peak intensty) remain the same. This effect is known as soliton quantization (Grudinin *et al.*, 1992).

In this chapter, we present an analytical method to look for solutions of the cubic and quintic CGLE (Akhmediev *et al.*, 1996). We start our analysis with the cubic case, where the results are easily obtained and interpreted, and use them as a basis for analysis of the quintic case. Apart from the soliton with fixed amplitude (which we also call the 'plain pulse'), we demonstrate the existence of arbitrary-amplitude pulses, flat-top pulses and other special solutions.

13.1 Complex quintic Ginzburg–Landau equation

In the form used in nonlinear optics, the quintic CGLE is:

$$i\psi_\xi + \frac{D}{2}\psi_{\tau\tau} + |\psi|^2\psi = i\delta\psi + i\epsilon|\psi|^2\psi + i\beta\psi_{\tau\tau} + i\mu|\psi|^4\psi - \nu|\psi|^4\psi, \quad (13.1)$$

where τ is the retarded time, ξ is the normalized propagation distance, δ, β, ϵ, μ and ν are real constants (we do not require them to be small), and ψ is a complex field. For the specific case of the optical fiber mentioned above, the physical meaning of these quantities is the following: ψ is the complex envelope of the electric field, δ is the linear excess gain at the carrier frequency, β describes spectral filtering ($\beta > 0$), ϵ accounts for nonlinear gain/absorption processes and μ represents a higher-order correction to the nonlinear amplification/absorption. We note that ν is a parabolic correction term (see Chapter 4) to the nonlinear refractive index. This means that the nonlinearity law exhibits a parabolic dependence on intensity. The parameter D determines the sign of the dispersion: $D = +1$ corresponds to anomalous dispersion and $D = -1$ to normal dispersion. While the former case does not require futher explanation, the latter is related, for example, to soliton propagation in fiber lasers with neodymium-doped amplifiers (Hofer *et al.*, 1991). Equation (13.1) has been written in such a way that if the right-hand side of it is set to zero, we obtain the standard NLSE. Many other non-equilibrium phenomena, such as processes in lasers (Haken, 1983; Jakobsen, 1992; Harkness *et al.*, 1994), binary fluid convection (Kolodner, 1991) and phase transitions (Graham, 1975) can also be described by this equation.

If the coefficients δ, β, ϵ, μ on the right-hand side are small, and $\nu = 0$, then soliton-like solutions of (13.1) can be studied by applying perturbative theory to the soliton solutions of the NLSE. This approach, however, cannot give all the relevant properties of soliton-like pulses and the regions in the parameter space where they exist. Finding exact solutions is an important step for understanding the full range of properties of the complex CGLE, thus helping to predict the behaviour resulting from an arbitrary initial condition. We consider both the cubic and the quintic CGLE, and derive all soliton solutions for both cases following the same procedure.

The cubic CGLE has been studied extensively (e.g. Hocking and Stewartson, 1972; Pereira and Stenflo, 1977; Nozaki and Bekki, 1984). The general pulse solution of this equation has a fixed amplitude for a given set of parameters. Careful investigation of the solutions with fixed amplitude shows that they become singular at certain values of the parameters, corresponding to a special line on the (β, ϵ) plane (Afanasjev, 1995a). Although the solution with fixed amplitude does not apply in this case, a new class of solutions arises, namely, the class of arbitrary-amplitude solitons.

The case of the quintic CGLE has been considered in a number of publications using numerical simulations, perturbative analysis and analytic solutions. The existence of soliton-like solutions of the quintic CGLE in the case of subcritical bifurcations ($\epsilon > 0$) has been shown numerically (Thual and Fauve, 1988; Brand and Deissler, 1989). A qualitative analysis of the transformation of the regions of existence of the pulse-like solutions, when the coefficients on the right-hand side change from zero to infinity, has been made by Hakim *et al.* (1990). An analytic approach, based on the reduction of (13.1) to a three-variable dynamical system, which allows us to obtain exact solutions for the quintic equation, has been developed by van Saarloos and Honhenberg (1990,1992).

The most comprehensive mathematical treatment of the exact solutions of the quintic CGLE, using Painlevé analysis and symbolic computations, is given by Marcq *et al.* (1994). The general approach used in that work is the reduction of the differential equation to a purely algebraic problem. The solutions include pulses, sinks, fronts and sources. The great diversity of possible types of solutions requires a careful analysis of each class of solutions separately. This is the reason why we have concentrated here solely on pulse-like solutions. We present in explicit form, and classify, all the solutions of this restricted class.

13.2 Perturbative approach

Perturbative analysis of the solitons of the cubic–quintic CGLE in the NLSE limit has been used by Pereira and Stenflo, (1977), Malomed (1987), Fauve and Thual (1990), Kodama *et al.*, (1992) and Afanasjev (1995b). Following this approach, let us consider the right-hand side of equation (13.1), with $D = +1$, as a small perturbation and write the solution as a soliton of the NLSE

$$\psi(\tau, \xi) = \frac{\eta}{\cosh[\eta(\tau + \Omega\xi)]} \exp[-i\Omega\tau + i(\eta^2 - \Omega^2)\xi/2]. \tag{13.2}$$

In the presence of the perturbation, the parameters of the pulse, viz. the amplitude η and frequency (or velocity) Ω, change adiabatically. The equations for them can be obtained from the equations for the energy and momentum.

Multiplying (13.1) by ψ^*, taking the complex conjugate, subtracting and integrating over τ leads to the evolution equation for the energy:

$$\frac{d}{d\xi} \int_{-\infty}^{\infty} |\psi|^2 d\tau = 2 \int_{-\infty}^{\infty} \left[\delta|\psi|^2 - \beta \left| \frac{\partial\psi}{\partial\tau} \right|^2 + \epsilon|\psi|^4 + \mu|\psi|^6 \right] d\tau.$$

Now, using (13.2) we have the equation for the evolution of $\eta(\xi)$, which is proportional to the pulse energy:

$$\frac{d\eta}{d\xi} = 2\eta \left[\delta - \beta\Omega^2 + \frac{1}{3}(2\epsilon - \beta)\eta^2 + \frac{8}{15}\mu\eta^4 \right]. \tag{13.3}$$

Similarly, using the equation for the momentum, we can derive the equation for $\Omega(z)$:

$$\frac{d\Omega}{d\xi} = -\frac{4}{3} \beta\Omega \, \eta^2. \tag{13.4}$$

The dynamical system of equations (13.3) and (13.4) for two real dependent variables has three fixed points at $\Omega = 0$, $\eta \geq 0$ including the origin ($\eta = 0$, $\Omega = 0$). The values of η^2 for the two other points are defined by finding the roots of a quadratic polynomial. The roots are positive (η is real) when $\epsilon < \beta/2$. The stability of these fixed points requires $\beta > 0$. The point at the origin is stable when $\delta < 0$ and $\beta > 0$. This condition is needed for the background state $\psi = 0$ to be stable. If both roots of the quadratic polynomial (in η^2) in the square brackets in (13.3) are positive, then one of the stationary points with non-zero η is stable. This fixed point is then a sink which defines the parameters of a stable approximate pulse-like solution of the quintic CGLE. Note that the term with ν in CGLE does not influence the location of the sink. It only introduces an additional phase term, $\exp(i8\nu\eta^4\xi/15)$, to the solution (13.2).

In the case of cubic CGLE, $\mu = 0$ and $\nu = 0$. The stationary point is then

$$\eta = \sqrt{\frac{3\delta}{\beta - 2\epsilon}}, \qquad \Omega = 0. \tag{13.5}$$

It is stable provided that $\delta > 0$, $\beta > 0$ and $\epsilon < \beta/2$. Clearly, in this case the soliton and the background cannot be stable simultaneously. Hence, this simple approach shows that to have both the soliton and the background stable, we need to have quintic terms in the CGLE.

This perturbative approach shows that, in general, the CGL tionary pulse-like solutions with fixed parameters. The dissipa in (13.1) break the scale invariance associated with the conser tem. Any pulse-like initial condition in close proximity to a fixed point will converge to a stable stationary solution. This observation is important for all-optical communication lines, because it shows that all pulses in a system with gain and loss will approach the same amplitude, width and velocity (Kodama and Hasegawa, 1992b) and the information flow will not be subjected to severe distortion.

The perturbative analysis presented above has some major limitations. Firstly, it can be applied only if the coefficients on the right-hand side of (13.1) are small, and this is not always the case in practice. Correspondingly, it describes the convergence correctly only for initial conditions which are close to the stationary solution. Secondly, the standard perturbative analysis cannot be applied to the normal dispersion case $(D = -1)$, when the NLSE itself does not have bright soliton solutions. However, (13.1) has stable pulse solutions for the normal dispersion case as well. Despite this, perturbative analysis gives valuable results for both cubic and quintic cases.

13.3 Special ansatz

In this section, we present a relatively simple method, based on the work of Akhmediev *et al.* (1996), which allows us to obtain and classify the various types of pulse-like solutions described by a special ansatz. Thus we obtain the class of solutions with fixed amplitude, and then we reveal the singularities and isolate several special solutions, including an unusual class of arbitrary-amplitude solitons, a family of flat-top solutions, a class of rational solutions, and the chirp-free solutions. Due to the restrictions imposed by our ansatz, these solutions do not cover the whole range of parameters, but they can serve as a basis for further generalizations.

First, we consider the stationary solutions of (13.1) with zero transverse velocity. This occurs when $\beta \neq 0$. The case $\beta = 0$ is considered in a separate section. Hence, we look for a solution of the form:

$$\psi(\tau, \xi) = A(\tau) \exp(-i\omega\xi), \tag{13.6}$$

where ω is a real constant. The complex function $A(\tau)$ can always be written in an explicit form as:

$$A(\tau) = a(\tau) \exp[i\phi(\tau)], \tag{13.7}$$

where a and ϕ are real functions of τ. By inserting (13.6), (13.7) into (13.1) and separating real and imaginary terms, we obtain:

$$\left(\omega - \frac{1}{2}\phi'^2 + \beta\phi''\right)a + 2\beta\phi'a' + \frac{1}{2}a'' + a^3 + \nu a^5 = 0,$$

$$\left(-\delta + \beta\phi'^2 + \frac{1}{2}\phi''\right)a + \phi'a' - \beta a'' - \epsilon a^3 - \mu a^5 = 0, \qquad (13.8)$$

where each prime stands for differentiation with respect to τ.

Let us now assume that

$$\phi(\tau) = \phi_0 + d \ln[a(\tau)], \qquad (13.9)$$

where d is the chirp parameter and ϕ_0 is an arbitrary phase. We suppose $\phi_0 = 0$ for simplicity. For the cubic case, this ansatz covers all pulse-like solutions. In the quintic case, however, (13.9) is a restriction imposed on $\phi(\tau)$, because the chirp could have a more general functional dependence on τ. Nevertheless, this ansatz allows us to find some classes of solutions in analytical form. After substitution of (13.9), equations (13.8) become:

$$\omega a + \left(\frac{1}{2} + \beta d\right) a'' + \left(\beta d - \frac{d^2}{2}\right)\frac{a'^2}{a} + a^3 + \nu a^5 = 0,$$

$$(13.10)$$

$$-\delta a + \left(\frac{d}{2} - \beta\right)a'' + \left(\frac{d}{2} + \beta d^2\right)\frac{a'^2}{a} - \epsilon a^3 - \mu a^5 = 0.$$

Now, we have two second-order ordinary differential equations in the same dependent variable, $a(\tau)$. To have a common solution, the two equations must be compatible. In general, this is not the case. However, for this particular system, they can be made compatible with a proper choice of the parameters.

To find the conditions for compatibility, we apply the following procedure. To start, we eliminate the first derivatives from the set of equations (13.10) to obtain:

$$\frac{d}{4}(1 + d^2)(1 + 4\beta^2)\frac{a''}{a} + \left(\frac{d}{2} + \beta d^2 + \epsilon\beta d - \frac{\epsilon d^2}{2}\right)a^2 + \left[\nu\left(\frac{d}{2} + \beta d^2\right)\right.$$

$$\left. + \mu\left(\beta d - \frac{d^2}{2}\right)\right]a^4 + \frac{\omega d}{2}(1 + 2\beta d) + \delta\left(\beta d - \frac{d^2}{2}\right) = 0. \qquad (13.11)$$

After integrating (13.11) we have:

$$\frac{d}{4}(1 + d^2)(1 + 4\beta^2)\frac{a'^2}{a^2} + \frac{1}{2}\left(\frac{d}{2} + \beta d^2 + \epsilon\beta d - \frac{\epsilon d^2}{2}\right)a^2 + \frac{1}{3}\left[\nu\left(\frac{d}{2} + \beta d^2\right)\right.$$

$$\left. + \mu\left(\beta d - \frac{d^2}{2}\right)\right]a^4 + \frac{\omega d}{2}(1 + 2\beta d) + \delta\left(\beta d - \frac{d^2}{2}\right) = 0. \qquad (13.12)$$

The integration constant is zero for solutions decreasing to zero at infinity.

On the other hand, we can eliminate the second derivative from (13.10), obtaining:

$$\frac{d}{4}(1 + d^2)(1 + 4\beta^2)\frac{a'^2}{a^2} + \left(\beta - \frac{d}{2} - \frac{\epsilon}{2} - \epsilon\beta d\right)a^2 - \left[\nu\left(\frac{d}{2} - \beta\right)\right.$$

$$+\mu\left(\beta d + \frac{1}{2}\right)\right] a^4 + \omega\beta - \frac{\omega d}{2} - \frac{\delta}{2} - \delta\beta d = 0. \qquad (13.13)$$

Equations (13.12) and (13.13) must be identical. Hence, the following set of three algebraic equations must be satisfied:

$$\nu(4d + 2\beta d^2 - 6\beta) + \mu(8\beta d - d^2 + 3) = 0, \qquad (13.14)$$

$$3d + 2\beta d^2 - 4\beta + 6\epsilon\beta d + 2\epsilon - \epsilon d^2 = 0, \qquad (13.15)$$

$$2\omega(d - \beta + \beta d^2) + \delta(1 - d^2 + 4\beta d) = 0. \qquad (13.16)$$

Equations (13.14)-(13.16) are the conditions of compatibility for the equations (13.10).

If both coefficients μ and ν are non-zero, then (13.14) gives the relation between the four parameters ϵ, β, μ and ν when the solution exists in the form of (13.7) and (13.9). The parameter d can be found from equation (13.15):

$$d = d_{\pm} = \frac{3(1 + 2\epsilon\beta) \pm \sqrt{9(1 + 2\epsilon\beta)^2 + 8(\epsilon - 2\beta)^2}}{2(\epsilon - 2\beta)}. \qquad (13.17)$$

This result shows that: (i) d can be found in terms of β and ϵ only, and (ii) the expression for d is the same for both the cubic and the quintic CGLE.

From (13.16), we obtain for ω:

$$\omega = -\frac{\delta(1 - d^2 + 4\beta d)}{2(d - \beta + \beta d^2)}. \qquad (13.18)$$

Now, taking into account (13.14) to (13.18), and after some cumbersome transformations, we can rewrite (13.13) (or (13.12)):

$$(a')^2 + \frac{2\nu}{8\beta d - d^2 + 3}a^6 + \frac{2(2\beta - \epsilon)}{3d(1 + 4\beta^2)}a^4 - \frac{\delta}{d - \beta + \beta d^2}a^2 = 0. \quad (13.19)$$

The coefficient of a^4 can equally be written in another way:

$$\frac{2\nu}{8\beta d - d^2 + 3} = \frac{\mu}{3\beta - 2d - \beta d^2}. \qquad (13.20)$$

It is important to note that (13.19) is a consequence of the set (13.10), and its solutions are equivalent to the solutions of (13.10). Equation (13.19) is an elliptic equation, and its solutions can be found relatively easily. The most important point for us is the form of the coefficients in (13.19). They have been reduced to their simplest forms, as this allows us to classify the solutions mainly in terms of ϵ and β. In what follows, we consider the solitons of the cubic and the quintic CGLE separately. In each section, we derive the analytical solution and then look for special cases.

13.4 Solitons of the cubic CGLE

13.4.1 Solitons with fixed amplitude

First, we concentrate on the cubic CGLE, that is (13.1) with $\nu = \mu = 0$.
Then (13.19) reduces to

$$(a')^2 + \frac{2(2\beta - \epsilon)}{3d(1 + 4\beta^2)} a^4 - \frac{\delta}{d - \beta + \beta d^2} a^2 = 0, \qquad (13.21)$$

which has the solution

$$a(\tau) = BC \operatorname{sech}(B\tau), \qquad (13.22)$$

where

$$C = \sqrt{\frac{3d(1 + 4\beta^2)}{2(2\beta - \epsilon)}}, \quad B = \sqrt{\frac{\delta}{d - \beta + \beta d^2}}, \qquad (13.23)$$

and d is given by (13.17), where the minus sign is chosen in front of the square root. The other value of d leads to an unphysical solution, as the expression under the square root for C then becomes negative. Solution (13.22) has been found by Hocking and Stewartson (1972), Pereira and Stenflo (1977) and others. An important feature of solution (13.22) is that its amplitude and width depend specifically on the parameters of the equation. This is a common property of solutions in non-conservative systems. In other words, (13.22) is a solution with fixed amplitude.

Solution (13.22) depends on three parameters: δ, β and ϵ. However, the parameter δ appears only in the expression for B. It can be seen that the variation of δ leads only to a rescaling of the soliton amplitude and width. So, in what follows, we analyse solution (13.22) on the (β, ϵ) plane.

To find the range of existence of solution (13.22), we note that, on the (β, ϵ) plane, the denominator in the expression for B is positive below the curve S given by

$$\epsilon_S = \beta \frac{3\sqrt{1 + 4\beta^2} - 1}{4 + 18\beta^2}, \qquad (13.24)$$

and negative above it (see Fig. 13.1). Hence, for solution (13.22) to exist, the value δ must be positive below S and negative above it. As this solution exists almost everywhere on the (β, ϵ) plane, we call it the general solution. S itself is the line where this solution becomes singular, i.e. its amplitude BC tends to infinity, while the width $1/B$ vanishes.

From numerical simulations (Akhmediev et al., 1995), it follows that S separates the regions of stable and unstable solitons in the (β, ϵ) plane. That is to say, solution (13.22) exists and is stable below S for $\delta > 0$. This is presented schematically in Fig. 13.1. We recall from perturbation theory that the solution is stable provided $\delta > 0$ and $\epsilon < \beta/2$ (Afanasjev, 1995b). The perturbation theory can be applied only for $|\delta|, |\beta|, |\epsilon| \ll 1$. S has two

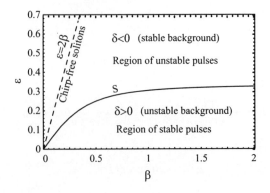

Figure 13.1 *The curve S (13.24) on the (β, ϵ) plane where the solutions with fixed amplitude, (13.22) and (13.40), become singular, and where the classes of special solutions with arbitrary amplitude, (13.26) and (13.46) exist. This plot applies to both the cubic and the quintic cases.*

limits

$$\epsilon \approx \beta/2 \text{ for } \beta \ll 1, \quad \epsilon \to 1/3 \text{ for } \beta \gg 1, \tag{13.25}$$

so, at small β, it coincides with the stability threshold given by perturbation theory (13.5).

For positive linear amplification ($\delta > 0$), the background state ($\psi = 0$) is unstable. If the initial conditions are close to the exact solution (13.22), and $\delta \ll 1$, this instability develops slowly and the soliton can propagate for distances up to $\xi_0 \sim \delta^{-1}$. Beyond that, radiation waves, growing linearly from the noise, become appreciable and can distort the soliton itself. The distance ξ_0 can be large enough to observe soliton interactions (Afanasjev, 1993). This situation is of interest for soliton-based communication lines (Mecozzi et al., 1991; Kodama and Hasegawa, 1992a). However, in other problems δ can be large, so ξ_0 is then small. The general conclusion is that either the soliton itself, or the background state, is unstable at each point in the (β, ϵ) plane. This means that the total solution is always unstable.

We have to emphasize the importance of the curve S. For the solution with fixed amplitude, it gives the range of existence, singularity and stability. Moreover, another significant class of solutions exists on this line.

13.4.2 The solution with arbitrary amplitude

Solution (13.22) does not exist on the curve given by (13.24). However, if we also impose the condition $\delta = 0$, then a new solution, valid only for (13.24), can be found:

$$a(\tau) = GF \operatorname{sech}(G\tau). \tag{13.26}$$

Here G is an arbitrary positive parameter, and d, ω and F are given by

$$d = \frac{\sqrt{1 + 4\beta^2} - 1}{2\beta},$$ (13.27)

$$\omega = -\, d \, \frac{1 + 4\beta^2}{2\beta} \, G^2,$$ (13.28)

$$F = \left(\frac{d\sqrt{1 + 4\beta^2}}{2\epsilon} \right)^{1/2}$$

$$= \frac{1}{\beta} \left[\frac{(2 + 9\beta^2)\sqrt{1 + 4\beta^2} \left(\sqrt{1 + 4\beta^2} - 1 \right)}{2(3\sqrt{1 + 4\beta^2} - 1)} \right]^{1/2}.$$ (13.29)

Solution (13.26) represents the arbitrary-amplitude soliton.

Arbitrary-amplitude solutions exist because, when $\delta = 0$, the cubic CGLE becomes invariant relative to the scaling transformation $\psi \to G\psi$, $\tau \to G\tau$, $\xi \to G^2\xi$. Hence, if we know a particular solution of this equation, then the whole family can be generated using this transformation. The singularity of solution (13.22) and existence of arbitrary-amplitude solutions were discovered numerically by Afanasjev (1995b). Note that all the parameters of solution (13.26) (except G) and the coefficient ϵ are expressed in terms of β.

The class of solutions (13.26) can be considered as a limiting case of the solution with fixed amplitude (13.22) when $\delta \to 0$ and, simultaneously, ϵ and β are moving towards S. At the line of singularity, S, the amplitude–width product C remains finite for the general solution (13.22) and has a finite limit on the line. The values C and F coincide on S. A plot of F versus β is given in Fig. 13.2.

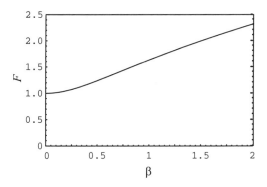

Figure 13.2 *Amplitude–width product F versus β for the solution with arbitrary amplitude of the cubic CGLE.*

It has been found, from numerical simulations (Afanasjev, 1995b), that the class of arbitrary-amplitude solutions is stable relative to small perturbations at any point of S. If we take an initial condition in the form of a superposition of the exact solution (13.26) and a small perturbation, eventually a stationary solution with some new value of G will be formed (Akhmediev and Afanasjev, 1995). For small β, the stationary solution (13.26) can be formed from the chirp-free initial condition, $\psi_0(\tau) = \eta \, \text{sech}(\eta\tau)$, as well. Figure 13.3 demonstrates stable propagation of three well-separated solitons (13.26) with G=0.75, 1, and 1.5, respectively.

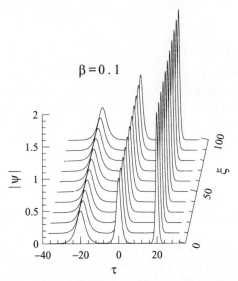

Figure 13.3 *Simultaneous propagation of three soliton solutions of the cubic CGLE, where each has a different amplitude.*

The most important feature of these solutions is that the background state, $\psi = 0$, is also stable, because $\delta = 0$. This means that, by removing the linear gain from the system, and imposing a particular relation between the coefficients ϵ and β, we can achieve stable propagation of these solitons on a stable background. It is remarkable that this class of solutions is the only family of stable pulses in the cubic model.

13.4.3 Chirp-free soliton

Apart from the singularity on S, solution (13.22) does not apply on the line $\epsilon = 2\beta$, as C then becomes indeterminate ($d \to 0$ when $\epsilon \to 2\beta$). However, the soliton amplitude remains finite in the vicinity of this line on the (β, ϵ) plane for finite fixed δ. It follows from (13.17) that, for $\epsilon = 2\beta$, the chirp parameter $d = 0$, and, from (13.18), that $\omega = \delta/2\beta$. Equation (13.19) becomes:

$$(a')^2 + a^4 + \frac{\delta}{\beta} a^2 = 0. \tag{13.30}$$

Its solution is:

$$a(\tau) = \sqrt{-\frac{\delta}{\beta}} \, \text{sech} \left(\sqrt{-\frac{\delta}{\beta}} \, \tau \right). \tag{13.31}$$

The coefficients δ and β must have opposite signs for this solution to exist. As $d = 0$, solution (13.31) has no chirp, in contrast to other soliton solutions of the CGLE. This occurs because of the special choice of the coefficients. In this case the complex constant $(1 - 2i\beta)$ can be factored out of (13.1) when it is reduced to an ordinary differential equation in terms of $a(\tau)$. The pulse itself is unstable, as these solutions are located on the (β, ϵ) plane above S (see Fig. 13.1).

13.5 Solitons of the quintic CGLE

13.5.1 Relation between coefficients

The soliton solutions of the quintic CGLE exist for a wide range of values of the coefficients β, ϵ, μ and ν. The ansatz (13.9) is the condition that restricts this range by imposing a relation (equation (13.14)) on them. Using (13.17), this relation can be rewritten as a linear equation in d:

$$\nu \left(\frac{12\epsilon\beta^2 + 4\epsilon - 2\beta}{\epsilon - 2\beta} d - 2\beta \right) + \mu \left(\frac{2\epsilon\beta - 16\beta^2 - 3}{\epsilon - 2\beta} d + 1 \right) = 0. \tag{13.32}$$

We can also eliminate d completely from equations (13.14) and (13.15) to obtain the following relation between the four coefficients β, ϵ, μ and ν:

$$\frac{27(\mu - 2\beta\nu)(1 + 2\epsilon\beta)^2}{(\epsilon - 2\beta)^2} - \frac{32(\nu + 2\beta\mu)^2}{2\beta\nu - \mu}$$

$$- \frac{60(1 + 2\epsilon\beta)(\nu + 2\beta\mu)}{\epsilon - 2\beta} - \mu + 2\beta\nu = 0. \tag{13.33}$$

Solving (13.33) for ϵ, we obtain

$$\epsilon = \frac{4\beta\mu^2 + 30\mu\nu + 120\beta^2\mu\nu + 4\beta\nu^2 \pm 3U}{-\mu^2 + 12\beta\mu\nu + 32\nu^2 + 108\beta^2\nu^2}, \tag{13.34}$$

where

$$U = (\mu - 2\beta\nu) \sqrt{3\mu^2 + 16\beta^2\mu^2 + 4\beta\mu\nu + 4\nu^2 + 12\beta^2\nu^2}. \tag{13.35}$$

This expression is the relation between the coefficients in explicit form. In contrast to the cubic equation, the general solution exists for both signs in the expression (13.17) for d. Equation (13.33) also applies to both cases. Hence, four different cases have to be considered.

Now we consider the zeros of μ and ν in the (β, ϵ) plane; these result from (13.34). If, in the expression (13.17) for d, we choose the negative sign $(d = d_-)$, then μ has to be zero on

$$\epsilon = \beta \frac{1 - 3\sqrt{1 + 3\beta^2}}{8 + 27\beta^2}. \qquad (13.36)$$

which is the solid curve in Fig. 13.4a. The values μ and ν have the same sign in the region above this line and opposite signs below it.

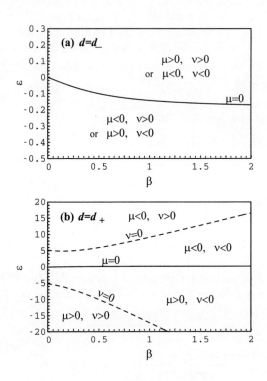

Figure 13.4 *The relation between the parameters μ and ν on the half-plane (β, ϵ) for which the quintic CGLE has analytic solutions. (a) $d = d_-$ in (13.17), (b) $d = d_+$.*

If we choose the positive sign in expression (13.17) for d (i.e. $d = d_+$), then μ becomes zero on the solid line in Fig. 13.4b

$$\epsilon = \beta \frac{1 + 3\sqrt{1 + 3\beta^2}}{8 + 27\beta^2}, \qquad (13.37)$$

and ν becomes zero on the two dashed curves in Fig. 13.4b defined by

$$\epsilon = \pm 3\sqrt{16\beta^2 + 3} - 4\beta. \qquad (13.38)$$

The value of ν changes sign on these curves (see Fig. 13.4b). These conclusions can be made more specific when we consider regions of existence for solutions.

In what follows, we consider solutions which exist when at least one of the coefficients μ or ν is non-zero, and express the solutions in terms of β, ϵ and ν. Using (13.20), the solutions can alternatively be expressed in terms of β, ϵ and μ.

13.5.2 Solutions with fixed amplitude

By using the substitution $f = a^2$, we can rewrite (13.19) in the form

$$f'^2 + \frac{8\nu}{8\beta d - d^2 + 3}f^4 + \frac{8(2\beta - \epsilon)}{3d(1 + 4\beta^2)}f^3 - \frac{4\delta}{d - \beta + \beta d^2}f^2 = 0. \quad (13.39)$$

This is again an elliptic-type differential equation. Bounded soliton-like solutions exist only if $4\delta/(d - \beta + \beta d^2) > 0$. The positive solution of (13.39) is:

$$f(\tau) = \frac{2f_1 f_2}{(f_1 + f_2) - (f_1 - f_2)\cosh(2\alpha\sqrt{f_1|f_2|}\,\tau)}, \quad (13.40)$$

where

$$\alpha = \sqrt{\left|\frac{2\nu}{8\beta d - d^2 + 3}\right|} = \sqrt{\left|\frac{\mu}{3\beta - 2d - \beta d^2}\right|}, \quad (13.41)$$

and f_1 and f_2 are the roots of the quadratic equation:

$$\frac{2\nu}{8\beta d - d^2 + 3}f^2 + \frac{2(2\beta - \epsilon)}{3d(1 + 4\beta^2)}f - \frac{\delta}{d - \beta + \beta d^2} = 0, \quad (13.42)$$

viz.

$$f_{1,2} = \frac{-(2\beta - \epsilon) \pm \sqrt{(2\beta - \epsilon)^2 + \frac{18\,\delta\,d^2\nu(1+4\beta^2)^2}{(8\beta d - d^2 + 3)(d - \beta + \beta d^2)}}}{6d\nu(1 + 4\beta^2)}$$
$$\times(8\beta d - d^2 + 3). \quad (13.43)$$

On the curve given by expression (13.38), this result must be replaced by:

$$f_{1,2} = \frac{-(2\beta - \epsilon) \pm \sqrt{(2\beta - \epsilon)^2 + \frac{9\,\delta\,d^2\mu(1+4\beta^2)^2}{(3\beta - 2d - \beta d^2)(d - \beta + \beta d^2)}}}{3d\mu(1 + 4\beta^2)}$$
$$\times(3\beta - 2d - \beta d^2). \quad (13.44)$$

We now discuss the conditions under which the soliton solution (13.40) exists. Clearly, one of the roots (we choose f_1) must be positive. The second one can have either sign. If it is also positive, we choose $f_1 < f_2$. Two situations arise:

(1) $2\nu/(8\beta d - d^2 + 3) > 0$. Here f_1 is positive, f_2 is negative, but $(2\beta - \epsilon)/d$ can have either sign. Hence, both values of d are suitable. For any ϵ and at

any $\beta > 0$, $8\beta d - d^2 + 3$ is positive when we use $d = d_-$. If we use $d = d_+$, then $8\beta d - d^2 + 3$ is negative in the area between the two dashed curves in Fig. 13.4b, and positive otherwise. Therefore ν must be positive in the former case, and its sign changes on the dashed curves as in Fig. 13.4b in the latter.

(2) $2\nu/(8\beta d - d^2 + 3) < 0$. Both roots f_1 and f_2 are positive, so $(2\beta - \epsilon)/d$ must be positive. Only $d = d_-$ satisfies this criterion. Now $8\beta d - d^2 + 3$ is always positive in this case, and ν can be only negative. This case is shown in Fig. 13.4a.

In both cases, the solution is defined by (13.40). The above analysis shows that, given a set of parameters, ϵ, β, δ and ν, in the area between the two dashed curves in Fig. 13.4b, there are two solutions when ν is negative, but only one when ν is positive. Conversely, outside this area, there are two solutions when ν is positive and only one when ν is negative. Besides, as $\frac{\delta}{d - \beta + \beta d^2}$ must be always positive, the restrictions on the sign of δ are the same as in the cubic case (see Fig. 13.1).

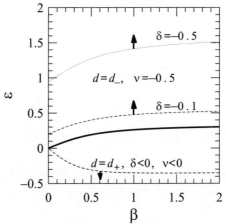

Figure 13.5 *Curves delimiting the regions on the (β, ϵ) plane where the stationary solutions (13.44) exist. The dot-dashed and the dotted curves are obtained by taking $d = d_-$, $\nu = -0.5$ with $\delta = -0.1$ and -0.5, respectively. The region where the solution exists is the area above the curve. The dashed curve is for $d = d_+$ and for negative values of δ and ν. In this last case, the allowed region is the area below the curve. The solid line represents curve S.*

As an illustrative example, Fig. 13.5 shows three curves which delimit the region of allowed values of the parameters (β, ϵ) where the solution exists for given values of ν and δ. The dot-dashed and the dotted curves have been obtained using $\nu = -0.5$, $d = d_-$, with $\delta = -0.1$ and -0.5, respectively. The region above each curve represents the allowed values of the parameters (β, ϵ) where the solution exists in each case. If both δ and

ν are negative, the corresponding curve is the same if the product $\delta\nu$ is kept constant. The dashed curve is for $d = d_+$. In this case the region of allowed values of (β, ϵ) is the area located below the curve, and as long as the signs of ν or δ do not change, it depends only on the product $\delta\nu$.

Given an arbitrary choice of three parameters, let us specify β, ν and δ. Then the stationary solution (13.44) exists for a certain range of values of ϵ. It can be seen from equation (13.14) that the parameter μ has two possible values, corresponding to the two possible values of $d = d_\pm$. Figure 13.6 shows the dependence of μ on ϵ, as given by (13.14), for $\beta = 0.5$ and $\nu = -0.5$. The solid curve is obtained by choosing $d = d_+$ in (13.17), and the dotted curve is for $d = d_-$. The horizontal lines mark the allowed values of ϵ where the solution exists for a fixed value of δ. These values are stated on the lines, together with the corresponding choice for d. Similar pairs of curves, $\mu(\epsilon)$, for $d = d_-$ and $d = d_+$, can be plotted for other values of β and ν. Then the value of δ delimits the values of ϵ where the analytic solution exists.

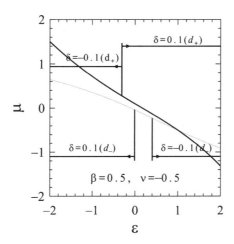

Figure 13.6 *Dependence of μ on ϵ (13.14) that must be satisfied as a necessary condition for the general solution to exist when $\beta = 0.5$ and $\nu = -0.5$. The solid curve is for $d = d_+$ in (13.17), while the dotted curve is for $d = d_-$. The horizontal lines mark the intervals where the solution given by (13.6)–(13.40) exists for a given value of δ. The value of δ is given above each line.*

13.5.3 Singularity at $\nu \to 0^-$

We can see that solution (13.40) has two different branches for the same set of parameters. When ν is negative, one of the solutions has a singularity at $\nu \to 0^-$. Thus $(2\beta - \epsilon)/d$ must be positive and finite. Then f_2 has the limit

$$\frac{3\delta d(1 + 4\beta^2)}{2(d - \beta + \beta d^2)(2\beta - \epsilon)^2},$$

and f_1 goes to infinity as

$$\frac{(\epsilon - 2\beta)(8\beta d - d^2 + 3)}{3d\nu(1 + 4\beta^2)},$$

and so the soliton amplitude goes to infinity. The singularity does not occur when $\nu \to 0^+$. The second solution in the limit $\nu \to 0^-$ coincides with solution (13.22) which applies in the case of the cubic CGLE. This singularity is trivial and is not related to any new solution.

13.5.4 The solution with arbitrary amplitude

Another singularity appears at

$$d - \beta + \beta d^2 = 0. \tag{13.45}$$

This occurs on the same curve S in the (β, ϵ) plane as in the cubic case (13.24). The singularity exists when the roots f_1 and f_2 have opposite signs. If β and ϵ satisfy (13.24) and we have $\delta = 0$, a class of soliton solutions with arbitrary amplitude exists. It is described by:

$$f(\tau) = \frac{3d(1 + 4\beta^2)P}{(2\beta - \epsilon) + D\cosh(2\sqrt{P}\,\tau)}, \tag{13.46}$$

where P is an arbitrary positive parameter and

$$D = \sqrt{(2\beta - \epsilon)^2 + \frac{18\,d^2\nu(1 + 4\beta^2)^2}{8\beta d - d^2 + 3}\,P}. \tag{13.47}$$

The values d and ω are given by

$$d = \frac{\sqrt{1 + 4\beta^2} - 1}{2\beta}, \tag{13.48}$$

$$\omega = -d\,\frac{1 + 4\beta^2}{2\beta}\,P. \tag{13.49}$$

This class of solutions is stable at any point on the special curve S and for any P in (13.46). The background state is also stable, as $\delta = 0$. Figure 13.7 shows the pulse profiles of three of these solutions, with $P = 2$, 1 and 0.5 respectively, as they propagate. No changes in the profiles are observed after propagating a very long distance. These pulses are the only analytic solutions which are stable in the whole region of parameters where they exist.

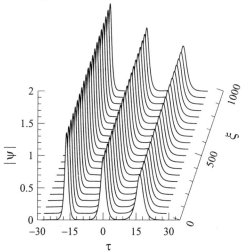

Figure 13.7 *Simultaneous propagation of three soliton solutions, with different amplitudes, of the quintic CGL with $\beta = \nu = 0.1$.*

13.5.5 Flat-top solitons

The soliton (13.40) becomes wider and flatter as the two positive roots approach each other. When $f_1 = f_2$, the soliton splits into two fronts with zero velocity. Each of them can be written in the form (we ignore the translations along τ):

$$f(\tau) = \frac{f_1}{1 + \exp(\pm \alpha f_1 \tau)}, \tag{13.50}$$

where

$$f_1 = \frac{(\epsilon - 2\beta)(8\beta d - d^2 + 3)}{6d\nu(1 + 4\beta^2)}, \tag{13.51}$$

and the sign in (13.50) determines the orientation of the front. The two roots f_1 and f_2 become identical when:

$$(2\beta - \epsilon)^2 = -\frac{18\,\delta\,\nu d^2\,(1 + 4\beta^2)^2}{(d - \beta + \beta d^2)(8\beta d - d^2 + 3)}. \tag{13.52}$$

This condition involves all the parameters of the equation. Depending on δ and ν, it can exist at any point in the (β, ϵ) plane.

The transition from the general solution, (13.40), to the flat-top solution (13.50), for $f_1 \to f_2$ is shown in Fig. 13.8. The top of the soliton becomes flatter as the roots approach each other.

If $f_1 = f_2$ exactly, the width of the pulse goes to infinity and the pulse decomposes into two fronts. In the region of non-zero intensity, the soliton phase $\phi(\tau)$ tends to a constant value exponentially. So, if we combine the

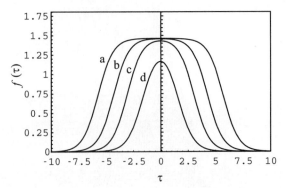

Figure 13.8 *The shapes of solutions (13.40) when the two roots f_1 and f_2 are close to each other. Separation into two fronts is a result of this proximity. The parameters chosen are: $\beta = 0.5$, $\nu = -0.5$, $\delta = -0.1$ and $d = d_-$, which gives $\mu = -0.227$, while ϵ is: (a) 0.4005969, (b) 0.4006, (c) 0.4007, (d) 0.41.*

two fronts (13.50) with opposite orientations, to form a wide, rectangular pulse of finite width, the influence of each front on the other is exponentially small. In other words, the two fronts (13.50) can join together without any domain boundary between them (cf. van Saarloos and Hohenberg, 1992).

Pulses and fronts have usually been considered as different solutions of the CGLE (van Saarloos and Hohenberg, 1990, 1992; Marcq *et al.*, 1994). The above results show that they can be transformed into each other by changing the parameters of the system. Moreover, these results give an indication of the range of parameters where we can expect smooth transitions from solitons to fronts. Stable stationary flat-top pulses have been observed experimentally in binary fluid convection (Kolodner *et al.*, 1988).

13.5.6 Rational solution

If $\delta = 0$ and (β, ϵ) is not located on the curve (13.24), then $\omega = 0$, and one of the roots of (13.42) (say, f_2) becomes zero. The other root is:

$$f_1 = \frac{(\epsilon - 2\beta)(8\beta d - d^2 + 3)}{3d\nu(1 + 4\beta^2)}. \tag{13.53}$$

Equation (13.39) can then be written in the form:

$$f'^2 + 4k_0[f - f_1]f^3 = 0, \tag{13.54}$$

where $k_0 = 2\nu/(8\beta d - d^2 + 3)$. The solution to this equation is a Lorentz function:

$$f(\tau) = \frac{f_1}{1 + k_0 f_1^2 \tau^2}. \tag{13.55}$$

The values f_1 and k_0 must be positive, which restricts the allowed values of the coefficients of the equation for this solution to exist.

The rational soliton is unstable for the whole range of the parameters where it exists. The rational solution represents a special, weakly localized limit of the solution with fixed amplitude (13.40). Rational solitons exist, and play an important role, in other integrable and non-integrable systems, including the NLSE and its generalizations.

13.5.7 Chirp-free soliton

Another degenerate case occurs when $\epsilon = 2\beta$ and $\mu = 2\beta\nu$. Equation (13.39) then reduces to:

$$f'^2 + \frac{8}{3}\nu f^4 + 4f^3 + \frac{4\delta}{\beta}f^2 = 0. \qquad (13.56)$$

The solution of this equation is:

$$f(\tau) = \frac{-2\delta}{\beta\left[1 - \sqrt{1 - \frac{8\delta\nu}{3\beta}}\cosh\left(2\sqrt{-\frac{\delta}{\beta}}\tau\right)\right]}. \qquad (13.57)$$

Clearly, this solution exists when δ/β is negative and ν positive. Here $d = 0$, so this solution to the CGLE has no phase chirp. This solution arises because the coefficients of the equation are chosen in such a way that a complex constant can be factored out from (13.1) when it is reduced to an ordinary differential equation in terms of $f = a^2$. The latter then becomes purely real.

Numerical simulations show that the chirp-free pulses given by our exact solutions are unstable for all values of the parameters. Nevertheless, we may expect the existence of chirp-free pulses beyond the limitations imposed by (13.34). Chirp-free solutions are important in various applications. One of them is the problem of obtaining a chirp-free pulse at the output of a laser. Such pulses could be used in transmission lines without additional modifications. While the pulse amplitude can be adjusted with relative ease by changing the values of the amplification or damping, δ, the control of chirp is a much more difficult task, as it requires knowledge of other system parameters. Our analysis reveals, to some extent, the range of these parameters where we can obtain chirp-free pulses.

13.6 Systems without spectral filtering

If $\beta = 0$, then it is possible to have solitons with non-zero velocity (i.e. travelling solitons). These solutions can be obtained using a simple transformation, because, for $\beta = 0$, (13.1) has an additional symmetry – it is invariant relative to a Galilean transformation, as explained in section 2.3. As a result, travelling pulse-like solutions can be obtained from zero-velocity ones using transformation (2.5). Hence, we can use the fixed-amplitude solution of section 13.5.2, put $\beta = 0$, and use transformation (2.5) to obtain the whole family of travelling pulses. Note that all the analysis of section 13.5.2 is valid in this case. The critical points in ϵ are the intersections of the special lines in Figs 13.1 and 13.4 with the vertical axis, $\beta = 0$. This last example completes the classification of possible pulse-like analytic solutions for (13.1).

Traditionally, it has been supposed that, if the coefficient β of the second-order derivative term on the right-hand side of (13.1) is non-zero, then only motionless and symmetric localized stationary solutions could exist. This follows, for example, from adiabatic perturbation theory – see, e.g. Elphick and Meron (1990) and references therein. However, perturbation theory cannot be applied if one or more of the coefficients on the right-hand side of (13.1) is not small.

13.7 Stability of solutions with fixed amplitude

An important issue is the stability of the exact solutions. As the system described by (13.1) is non-conservative, the stability can only be analysed numerically. Such an analysis includes the solution of the linearized problem, i.e. the calculation of the perturbation eigenmodes and their growth rates. A perturbed solution has to be written in the form:

$$\psi(\tau, \xi) = [A_0(\tau) + \gamma g(\tau, \xi)] \exp(-i\omega\xi), \qquad (13.58)$$

where $A_0(\tau)$ is the stationary solution under study (13.7), γ is a small parameter and $g(\xi, \tau)$ is a perturbation function. Inserting (13.58) into (13.1), and linearizing in the small parameter γ, we obtain:

$$ig_\xi + \omega g + \left(\frac{1}{2} - i\beta\right) g_{\tau\tau} + 2|A_0|^2(1 - i\epsilon)g + A_0^2(1 - i\epsilon)g^*$$

$$+ (\nu - i\mu)(3|A_0|^4 g + 2|A_0|^2 A_0^2 g^*) = 0. \qquad (13.59)$$

This equation for the perturbation function g may have many possible types of solutions. Moreover, because the linear operator in (13.59) is not Hermitian, its eigenvalues are, in general, complex.

Consider the case with negative ν. Figure 13.9 shows, on a logarithmic scale, the perturbation growth rate as a function of ϵ, for $(\beta, \nu) = (0.5, -0.5)$, with $\delta = \pm 0.1$ and ± 0.001. Figure 13.9a is for the cases with negative δ while Fig. 13.9b is for the positive ones. Each curve is labelled with its value of δ. For each δ, there are two curves, corresponding to the two possible solutions associated with the values of d. For $|\delta| = 0.001$, we have a solid curve ($d = d_-$) and a dotted one ($d = d_+$), while for $|\delta| = 0.1$, the dot-dashed and the dashed curves correspond to $d = d_-$ and $d = d_+$, respectively. Clearly, in all cases the solutions obtained for $d = d_+$ have much higher instability growth rates than those associated with d_-. Consequently, these latter solutions should be of more interest from a practical point of view. Moreover, as we will see below, those for negative δ are related to the stable solutions.

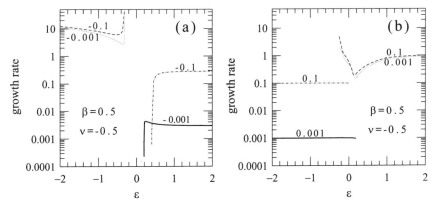

Figure 13.9 *The growth rate of the predominant perturbation eigenmode associated with the soliton solution, as a function of ϵ for $\beta = 0.5$, $\nu = -0.5$ and (a) $\delta < 0$ (b) $\delta > 0$. In both cases the solid curve is for $|\delta| = 0.001$ and $d = d_-$, the dot-dashed curve is for $|\delta| = 0.1$ and $d = d_-$, the dotted curve is for $|\delta| = 0.001$ and $d = d_+$, and the dashed curve is for $|\delta| = 0.1$ and $d = d_+$. For clarity we have labelled each curve with its value of δ.*

For positive δ and for the solutions obtained using $d = d_-$, the figure shows that the perturbation growth rate exactly equals δ, i.e. their instability has its origin solely in the instability of the uniform background, $\psi = 0$. The result shows that the pulse itself is stable. This is exactly the case which occurs in optical transmission lines. The excess gain is usually positive in order to amplify the pulse itself, but it is kept small to avoid an appreciable growth of the background. What Fig. 13.9a indicates, then, is that we can have stable propagation of these states over great distances, as long as the excess linear gain, δ, is low enough. Figure 13.10 confirms this last assertion. The initial condition is the following:

$$\psi(\tau, 0) = A_0(\tau)[1 + \Gamma(\tau)] \tag{13.60}$$

where $A_o(\tau)$ is the corresponding stationary solution (13.7) for the following coefficients: $\beta = 0.5$, $\nu = -0.5$, $\epsilon = 0.1$, $\delta = 0.005$, and $d = d_-$ ($\Rightarrow \mu = -0.089$), and Γ is random noise obeying

$$< \Gamma(\tau) > = 0, \qquad \sqrt{< |\Gamma(\tau)|^2 >} = 0.2. \tag{13.61}$$

The noise term is intended to seed any latent instability of the system, as it surely contains all possible exponentially growing eigenfunctions of the linearized problem. Figure 13.10 shows that the solution very quickly approaches the profile corresponding to the stationary solution while eliminating the random fluctuations, and then maintains this shape for long distances. After propagating a distance of $\xi = 1000$, the profile has not experienced any significant change. On further propagation, the growth of the background starts to be perceptible.

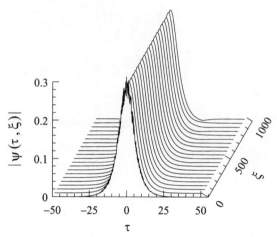

Figure 13.10 *The evolution of the solution for $\beta = 0.5$, $\epsilon = 0.1$, $\nu = -0.5$, $\delta = +0.005$ and $\mu = -0.09$. This stationary solution is initially perturbed, as indicated in (13.60) and (13.61).*

The solution corresponding to $d = d_+$ exists only for negative ϵ, and it is always unstable. The solution for $d = d_-$ exists for positive ϵ and negative μ, i.e. in the region where stable propagation is possible. The growth rate for this solution is always smaller, and it vanishes for $\epsilon \to 0.4006$. This is the closest point to the value of ϵ_* with $f_1 = f_2$. The solution does not exist for $\epsilon < \epsilon_*$. In reality, the region of existence of the flat-top solitons is larger than the value given by the analytic expression.

For negative δ (and $d = d_-$), at any given ϵ, the perturbation growth rate decreases as $|\delta|$ decreases. Another interesting feature of the growth rate curves in Fig. 13.9a is that, at $d = d_-$, the solution becomes more stable as we move to its smallest allowed value of ϵ. This happens when f_1 approaches f_2. Figure 13.11 shows an example of stable propagation for these cases. The initial conditions are (13.60) and (13.61), with $A_0(\tau)$ being the stationary solution for the following values of the coefficients: $\beta = 0.5$, $\nu = -0.5$, $\delta = -0.1$, $\epsilon = 0.4006$ and $d = d_-$ ($\Rightarrow$ $\mu = -0.227$). For this specific case, $f_1 = 1.458$, and $f_2 = 1.469$. At first the solution recovers its unperturbed shape and then it propagates for the whole distance that we have considered ($\xi_{max} = 10000$) without modifying its profile. We conclude that solutions obtained for negative δ and $d = d_-$ are stable when f_1 is close to f_2 (dot-dashed and dotted curves in Fig. 13.9). The transformation of the pulses into a pair of zero-velocity fronts occurs on these lines.

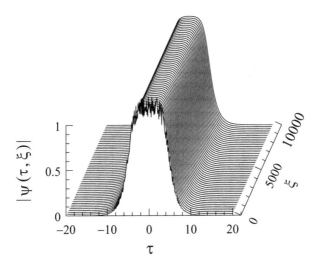

Figure 13.11 *Stable propagation of a flat-top soliton. The parameters chosen for this simulation are: $\beta = 0.5$, $\epsilon = 0.4006$, $\nu = -0.5$, $\delta = -0.1$ and $d = d_-$, which gives $\mu = -0.227$. The perturbation added initially to the stationary solution is a random function.*

Similar results on the stability of these solutions are valid for other values of β and ν. Figure 13.12 shows the perturbation growth rate for different solutions with $\nu = -0.1$, $\delta = \pm 0.01$, and $\beta = 0.1$ and 0.5. The corresponding value of β is given on each curve. The curves for the solution obtained using $d = d_+$ are plotted with dotted and dashed lines, while those for $d = d_-$ are represented by continuous and in dot-dashed lines. The curves exhibit the same qualitative features as those of Fig. 13.9. The general behaviour of the perturbation growth rate for other values of β and ν (< 0) is similar to that shown in Figs 13.9 and 13.12.

Similar analysis shows that the solutions for positive ν are always unstable both for $d = d_-$ and $d = d_+$. For negative δ and $d = d_-$, the solution exists only for positive values of ϵ, which then necessitates positive values for μ. Thus the nonlinear gain compensates for the linear losses. If $d = d_+$, the solution exists for very high values of $|\epsilon|$, which also produce high values for ν. For instance, for $\beta = 0.5$, $\nu = 0.5$, and $\delta = -0.01$, the solution obtained taking $d = d_+$ exists in the interval $(-\infty, -9.93)$, while for $\epsilon = -12$, μ becomes 32.97. The corresponding solutions are highly unstable.

The general conclusion from the stability analysis is that although exact solutions (13.40) to the quintic CGLE can be found when a specific relation

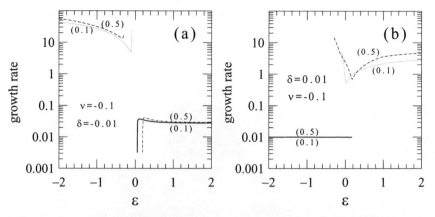

Figure 13.12 *The growth rate of the predominant perturbation eigenmode associated with the soliton solution, as a function of ϵ for $\nu = -0.1$, with $\beta = 0.5$ and 0.1, and with (a) $\delta = -0.01$ (b) $\delta = -0.05$. The different curves correspond to: (i) $\beta = 0.5$ and $d = d_-$ (solid), (ii) $\beta = 0.1$ and $d = d_-$ (dot-dashed), (iii) $\beta = 0.5$ and $d = d_+$ (dashed), (iv) $\beta = 0.1$ and $d = d_+$ (dotted). For clarity, each curve is labelled with its value of β (in parentheses).*

between the parameters is satisfied, almost all of them are unstable. An exception appears in the vicinity of the boundary that separates pulses from pairs of fronts. The perturbation growth rate of these soliton solutions falls to zero when we tend to this limit. These stable solutions have flat tops, indicative of the transition from pulses to fronts.

13.8 Regions in the parameter space where stable pulses exist

It is crucially important to know the values of the coefficients where the quintic CGLE has stable pulse-like solutions. Stable pulses are the most interesting objects in optics. They are produced by laser systems and they constitute bits of information, transmitted in optical fiber systems. Hence, it is important to know where we can expect stable pulses in the parameter space. This is the region of global stable pulse propagation, i.e. the region in the parameter space where a broad class of initial conditions converge to a stationary pulse, which therefore represents a stable pulse-like solution of the quintic CGLE.

The first observation of strictly stable pulse-like solutions was reported by Thual and Fauve (1988). They found some points in the parameter space where stable pulses exist. Rough estimates of the location of the boundaries between fronts and pulse-like solutions of the CGLE have been made by Hakim et al. (1990). However, there is no sharp boundary between the two classes of solutions. It has been found by van Saarloos and Hohenberg

(1992) that, for some values of the parameters, a variety of fronts and pulse-like solutions exist.

In this section, we present numerical results obtained by Soto-Crespo *et al.* (1996), which give the values of the coefficients $(\delta, \beta, \epsilon, \mu, \nu)$ (i.e. the subspace of the parameter space) of the quintic CGLE where stable pulses exist. Stable pulses exist in a certain region, and it is interesting to compare it with that of a lower dimensionality where the analytical solutions given by (13.40) exist.

Let us first fix some limits on the parameter space where we seek stable pulses. The parameter β clearly must be non-negative, in order to stabilize the soliton in frequency domain. The linear gain coefficient δ must be zero or negative to provide stability of the background. In this case, for $\mu = 0$, stable pulses can exist only for ϵ above the curve S. We choose $\mu < 0$ to stabilize the pulse against the collapse. The parameter ν can have either sign.

Stable pulses can be found numerically from the propagation equation (13.1), taking a Gaussian pulse of arbitrary amplitude and width as the initial condition (Soto-Crespo *et al.*, 1996). The shape of the initial pulse appears to be of little importance. If the solution converges to a stationary one, it can be viewed as a stable solution, and the chosen set of the parameters can be deemed to belong to the class of those which permit the existence of solitons.

Figure 13.13 shows three examples of the soliton solutions found with this method. The corresponding values of the coefficients are: $\beta = 0.5$, $\delta = \nu = \mu = -0.1$, with ϵ being 0.38 (solid curve), 0.52 (dotted curve) and 0.66 (dashed curve). By repeating these calculations systematically for other sets of parameters, it is possible to construct the regions in the parameter space where stable propagation of bounded solutions is possible.

Figure 13.14 shows the areas in the (β, ϵ) plane were soliton solutions were found numerically. The differently hatched areas correspond to different values of the parameter μ. The lower curve (dashed) which represents the curve S, is plotted to allow us to make some comparisons with the conclusions obtained from the analytic solutions. First of all, we note that the region of stable pulses is always above S, and that the lower boundary of the stability region (solid line) is roughly parallel to S. The distance between this lower boundary and S depends on δ, μ and ν. For small μ, ν and δ, this distance is small. For given values of ν and δ, the hatched regions become wider as $|\mu|$ increases, and the lower boundary becomes higher. For fixed ν and μ, the lower boundary approaches S as δ goes to zero. We would expect that, at zero δ, S would denote the onset of instability.

Figure 13.14 gives a rough idea of how the regions of existence of stable pulses in the plane (β, ϵ) change when μ and δ are changed. Stationary pulses must balance loss and gain. For systems whose parameters are located below the lower boundary of the hatched regions, pulses attenuate as

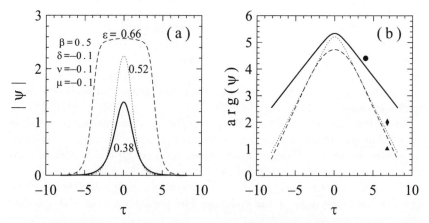

Figure 13.13 *Soliton solutions found numerically for $\beta = 0.5$, $\nu = \mu = \delta = -0.1$, with ϵ being 0.38 (solid curve), 0.52 (dotted) and 0.66 (dashed). (a) amplitude profile, $|\psi|$. (b) Phase profile, $arg(\psi)$. The circle, diamond and triangle, labelling the cases $\epsilon = 0.38$, 0.52, 0.66 respectively, will be used in later figures to locate these solutions in the parameter space.*

they propagate. The energy flux added to the initial pulse due to positive ϵ is less than the energy decrease due to linear ($\beta < 0$, $\delta > 0$) and nonlinear ($\mu < 0$) losses. The physical processes on the upper boundary are different. In general, the upper boundary in the (β,ϵ) diagram coincides with the curve where fronts have zero velocity. Above that curve, two fronts of the wide pulse diverge, while below it two fronts converge, forming a stable pulse at the end of this process. Thus, stable pulses can exist only below that line (see sections 13.10 and 13.11 for details).

Now, consider other planes in the parameter space where we have found stable pulse-like solutions. Figure 13.15a shows the region of stable pulses in the (ν, ϵ) plane for fixed values of μ, δ and β, as indicated in the figure. The plot shows that the width of the strip in Fig. 13.14 increases greatly as ν increases. The dashed curve in Fig. 13.15 shows where the exact analytical solutions are located for the same set of parameters. Interestingly enough, this line is also almost parallel to the upper border of the area of stable pulses, but is located some distance from it. This shows that the analytical solutions are beyond that region and therefore are unstable.

Figure 13.15b shows the area of stable pulses in the plane (μ, ϵ) for fixed values of ν, δ and β. As $|\mu|$ increases, the interval of allowed values of ϵ becomes wider, and its central value larger. This last observation is also expected, as it indicates that larger fifth-order nonlinear losses must be compensated for by increasing the third-order nonlinear gain. The width of the strip becomes infinitesimally small at $\mu \approx -0.04$. The dashed line

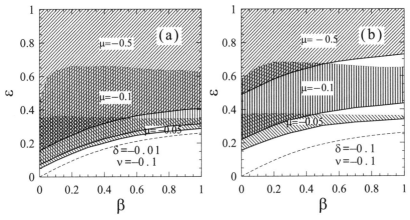

Figure 13.14 *Regions in the* (β,ϵ) *plane where stable pulse-like solutions are found. Differently hatched areas are for different values of μ, as indicated in each area. All these areas are located above S (dashed curve). We have set $\nu = -0.1$ with (a) $\delta = -0.01$, (b) $\delta = -0.1$.*

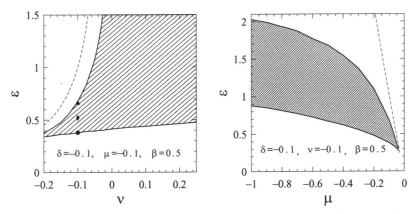

Figure 13.15 *Regions in the planes (ν,ϵ) (left) and (μ,ϵ) (right) where stable pulses exist. Here $\beta = 0.5$ and $\delta = -0.1$. The dashed lines represent the points where the analytical solution given by (13.40) exists. The three symbols – circle, diamond and triangle – show the location of the solutions represented in Fig. 13.13.*

represents the points where the exact analytical solutions are located for the given values of $(\delta,\nu,\beta) = (-0.1,-0.1,0.5)$. Again, it can be seen that they are not in the area of stable pulses. However, in this case the distance between the region of stable pulses and the exact analytical solutions increases with μ and goes to zero at $\mu \to -0.04$. The instability growth rate of the corresponding analytical solution for this value of μ becomes negligible.

What is the dependence on δ of the region where stable pulses exist? Figure 13.16 shows this dependence for fixed values of μ, ν and β. In this figure, $\nu = \mu = -0.1$ and $\beta = -0.5$. As the linear excess gain decreases, the interval of allowed values of ϵ increases, while its central value increases as $|\delta|$ increases. This logically means that larger linear losses must be compensated for (if the remaining parameters are constant) by increasing the third-order nonlinear gain. For the above values of ν, μ and β, (13.34) gives $\epsilon = 1$, which is above the hatched region.

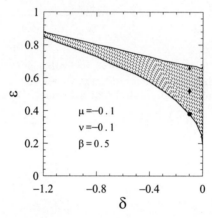

Figure 13.16 *Region in the (δ, ϵ) plane where stable pulses are possible. Here $\beta = 0.5$, and $\nu = \mu = -0.1$. For these parameters, the analytical solution exists at $\epsilon = 1$. The three symbols – circle, diamond and triangle – show the location of the solutions represented in Fig. 13.13.*

13.9 Reduction to a set of ordinary differential equations

Analytic solutions can be found only for certain combinations of parameters. One way to find stationary solutions of (13.1) numerically is to reduce it to a set of ODEs. Thus, we seek solutions in the form:

$$\psi = a(t) \, \exp[i\phi(t) - i\omega\xi], \tag{13.62}$$

where a and ϕ are real functions of $t = \tau - v\xi$, v is the pulse velocity and ω is the nonlinear shift of the propagation constant. Substituting (13.62) into (13.1), we obtain an equation for two coupled functions, a and ϕ. Separating real and imaginary parts, we get the following set of two ODEs:

$$[\omega - \tfrac{1}{2}D\phi'^2 + \beta\phi'' + v\phi']a + 2\beta\phi'a' + \tfrac{1}{2}Da'' + a^3 + va^5 = 0,$$

$$(-\delta + \beta\phi'^2 + \tfrac{1}{2}D\phi'')a + (D\phi' - v)a' - \beta a'' - \epsilon a^3 - \mu a^5 = 0, \tag{13.63}$$

where each prime denotes a derivative with respect to t. It can be transformed into:

$$\omega a + v\frac{M}{a} - \tfrac{1}{2}DM^2/a^3 + \beta M'/a + \tfrac{1}{2}Da'' + a^3 + \nu a^5 = 0,$$

$$-\delta a - va' + \beta M^2/a^3 + \tfrac{1}{2}DM'/a - \beta a'' - \epsilon a^3 - \mu a^5 = 0,$$

(13.64)

where $M = a^2\phi'$.

Separating derivatives, we obtain:

$$M' = \frac{2(D\delta-2\beta\omega)}{1+4\beta^2}a^2 + \frac{2(D\epsilon-2\beta)}{1+4\beta^2}a^4 + \frac{2(D\mu-2\beta\nu)}{1+4\beta^2}a^6$$

$$- \frac{4\beta v}{1+4\beta^2}M + \frac{2Dv}{1+4\beta^2}ay,$$

(13.65)

$$y' = \frac{M^2}{a^3} - \frac{2(D\omega+2\beta\delta)}{1+4\beta^2}a - \frac{2(D+2\beta\epsilon)}{1+4\beta^2}a^3 - \frac{2(D\nu+2\beta\mu)}{1+4\beta^2}a^5$$

$$- \frac{4\beta v}{1+4\beta^2}y - \frac{2Dv}{1+4\beta^2}\frac{M}{a}, a' = y.$$

This set contains all stationary and uniformly translating solutions. The parameters v and ω are the eigenvalues of (13.65). In the (M, a) plane, the solutions corresponding to pulses are closed loops starting and ending at the origin.

If we are only interested in zero-velocity ($v = 0$) solutions, (13.65) can be simplified:

$$M' = \frac{2(D\delta - 2\beta\omega)}{1 + 4\beta^2}a^2 + \frac{2(D\epsilon - 2\beta)}{1 + 4\beta^2}a^4 + \frac{2(D\mu - 2\beta\nu)}{1 + 4\beta^2}a^6,$$

$$y' = \frac{M^2}{a^3} - \frac{2(D\omega + 2\beta\delta)}{1 + 4\beta^2}a - \frac{2(D + 2\beta\epsilon)}{1 + 4\beta^2}a^3 - \frac{2(D\nu + 2\beta\mu)}{1 + 4\beta^2}a^5,$$

$$a' = y.$$

(13.66)

This set of first-order ODEs can be solved numerically. The asymptotic behaviour of (13.66) at small a is given by

$$a = a_0 \exp(g\tau),$$

$$M = \frac{D\delta - 2\beta\omega}{g(1 + 4\beta^2)}a_0^2 \exp(2g\tau),$$

where the soliton tail exponent g can be found from the biquadratic equation

$$g^4 + \frac{2(D\omega + 2\beta\delta)}{1 + 4\beta^2}g^2 - \frac{(D\delta - 2\beta\omega)^2}{(1 + 4\beta^2)^2} = 0.$$

(13.67)

Thus

$$g^2 = \pm\sqrt{\frac{\omega^2 + \delta^2}{1 + 4\beta^2} - \frac{D\omega + 2\beta\delta}{1 + 4\beta^2}}$$

(13.68)

Using this approximation for the tails, it is possible to find the rest of the pulse solution with a shooting method. Examples are given below.

13.10 Composite pulses

Stable plain pulses exist only when δ and μ are negative, and β and ϵ are positive. The parameter ν can be either positive or negative. These conditions mean that one term (that with coefficient ϵ) provides gain for the pulse, while three other terms (those with coefficients δ, β and μ) produce losses. At certain values of the parameters, the pulse-like initial condition converges to a new type of solution, namely, a composite pulse (CP). We may view this as a combination of a source and two fronts.

The CP profile, phase profile and Fourier transform are shown in Fig. 13.17. The corresponding curves for the plain pulse solution (for the same set of parameters) are also given for comparison. The CP consists of two fronts with a small 'hill' between them. This hill is the domain boundary between the two fronts, as they have non-zero wavevectors. This hill should be counted as a source, because it follows from the phase profile that energy flows from the centre to the CP wings. Note that the flat regions between the source and the fronts are relatively small. The typical width of the source is the same as the typical width of the front.

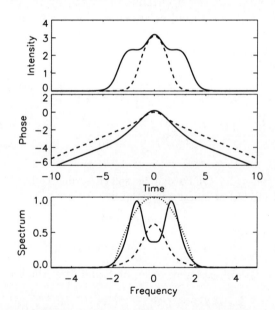

Figure 13.17 *The amplitude, phase profiles, and the spectra of the composite pulse (solid curves) and plain pulse (dashed curves) having $\delta = -0.1$, $\beta = 0.5$, $\epsilon = 1.75$, $\mu = -0.6$ and $\nu = -0.1$. The dotted curve gives the width of the spectral filtering $(1 - \beta^2\omega^2)$.*

To compare the CP with a plain pulse, we use the (a, M) plane. From Fig. 13.18 we see that that there are many similarities between the two solutions. In particular, the top part of the CP (i.e. source) has the same shape as the top part of the pulse. The wings of the plain pulse and the CP are also very similar. The difference between the nonlinear propagation constants ω of two structures is around 10%. The similarity between the asymptotes follows from the linearized version of (13.1).

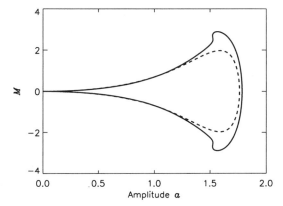

Figure 13.18 *The plain pulse (dashed curve) and the composite pulse (solid curve) on the (a, M) plane. The parameters are the same as in Fig. 13.17.*

The spectrum of the CP is shown in Fig. 13.17. It has a central dip and two well-separated peaks. To explain this structure in the spectrum, we note that the CP consists of two short plane waves located between the source and the two fronts. These plane waves have wavevectors in opposite directions, and the spectrum of this structure consists of two peaks. The spectral separation between the peaks is determined by the difference between the wavevectors of the two plane waves.

Figure 13.19 shows the ranges of parameters where the stable plain pulses and the CPs exist. The range of existence of the CPs is limited by the positive front velocity threshold. The upper boundary on this plot corresponds to the positive front velocity threshold and, hence, to the transition into a pair of fronts. The lower boundary corresponds, in general, to the transition from the composite pulse to the plain pulse – this can be quite complicated.

13.11 Moving pulses

Moving pulses (MPs) can be observed as a result of the instability of the CP at the lower boundary of the region of stable CP on the (μ, ϵ) plane. If an anti-symmetric perturbation has a large growth rate, then instead of a transformation into a plain pulse, we obtain a spontaneous transformation into an asymmetric MP (Fig. 13.20). The moving pulse can also be formed by choosing a moving initial condition. In this case, both left- and right-

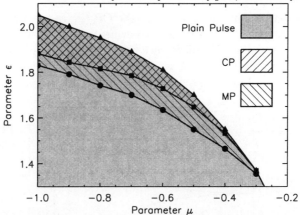

Figure 13.19 *The range of existence of stationary stable pulses on the (μ, ϵ) plane. The solid curve marked by triangles gives the zero front velocity threshold, and filled areas give the ranges of existence of the plain pulse, the composite pulse and the moving pulse. Parameters are $\delta = -0.1$, $\beta = 0.5$ and $\nu = -0.1$.*

moving pulses can be created. They are mirror images, due to the time-reversal symmetry of (13.1).

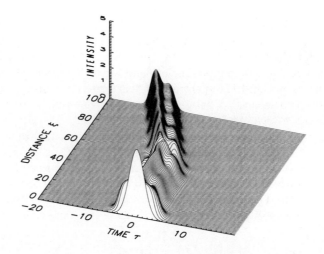

Figure 13.20 *Spontaneous transformation of a composite pulse into a moving pulse for $\delta = -0.1$, $\beta = 0.5$, $\nu = -0.1$ $\epsilon = 1.8$ and $\mu = -0.8$.*

The amplitude, phase profile and spectra of the MPs are given in Fig. 13.21. The amplitude profile is indeed very close to the profiles of the plain pulse and the composite pulse. In other words, an MP can be considered as the bound state (i.e. nonlinear superposition) of the plain pulse and the front, or as a CP in which one of the fronts is missing. The spectrum of the MP is asymmetric. There are two unequal peaks separated by a valley.

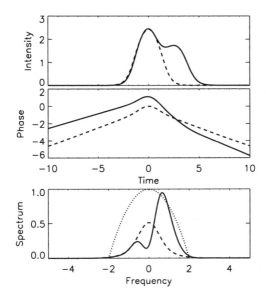

Figure 13.21 *The amplitude, phase profiles and spectra of the moving pulse (solid curves) and the plain pulse (dashed curves). The parameters are the same as those in Fig. 13.20. The dotted curve gives the width of the spectral filtering $(1-\beta^2\omega^2)$.*

It follows from the representation of the MP as the bound state of a pulse and a front, that its energy can be roughly written as $E_{\mathrm{MP}} = E_{\mathrm{P}} + E_{\mathrm{F}}$, while the energy of the composite pulse is roughly $E_{\mathrm{CP}} = E_{\mathrm{P}} + 2E_{\mathrm{F}}$, where E_{P} and E_{F} are the energies of the plain pulse and front, respectively. In other words, the difference between the MP energy and the energy of the plain pulse (which exists for the same set of parameters) is half of the difference between the energy of the CP and the plain pulse. Numerical calculations show qualitative agreement with this prediction, as Fig. 13.22 shows.

The MP always moves with the pulse leading (Fig. 13.20). This again shows that the MP is a nonlinear combination of a pulse and a front, because it exists in the region of parameters where the front velocity is negative. The front pushes the pulse from one side. Another important feature is that the velocity of an MP is always smaller than the velocity of the front for the same set of parameters. For the MP, the front tends to move with its own velocity but the pulse tends to be stationary, due to the spectral filtering. The resulting velocity of the MP is determined by competition between these two processes. The distance between the centre of the pulses and the front in the MP is slightly larger that the corresponding distance in the CP, because the pulse offers less resistance to the pushing force of the front. The difference between the predicted and the observed energy of the MP follows from this observation (Fig. 13.22).

Note the analogy between the MP of the CGLE and the asymmetric stationary solution in the case of nonlinear guided waves. It has been shown that asymmetric solutions of the symmetric nonlinear problem can exist in the conservative case (Akhmediev, 1982). We may expect that this will also be true for the non-conservative problem. In the case of the CGLE, the medium is homogeneous, while for nonlinear planar guided waves there is a spatially symmetric inhomogeneity. On the other hand, the presence of spectral filtering can be considered as an inhomogeneity in the frequency domain, so an asymmetric pulse can appear. This can happen equally well in the spatial domain (for nonlinear guided waves) and in the frequency domain (for the CGLE).

The range of existence of the MP is even larger than that of the CP, as can be seen from Figs 13.19 and 13.22. The upper boundary in ϵ is almost the same for plain pulses, the CP and the MP, while the range of existence of the MP is approximately twice as wide as that of the CP. However, neither the CP nor the MP exists when $\mu > -0.3$. Apparently, this is the threshold where double matching (between amplitudes and wavevectors) of the pulse and front becomes possible. As we decrease μ below -0.3, the range of this matching increases. This again proves that the CP and the MP have a common nature. The MP exists for a continuous range of parameters, and this range is comparable with the range of existence of the plain pulses.

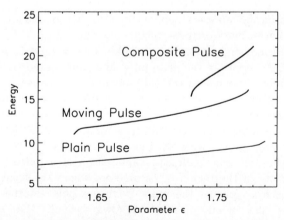

Figure 13.22 *The energies of the plain pulse, the composite pulse and the moving pulse versus ϵ. Here $\delta = -0.1$, $\beta = 0.5$, $\mu = -0.6$ and $\nu = -0.1$.*

At the lower boundary of their existence in ϵ, the MPs are transformed into plain pulses. This process can be explained by noting that, as the front velocity increases, resistance from the pulse also increases. At some point, the front velocity overcomes the repulsion and the front is absorbed by the pulse. In this sense, the lower boundaries for the CP and the MP on the (μ, ϵ) plane have the same meaning.

13.12 Coexistence of pulses

Pulses, fronts and sources can be considered as elementary building blocks which can be combined to form more complicated structures. Stable structures arise as a result of the strong interaction between pulses and fronts. However, such structures require the matching conditions for the amplitude and the wavevector to be fulfilled. This is the reason why they only exist in a relatively narrow range of parameters. Two different stable stationary solutions of the quintic CGLE, for the same set of parameters, can coexist. This result is quite natural if we consider fronts, pulses and sources as elementary building blocks which can be combined to form more complicated structures.

The CP solutions exist in the range of parameters where the front has a small negative velocity. Hence, the CP exists due to repulsion between the source and the front. When the front velocity is small, fronts would move toward each other, but they encounter repulsion from the source which keeps them at a fixed distance. For parameters where the front velocity is large enough, they overcome this repulsion and a plain pulse is formed. For the range of parameters where the front has zero wavevector, the plain pulse and the CP become indistinguishable.

Composite and moving pulses, although they seem exotic, may play a crucial role in the dynamics of some particular systems. We have seen that the exact solution of the quintic CGLE becomes stable in the vicinity of transition from pulses to fronts. At the same time, this is the very range of existence of the CPs and MPs. So, as the system parameters are changed toward the stability region, the appearance of CPs and MPs is possible. For such interesting forms of solution, we have neither an exact solution, nor even some approximate one.

13.13 Interaction of moving and stationary pulses

There are four stable pulse solutions in some regions of the parameter space, viz. a plain pulse, a composite pulse, and left- and right-moving pulses. This gives three possibilities for pair interaction, i.e. interaction of a moving pulse with a plain pulse, with a composite pulse, and with another moving pulse. The result of the collision also depends on the relative phase of the interacting pulses. If a moving pulse collides with a stationary pulse, and the propagation constants are unequal, then the relative phase at the collision point depends on both the initial phase difference and the initial separation, so it is difficult to control the relative phase in numerical simulations. If two moving pulses collide, the relative phase between them is determined only by the initial phase difference, so it can easily be controlled.

Figure 13.23 shows different scenarios of interaction. In the case of the interaction between MPs, the formation of either a CP (for in-phase MPs, see Fig. 13.23a), or a plain pulse, or another MP (for out-of-phase pulses) can be observed. Collisions between stationary CPs and MPs also result in several possibilities. Figure 13.23b shows one of them, namely, absorption of an MP by a CP. The only difference between the CP before and after the collision is a small shift in τ.

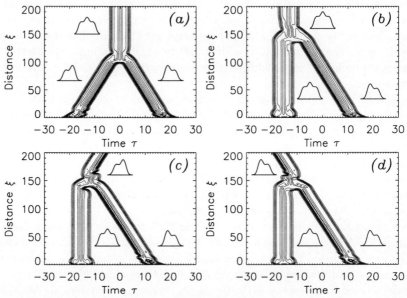

Figure 13.23 *Interaction between moving and composite pulses; here* $\delta = -0.1$, $\beta = 0.5$, $\nu = -0.1$ *and* $\mu = -0.6$. *(a) Interaction between in-phase MPs for* $\epsilon = 1.75$. *(b) Interaction between MP and CP for* $\epsilon = 1.75$. *(c) Interaction between MP and CP for* $\epsilon = 1.73$ *(note that the orientation of the MP reverses after the collision). (d) Interaction between MP and CP for* $\epsilon = 1.73$. *The relative phase is changed by* π *in comparison with (c). Profiles near the contour plots show the type of the pulse (i.e. CP or MP) and orientation of the MP.*

Figure 13.23a,b is plotted for $\epsilon = 1.75$. For smaller ϵ, say $\epsilon \approx 1.73$, the CP still exists, but it is less stable. For this set of parameters, an MP is formed after the collision, and the direction of propagation of this MP depends on the phase difference between the pulses (Fig. 13.23c,d). If the MP changes its direction of motion after collision, as in Fig. 13.23c, it also changes its orientation, because an MP always moves with the pulse leading and the front trailing. Potentially, there are other possible types of interaction, including complete annihilation of interacting pulses, and the tunnelling of one pulse through the other.

13.14 Soliton bound states

Separate plain pulses in the CGLE model can form bound states (Brand and Deissler, 1989; Afanasjev and Akhmediev, 1995) which are also stationary solutions of the CGLE. The pulses in such bound states only overlap weakly. Bound states of two pulses exist for a relatively wide range of parameters. The physical nature of these bound states is still unclear, although a few models have been considered (Malomed, 1991). In this section we briefly describe the results of Afanasjev and Akhmediev (1995, 1996). These simulations demonstrate the existence of in-phase and out-of-phase bound states.

If the conservative effects dominate over the dissipative ones (i.e. δ, β, ϵ, $\mu \ll 1$), the soliton interaction in the CGLE model is qualitatively similar to the soliton interaction in the NLSE (Afanasjev, 1993). This means that in-phase solitons attract each other and out-of-phase solitons interact repulsively, although the strength of the interaction can be severely affected by the presence of spectral filtering.

If the dissipative effects are of the same order as the conservative ones, then bound states may be formed. Figure 13.24 gives examples of bound states of in-phase ($\phi = 0$) and out-of-phase ($\phi = \pi$) solitons which exist in anomalous ($D = +1$) and normal ($D = -1$) dispersion regions. The separation between the pulses in these bound states is of the same order as the pulse width. Note also that the out-of-phase pulses are closer to each other than the in-phase pulses. Bound states of two solitons do exist, but, as numerical simulations show, all of them are unstable.

13.15 Soliton interactions

The best way to study the possibility of the formation of stable bound states is to study numerically the dynamics of the interaction between two solitons. If we choose an initial condition in the form of two solitons,

$$\psi_0(\tau) = F(\tau - \tau_0/2) + F(\tau + \tau_0/2) \exp(i\phi_0) \qquad (13.69)$$

where $F(\tau)$ is the soliton shape found from numerical simulations, τ_0 is the separation between the solitons and ϕ_0 is the initial phase difference, then we can observe the behaviour of the two-soliton solution in the (τ_0, ϕ_0) plane. If the bound state is stable, then these two parameters will converge to a stationary point corresponding to a bound state. We suppose that the solitons are well separated and that their shapes hardly change.

The amplitudes of two interacting solitons of the CGLE, to a first ap-

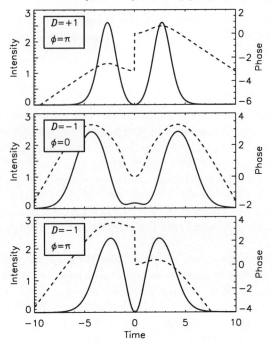

Figure 13.24 *Bound state intensity (solid curves) and phase (dashed curves) profiles*

proximation, are fixed for a given set of parameters. This means that the two main parameters may change when the pulses interact: the relative distance and relative phase between the solitons. Hence, a convenient representation of the dynamics of interaction is a plot in polar coordinates ($\tau \cos \phi$, $\tau \sin \phi$) where τ and ϕ are the separation and phase difference between the solitons at arbitrary ξ. Some trajectories of solutions in this plane, found from numerical simulations (Afanasjev and Akhmediev, 1996), are shown in Fig. 13.25. Points on the horizontal axis correspond to symmetric and anti-symmetric bound states. Trajectories around them behave like those around a saddle point – this shows that both of them are unstable. Although all the bound states considered are unstable, they clearly play a pivotal role in the overall dynamics of two-soliton interaction. If we choose suitable initial conditions, they can be observed propagating together for long distances. The existence of separatrices in Fig. 13.25 shows that the two solitons can remain at a fixed distance from each other, unless the phase difference between them evolves from π to 0. Moreover, if the initial phase difference between the solitons is exactly π, the out-of-phase bound state serves as an attractor.

In the case of normal dispersion, both in-phase and out-of-phase bound

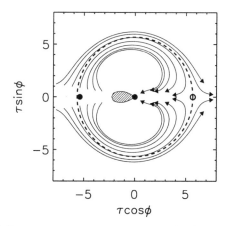

Figure 13.25 *Soliton interaction trajectories in the* $(\tau \sin \phi, \tau \cos \phi)$ *plane for* $\beta = 1$ *(solid curves). The arrows show the direction of motion. The closed circle shows the centre of the frame, and bound state of out-of-phase solitons. The open circle shows the threshold between attraction and repulsion for in-phase solitons. The dashed curve is a 'separatrix'. The shaded area is a 'forbidden' zone.*

states of solitons also exist. An example of a trajectory in the (τ, ϕ) plane is shown in Fig. 13.26. Other trajectories are similar to this one, or mirror images of it relative to the horizontal axis. The behaviour of the trajectories is similar to the previous case. Both points corresponding to bound states repel the trajectories, so that both of them are unstable.

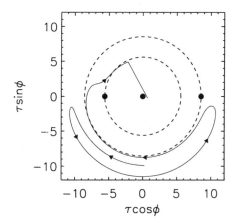

Figure 13.26 *Soliton interaction trajectories in the* $(\tau \sin \phi, \tau \cos \phi)$ *plane for the normal dispersion case. The closed circles show the centre of the frame, the bound state of out-of-phase solitons and the bound state of in-phase solitons.*

Another difference is that the trajectories move parallel to the separatrix in opposite directions relative to the previous case. One more qualitative difference is that there is a point in the plane (an unstable focus) which can correspond to the bound state of two solitons in quadrature. The dynamics of the soliton interaction close to this bound state is shown in Fig. 13.27. This bound state is, again, unstable. The behaviour is qualitatively similar for other values of the parameters of the equation. The general conclusion is that, even if two-soliton bound states do exist in the model of the CGLE, they are always unstable.

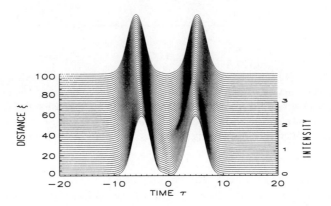

Figure 13.27 *Dynamics of a pair of solitons with phase difference of $\pi/2$ between them.*

Note that bound states of two solitons may exist in conservative problems as well (see Chapter 10). They are also unstable to variations of the relative phase (in addition to being unstable to variations of the relative amplitude). As far as we know, no stable bound states of two bright solitons in one-dimensional homogenous media have been found so far. The formation and stability of bound states of two solitons merits further investigation.

13.16 Concluding remarks

The cubic CGLE is an important model for describing optical transmission systems with guiding filters. The use of nonlinear gain ($\epsilon > 0$) in these systems allows the reduction or suppression of the growth of linear radiation and the use of stronger spectral filtering ($\beta \sim 1$). In this case, (13.24) gives the instability threshold. By removing the linear gain from the system, stable propagation of both the soliton and the background can be achieved.

The actual range of parameters where the special solutions exist is of great importance, and many experimental and numerical observations have been performed in this range. This can be illustrated using arbitrary-amplitude pulses as an example. Indeed, for the systems described by the cubic CGLE (e.g. soliton transmission systems), the main limitation is due to the growth of linear radiation (due to the instability of the background state). Emission of spontaneous noise by amplifiers also contributes to this effect. Thus, it is desirable to reduce this instability, and, in order to do this, the excess linear gain δ should be kept as low as possible. At the same time, soliton collapse (which occurs if $\epsilon > \epsilon_S$), also should be avoided. Hence, the optimal regime lies near the curve S, where the nonlinear gain and the spectral filtering balance each other, so a significant contribution from the linear gain is not necessary.

Another example of a soliton fiber system which has its working regime near S is the soliton fiber laser with saturable absorption (Haus *et al.*, 1991; Chen *et al.*, 1994). Such a laser is described by the cubic CGLE, but the difference from (13.1) is that the linear amplification coefficient, δ, depends on the total pulse energy, E. In the mode-locked regime, δ has a small negative value to stabilize the background state. If the pulse energy increases, the absorption also increases, and vice versa. Clearly, the stability of such a system depends on the slope of the $\delta(E)$ curve. However, to provide self-starting of the laser, δ should have a small positive value when E is much smaller than the energy of the stationary pulse. So, again, the optimum regime lies near S, where the nonlinear gain and spectral filtering partly cancel each other, and the absolute value of δ can be kept small.

One of the systems which is described by the quintic CGLE is a soliton fiber laser with fast saturable absorption. In this case, the soliton is supported by nonlinear gain, but loses energy due to three effects: spectral filtering, linear losses and the quintic stabilizing term. However, even a small linear loss is enough to keep the background state stable. Hence, the stationary state exists basically as a result of a balance between the nonlinear gain, spectral filtering, and the quintic stabilizing term, thus proving the importance of the study of arbitrary-amplitude pulses in the quintic model.

Also, we would like to outline some possible future work in this direction. The most important problem involving generalization is the exact solution of the quintic Ginzburg–Landau equation. This would necessitate the use of an ansatz more general than (13.6) and (13.7). Further work should be stimulated by the existence of stable solutions, for which no analytical expression is known.

The natural generalization of the Ginzburg–Landau equation is the vector Ginzburg–Landau equation. It can describe waves in birefringent media with gain and loss. Such an equation has already been studied in the context of fluid dynamics (Deissler and Brand, 1990), and now it appears to be an important model for soliton communication lines with polarization multiplexing (Wabnitz, 1995) and soliton fiber lasers (Haus *et al.*, 1995).

Recently, the effect of soliton polarization self-orientation has been discovered in this model (Afanasjev, 1995a).

References

Abdullaev F. Kh., Abrarov R. M. and Darmanyan S. A. (1989) Dynamics of solitons in coupled optical fibers, *Opt. Lett.*, **14**, No 2, 131–133.

Abdullaev F., Darmanyan S. and Khabibullaev P. (1993) *Optical Solitons*, Springer–Verlag, Berlin, Heidelberg.

Abdulloev K. O., Bogolubsky I. L. and Makhankov V. G. (1976) One more example of inelastic soliton interaction, *Phys. Lett. A*, **56**, 427–428.

Ablowitz M. J. and Clarkson P. A. (1991) *Solitons, Nonlinear Evolution Equations and Inverse Scattering*, London Mathematical Society Lecture Notes Series **149**, Cambridge University Press, Cambridge.

Ablowitz M. J., Kaup D. J., Newell A. C. and Segur H. (1974) The inverse scattering transform–Fourier analysis for nonlinear problems. *Stud. Appl. Math.*, **36**, 249–315.

Abramowitz M. and Stegun I. A. (1964) *Handbook of Mathematical Functions with Formulas, Graphs, and Mathematical Tables*, Dover, New York.

Aceves A. B. and Wabnitz S. (1989) Self-induced transparency solitons in nonlinear periodic media, *Phys. Lett. A*, **141**, No 1/2, 37.

Aceves A. B., De Angelis C. and Rubenchik A. M. (1994) Multidimensional solitons in fiber arrays, *Opt. Lett.*, **19**, No 5, 329.

Aceves A. B., Moloney J. V. and Newell A. C. (1989) Theory of light beam propagation at nonlinear interfaces. I. Equivalent–particle theory for a single interface, *Phys. Rev. A*, **39**, 1809–1827.

Afanasjev V. V. (1993) Interpretation of the effect of reduction of soliton interaction by bandwidth–limited amplification, *Opt. Lett.*, **18**, No 10, 790–792.

Afanasjev V. V. (1995a) Soliton polarization self–orientation, *Opt. Lett.*, **20**, 1967–1969.

Afanasjev V. V. (1995b) Soliton singularity in the system with nonlinear gain, *Opt. Lett.*, **20**, No 7, 704–706.

Afanasjev V. V. (1995c) Soliton polarization rotation in fiber lasers *Opt. Lett.*, **20**, No 3, 270–272; see also *Nonlinear Guided Waves and Their Applications*, **6**, OSA Technical digest series, Optical Society of America, Washington DC, 73–75.

Afanasjev V. V. and Akhmediev N. N. (1995) Soliton interaction and bound states in amplified–damped fiber systems, *Opt. Lett.*, **20**, No 19, 1970–1972.

Afanasjev V. V. and Akhmediev N. N. (1996) Soliton interaction in nonequilibrium dynamical systems, *Phys. Rev. E*, **53**, No 6, 6471.

Agranovich V. M. and Ginzburg V. L. (1966) *Spatial Dispersion in Crystal Optics and the Theory of Excitons*, Interscience, Innsbruck.

Agranovich V. M., Babichenko V. S., Chernyak V. Ya. (1980) Nonlinear surface polaritons, *JETP Lett.*, **32**, 512–515. Original (in Russian) *Pis'ma Zh. Eksp.*

Teor. Fiz., **32**, No 8, 532–535 (1980).

Agrawal G. P. (1991) Optical pulse propagation in doped fiber amplifier, *Phys. Rev. A* **44**, 7493–7501.

Agrawal G. P. (1993) Effect of two–photon absorption on the amplification of ultrashort optical pulses, *Phys. Rev. E* **48**, 2316–2318.

Agrawal G. P. (1995) *Nonlinear Fiber Optics*, 2nd edn, Academic Press Inc., San Diego, CA.

Ainslie B. J. and Day C. R. (1986) A review of single–mode fibers with modified dispersion characteristics, *J. Lightwave Technol.*, **LT–4**, No 8, 967–979.

Aitchison J. S., Weiner A. M., Silberberg Y., Leaird D. E., Oliver M. K., Jackel J. L. and Smith W. E. (1991) Experimental observation of spatial soliton interactions, *Opt. Lett.*, **16**, No 1, 15–17.

Akhmanov C. A. and Khokhlov R. V. (1964) *Problems of Nonlinear Optics: Electromagnetic Waves in Nonlinear Dispersive Media*, VINITI, Moscow.

Akhmediev N. N. (1982) Novel class of nonlinear surface waves: asymmetric modes in a symmetric layered structure, *Sov. Phys. JETP*, **56**, 299–303. Original (in Russian): *Zh. Eksp. Teor. Fiz.*, **83**, No 8, 545–553 (1982).

Akhmediev N. N. (1991) The problem of stability and excitation of nonlinear surface waves, in: *Modern Problems in Condensed Matter Sciences* Vol.XXIX, ed H.–E. Ponath and G. I. Stegeman, Chapter 3, 289–321, North–Holland, Amsterdam.

Akhmediev N. N. and Afanasjev V. V. (1995) Novel arbitrary–amplitude soliton solutions of the cubic–quintic complex Ginzburg–Landau equation, *Phys. Rev. Lett.*, **75**, 2320–2323.

Akhmediev N. and Ankiewicz A. (1993a) Novel soliton states and bifurcation phenomena in nonlinear fibre couplers, *Phys. Rev. Lett.*, **70**, No 16, 2395–2398.

Akhmediev N. and Ankiewicz A. (1993b) Spatial soliton X–junctions and couplers, *Opt. Commun.*, **100**, 186–192.

Akhmediev N. and Ankiewicz A. (1993c) First order exact solutions of the NLSE in the normal dispersion regime, *Phys. Rev. A*, **47**, 3213–3221.

Akhmediev N. and Ankiewicz A. (1993d) Theory of amplification of nonlinear guided waves, *Phys. Rev. A*, **47**, 2196–2204.

Akhmediev N. N. and Buryak A. V. (1994) Soliton states and bifurcation phenomena in three–core nonlinear fiber couplers, *JOSA B*, **11**, No 5, 804–809.

Akhmediev N. N. and Buryak A. V. (1995a) Soliton states and bifurcations in periodically twisted birefringent fiber filters, *Opt. Commun.*, **113**, 505–508.

Akhmediev N. N. and Buryak A. V. (1995b) Stationary pulse propagation in N–core nonlinear fiber arrays, *IEEE J. Quant. Electron.*, **QE–31**, No 4, 682–688.

Akhmediev N. N. and Karlsson M. (1995) Cherenkov radiation emitted by solitons in optical fibers, *Phys. Rev. A*, **51**, No 3, 2602–2607.

Akhmediev N. N., Korneev V. I. (1986) Modulation instability and periodic solutions of the nonlinear Schrödinger equation, *Theor. Math. Phys.*, **69**, 1089. Original (in Russian): *Teor. Mat. Fiz. (USSR)*, **69**, 189.

Akhmediev N.N. and Mitskevich N.V. (1991) Extremely high degree of N-soliton pulse compression in an optical fiber. *IEEE J. Quant. Electron.*, **QE-27**, No 3, 849–857.

Akhmediev N. N. and Ostrovskaya N. V. (1988) Stability of nonlinear waves in a symmetric planar waveguide, *Sov. Phys. Tech. Phys.*, **33**, 1333–1337.

Akhmediev N. N. and Soto–Crespo J. M. (1993) Generation of a train of three–dimensional optical solitons in a self–focussing medium, *Phys. Rev. A*, **47**, 1358–1364.

Akhmediev N. N. and Soto–Crespo J. M. (1994a) Propagation dynamics of ultrashort pulses in nonlinear fiber couplers, *Phys. Rev. E*, **49**, 4519–4529.

Akhmediev N. N. and Soto - Crespo J. M. (1994b) Dynamics of solitonlike pulse propagation in birefringent optical fibers, *Phys. Rev. E*, **49**, No 6, 5742–5754.

Akhmediev N. N. and Wabnitz S. (1992) Phase detecting of solitons by mixing with a continuous–wave background in an optical fiber, *JOSA B*, **9**, No 2, 236–242.

Akhmediev N. N., Eleonskii V. M. and Kulagin N. E. (1985a) Generation of periodic trains of picosecond pulses in an optical fiber: Exact solutions, *Sov. Phys. JETP*, **62**, No 5, 894–899.

Akhmediev N. N., Korneev V. I. and Kuz'menko Yu. V. (1985b) Excitation of nonlinear surface waves by Gaussian light beams, *Sov. Phys. JETP*, **61**, 62–67 (1985). Original (in Russian): *Zh. Eksp. Teor. Fiz.*, **88**, 107.

Akhmediev N. N., Eleonskii V. M. and Kulagin N. E. (1987) Exact first–order solutions of the nonlinear Schrödinger equation, *Theor. Math. Phys.*, **72**, 183–196 (1987). Original (in Russian): *Teor. Mat. Fiz. (USSR)* **72**, 809–818.

Akhmediev N. N., Eleonskii V. M., Kulagin N. E. and Shil'nikov L. P. (1989a) Steady state pulses in a birefringent nonlinear optical fiber: soliton multiplication processes, *Sov. Tech. Phys. Lett.*, **15**, No 8, 587–588. Original (in Russian): *Pis'ma Zh. Tekh. Fiz.* **15**, 19–23 (1989).

Akhmediev N. N., Nabiev R. F. and Popov Yu. M. (1989b) Three–dimensional modes of a symmetric nonlinear plane waveguide, *Opt. Commun.*, **69**, Nos 3/4, 247–252.

Akhmediev N. N., Heatley D. R., Stegeman G. I. and Wright E. M. (1990) Pseudorecurrence in two–dimensional modulation instability with a saturable self–focusing nonlinearity, *Phys. Rev. Lett.*, **65**, No 12, 1423–1426.

Akhmediev N. N., Eleonskii V. M. and Kulagin N. E. (1991) An example of a dynamical system connected with the nonlinear Schrödinger equation, *Selecta Mathematica Sovietica*, **10**, No 1, 55–59.

Akhmediev N. N., Korneev V. I. and Nabiev R. F. (1992) Modulation instability of the ground state of nonlinear wave equation. Optical machine–gun, *Opt. Lett.*, **17**, No 6, 393–395.

Akhmediev N. N., Ankiewicz A. and Soto-Crespo J. M. (1993) Does the nonlinear Schrödinger equation describe propagation in nonlinear waveguides? *Opt. Lett.*, **18**, 411–413.

Akhmediev N. N., Buryak A. V. and Karlsson M. (1994a) Radiationless optical solitons with oscillating tails, *Opt. Comm.*, **110**, 540–544.

Akhmediev N. N., Buryak A. V. and Soto - Crespo J. M. (1994b) Elliptically polarized solitons in birefringent optical fibers, *Optics Comm.*, **112**, 278–282.

Akhmediev N. N., Buryak A. V., Soto-Crespo J. M. and Andersen D. R. (1995) Phase-locked stationary soliton states in birefringent nonlinear optical fibers, *JOSA B*, **12**, No. 3, 434–439.

Akhmediev N. N., Afanasjev V. V. and Soto - Crespo J. M. (1996) Singularities and special soliton solutions of the cubic–quintic complex Ginzburg–Landau equation, *Phys. Rev. E*, **53**, No 1, 1190.

Allen L. and Eberly J. H. (1974) *Optical Resonance and Two Level Atoms*, Wiley, New York.

Anderson D. (1983) Variational approach to nonlinear pulse propagation in optical fibers, *Phys. Rev. A*, **27**, No 6, 3135–3145.

Anderson R. L. and Ibragimov N. H. (1979), *Lie–Bäcklund Transformations in Applications*, SIAM Studies in Applied Mathematics, Society for Industrial and Applied Mathematics, Philadelphia.

Ankiewicz A. (1988a) Novel effects in non–linear coupling, *Opt. Quant. Electron.*, **20**, 329–337.

Ankiewicz A. (1988b) Interaction coefficient in detuned nonlinear couplers, *Opt. Commun.*, **66**, 311–314.

Ankiewicz A. and Akhmediev N. N. (1996) Analysis of bifurcations for parabolic nonlinearity optical couplers, *Opt. Commun.*, **124**, 95–102.

Ankiewicz A. and Peng G. D. (1991a) Soliton interaction in optical fibre couplers, *Int. J. Optoelectronics*, **6**, 15–22.

Ankiewicz A. and Peng G. D. (1991b) Accurate variational method for nonlinear fibre devices, *Opt. Commun.*, **84**, 71–75.

Ankiewicz A. and Peng G. D. (1993) Generalized Gaussian analysis of fibres with non–Kerr law nonlinearities, *Opt. Quant. Electron.*, **25**, 147–155.

Ankiewicz A. and Tran H. T. (1991) A new class of nonlinear guided waves, *J. Mod. Opt.*, **38**, 1093–1106.

Ankiewicz A., Akhmediev N., Peng G. D. and Chu P. L. (1993) Limitations of the variational approach for soliton propagation in nonlinear couplers, *Opt. Commun.*, **103**, 410–416.

Ankiewicz A., Karlsson M. and Akhmediev N. (1994) Dark soliton pairs in fiber couplers, *Opt. Commun.*, **111**, 116–122.

Ankiewicz A., Akhmediev N. N., and Soto–Crespo J. M. (1995a) Novel bifurcation phenomena for solitons in nonlinear saturable couplers, *Opt. Commun.*, **116**, 411–415.

Ankiewicz A., Akhmediev N. N. and Peng G. D. (1995b) Stationary soliton states in couplers with saturable nonlinearity, *Opt. Quant. Electron.*, **27**, 193–200.

Askaryan G. A. (1962) Cherenkov radiation and transition radiation from electromagnetic waves, *Sov. Phys. JETP*, **15**, No 5, 943–946. Original (in Russian): *Zh. Teor. Eksp. Fiz.*, **42**, 1360–1364.

Atkinson D., Loh W., Afanasjev V. V., Grudinin A. B., Seeds A. J. and Payne D. N. (1994) Increased amplifier spacing in a soliton system with quantum–well saturable absorbers and spectral filtering, *Opt. Lett.* **19,** 1514–1516.

Bélanger P. A., Gagnon L. and Paré C. (1989) Solitary pulses in an amplified nonlinear dispersive medium, *Opt. Lett.*, **14**, No 17, 943–945.

Bélanger P. A. and Paré C. (1990) Soliton switching and energy coupling in two–mode fibers: analytical results, *Phys. Rev. A*, **41**, 5254–5256.

Bélanger N. and Bélanger P. A. (1996) Bright solitons on a c.w. background, *Opt. Commun.*,**124**, 301–308.

Bendow B., Gianino P., Tzoar N. and Jain M. (1980) Theory of nonlinear pulse propagation in optical waveguides, *JOSA B*, **70**, No 5, 539–546.

Benilov E. S., Grimshaw R. and Kuznetsova E. P. (1993) The generation of radiating waves in a singularly–perturbed Korteveg–de Vries equation, *Physica D*, **69**, 270.

Benjamin T. J. and Feir J. E. (1967) The disintegration of wave trains on deep water, *J. Fluid. Mech.*, **27**, 417–430.

Bennion I., Goodwin M. J. and Stewart W. J. (1985) Experimental nonlinear optical waveguide device, *Electron. Lett.*, **21**, No 1, 41–42.

Berkhoer A. L. and Zakharov V. E. (1970) Self excitation of waves with different polarizations in nonlinear media, *Sov. Phys. JETP*, **31**, No 3, 486–490. Original (in Russian): *Zh. Eksp. Teor. Fiz.*, **58**, 903–911 (1970).

Bernstein L. J., DeLong K. W. and Finlayson N.,(1993) Self–trapping transitions in a discrete NLS model with localized initial conditions, *Phys. Lett. A*, **181**, 135–141.

Bespalov V. I. and Talanov V. I. (1966) On the filament structure of a light beam in nonlinear liquids, *Sov. Phys. JETP Lett.* Original (in Russian): *Pis'ma Zh. Exp. Teor. Phys.*, **3**, 471.

Black R. J. and Ankiewicz A. (1985) Fiber optic analogies with mechanics, *Am. J. Phys.*, **53**, 554–563.

Blagoeva A. B., Dinev S. G., Dreischuh A. A. and Naidenov A. (1991) Light bullets formation in a bulk media, *IEEE J. Quant. Electron*, **QE–27**, No 8, 2060–2065.

Blair S., Wagner K. and McLeod R. (1994) Asymmetric spatial soliton dragging, *Opt. Lett.*, **19**, No 23, 1943–1945.

Bloembergen N. (1965) *Nonlinear Optics*, Benjamin, New York.

Blow K. J. and Doran N. J. (1985) Multiple dark soliton solutions of the nonlinear Schrödinger equation, *Phys. Lett. A*, **107**, No 2, 55–58.

Blow K. J. and Wood D. (1989) Theoretical description of transient stimulated Raman scattering in optical fibers, *IEEE J. Quant. Electron.*, **QE–25**, 2665–2673.

Blow K. J., Doran N. J. and Phoenix S. J. D. (1992) The soliton phase, *Opt. Commun.*, **88**, No 2/3, 137.

Blow K. J., Doran N. J. and Wood D. (1987) Polarization instabilities for solitons in birefringent fibers, *Opt. Lett.*, **12**, No 3, 202–204.

Boardman A. D. and Egan P. (1986) Optically nonlinear waves in thin films, *IEEE J. Quant. Electron.*, **QE–22**, 319–324.

Bolotovskii B. M. and Ginzburg V. L. (1972) The Vavilov–Cherenkov effect and the Doppler effect in the motion of sources with superluminal velocity in a vacuum, *Sov. Phys. Uspekhi*, **15**, No 2, 184–192.

Bona J. L., Pritchard W. G. and Scott L. R. (1980) Solitary-wave interaction, *Phys. Fluids.*, **23**, No 6, 438–441.

Born M. and Wolf E. (1980) *Principles of Optics*, 6th edn, Pergamon Press, New York, Section 1.4.

Boyer G. R. and Carlotti X. F. (1986) Nonlinear propagation in a single–mode optical fiber in case of small group velocity dispersion, *Opt. Comm.*, **60**, No 1/2, 18–22.

Brand H. R. and Deissler R. J. (1989) Interaction of localized solutions for subcritical bifurcations, *Phys. Rev. Lett.* **63**, No 26, 2801–2804.

Buckingham A.D. and Orr B. J. (1967) *Rev. Chem. Soc.*, **21**, 195.

Bullough, R.K. and Caudrey, P. J. (eds), (1980) *Solitons*, Springer–Verlag, Berlin.

Buryak A. V. and Akhmediev N. N. (1994a) Influence of radiation on soliton dynamics in nonlinear fibre couplers, *Opt. Commun.*, **110**, 287–292.

Buryak A. V. and Akhmediev N. N. (1994b) Internal friction between solitons in near–integrable systems, *Phys. Rev. E.*, **50**, No 4, 3126–3133.

Buryak A. V. and Akhmediev N. N. (1995) Stability criterion for stationary bound states of solitons with radiationless oscillating tails, *Phys. Rev. E*, **51**, No 4, 3572–3578.

Butcher P. N. (1965) Nonlinear optical phenomena, Bulletin 200, Engineering Experiment Station, Ohio State University, Columbus.

Cai D., Bishop A. R., Gronbech - Jensen N. and Malomed B. (1994) Bound solitons in the a.c.–driven damped nonlinear Schrödinger equation, *Phys. Rev. E* **49**, No 2, 1677–1679.

Cao X. D. and Meyerhofer D. D. (1994) Soliton collisions in optical birefringent fibers, *JOSA B*, **11**, No 2, 380–385.

Cao X. D. and Meyerhofer D. D. (1995) Optimization of pulse shaping using nonlinear polarization rotation, *Opt. Commun.*, **120**, 316–320.

Cavalcanti S. V., Cressoni J. C., da Cruz H. R. and Gouveia–Neto A. S. (1991) Modulation instability in the region of minimum group–velocity dispersion of single–mode optical fibers via an extended nonlinear Schrödinger equation, *Phys. Rev. A*, **43**, No 11, 6162–6165.

Champneys A. R. and Toland J. F., (1993) Bifurcation of a plethora of multi–modal homoclinic orbits for autonomous Hamiltonian systems, *Nonlinearity*, **6**, 665–721.

Chen C.-J., Wai P. K. A. and Menyuk C. R. (1994) Stability of passively mode–locked fiber lasers with fast saturable absorption, *Opt. Lett.*, **19**, No 3, 198–200.

Chen C.-J., Wai P. K. A., and Menyuk C. R. (1995) Self–starting of passively mode–locked lasers with fast saturable absorbers, *Opt. Lett.*, **20**, No 4, 350–352.

Chen W. and Mills D. L. (1987) Gap solitons and the nonlinear optical response of superlattices, *Phys. Rev. Lett.*, **58**, No 2, 160–163.

Chen Y., Snyder A. W. and Mitchell D. J. (1990a) Power flow in triple core couplers, *Electron. Lett.*, **26**, No 1, 76.

Chen Y., Snyder A. W. and Mitchell D. J. (1990b) Ideal optical switching by nonlinear multiple (parasitic) core couplers, *Electron. Lett.*, **26**, No 1, 77 – 78.

Chi S. and Guo Q. (1995) Vector theory of self–focusing of an optical beam in Kerr media. *Opt. Lett.*, **20**, No 15, 1598–1600.

Chiao R. Y., Garmire E. and Townes C. H. (1964) Self–trapping of optical beams, *Phys. Rev. Lett.*, **13**, No 15, 479–482.

Christodoulides D. N. (1988) Black and white vector solitons in weakly–birefringent optical fibers, *Phys. Lett. A*, **132**, 451–452.

Christodoulides D. N. and Joseph R. I. (1984) Exact radial dependence of the field in a nonlinear dispersive dielectric fiber: bright pulse solutions, *Opt. Lett.*, **9**, No 6, 229–231.

Christodoulides D. N. and Joseph R. I. (1988a) Vector solitons in birefringent nonlinear dispersive media, *Opt. Lett.*, **13**, No 1, 53–55.

Christodoulides D. N. and Joseph R. I. (1988b) Discrete self–focusing in nonlinear arrays of coupled waveguides, *Opt. Lett.*, **13**, No 9, 794–795.

Christodoulides D. N. and Joseph R. I. (1989) Slow Bragg solitons in nonlinear periodic structures, *Phys. Rev. Lett.*, **62**, No 15, 1746.

Christov I. P., Murnane M. M., Kapteyn H. C., Zhou J. and Huang C. - P. (1994) Fourth–order dispersion–limited solitary pulses, *Opt. Lett.*, **19**, No 18, 1465–1467.

Chu P. L. and Desem C. (1985) Effect of third order dispersion of optical fibre on soliton interaction, *Electron. Lett.*, **21**, 228–229.

Cohen L. G. and Mammel W. L. (1982) Low–loss quadruple–clad single–mode lightguides with dispersion below 2 ps/km nm over the 1.28 μm–1.65 μm wavelength range, *Electron. Lett.*, **18**, No 24, 1023–1024.

Conte R. and Musette M. (1993) Linearity inside nonlinearity: exact solutions to the complex Ginzburg–Landau equation, *Physica D* **69**, 1–17.

Cook P. A. (1986) *Nonlinear Dynamical Systems*, Prentice–Hall International.

Cowan S., Enns R. H., Rangnekar S. S. and Sunghera S. S. (1986) Quasi–soliton and other behaviour of the nonlinear cubic–quintic Schrödinger equation, *Can. J. Phys.*, **64**, 311–315.

Daino B., Gregori G. and Wabnitz S. (1985) Stability analysis of nonlinear coherent coupling, *J. Appl. Phys.*, **58**, 4512–4514.

Dauxois T. and Peyrard M. (1993) Energy localization in nonlinear lattices, *Phys. Rev. Lett.*, **70**, No 25, 3935–3938.

David D., Holm D. D. and Tratnik M. V. (1990) Hamiltonian chaos in nonlinear optical polarization dynamics, *Phys. Reports*, **187**, No 6, 281–370.

De La Fuente R., Barthelemy A. and Froehly C. (1991) Spatial soliton–induced guided waves in a homogeneous nonlinear Kerr medium, *Opt. Lett.*, **16**, No 11, 793–795.

De Sterke C. M. and Sipe J. E. (1991) Polarization instability in a waveguide geometry, *Opt. Lett.*, **16**, 202–204.

De Sterke C. M. and Sipe J. E. (1994) Gap solitons, in *Progress in Optics*, **33**, 205.

Deissler R. J. and Brand H. R. (1990) The effect of nonlinear gradient terms on localized states near a weakly inverted bifurcation, *Phys. Lett. A*, **146**, 252–255.

Deissler R. J. and Brand H. R. (1991) Interaction of two-dimensional localized solutions near a weakly inverted bifurcation, *Phys. Rev. A*, **44**, No 6, R3411–R3414.

Deissler R. J. and Brand H. R. (1994) Periodic, quasiperiodic and chaotic localized solitons of the quintic complex Ginzburg–Landau equation, *Phys. Rev. Lett.*, **72**, No 4, 478–481.

Deissler R. J. and Brand H. R. (1995) Interactions of breathing localized solutions for subcritical bifurcations, *Phys. Rev. Lett.*, **74**, No 24, 4847–4850.

Desem C. and Chu P. L. (1987) Reducing soliton interaction in single–mode optical fibres, *IEE Proc.*, **134**, Pt. J, No 3, 145–151.

Dodd R. K., Eilbeck J. C., Gibbon J. D. and Morris H. C. (1984) *Solitons and Nonlinear Wave Equations*, Academic Press, London.

Doerr C. R., Haus H. A. and Ippen E. P. (1994a) Additive–pulse limiting, *Opt. Lett.*, **19**, 31–33.

Doerr C. R., Haus H. A. and Ippen E. P. (1994b) Asynchronous soliton mode locking, *Opt. Lett.*, **19**, No 23, 1958–1960.

Doran N. J. and Blow K. J. (1983) Solitons in optical communications, *IEEE J. Quant. Electron.*, **QE–19**, 1883–1888.

Doran N. J. and Wood D. (1987) Soliton processing element for all–optical switching and logic, *JOSA B*, **11**, 1843–1846.

Dowling R. J. (1990) Stability of solitary waves in a nonlinear birefringent optical fiber, *Phys. Rev. A*, **42**, No 9, 5553–5560.

Drazin P. G. and Johnson R. S. (1989) *Solitons: An Introduction*, Cambridge University Press, New York.

Duling I. N. (1991) Subpicosecond all–fibre erbium laser, *Electron. Lett.*, **27**, No. 6, 544–545.

D'ychenko A. I., Zakharov V. E., Pushkarev A. N., Shvetz V. F. and Yan'kov V. V. (1989) Soliton turbulence in nonintegrable wave systems, *Sov. Phys. JETP*, **69**, No 6, 1144–1147. Original (in Russian): *Zh. Eksp. Teor. Fiz.*, **96**, 2026–2031.

Edmundson D. E. and Enns R. H. (1992) Robust bistable light bullets, *Opt. Lett.*, **17**, No 8, 586–588.

Eguchi M., Hayata K. and Koshiba M. (1991) Effect of birefringence on the interaction between adjacent nonlinear pulses, *Opt. Lett.*, **16**, 82.

Eleonskii V. M., Oganes'yants L. G. and Silin V. P. (1972a) Cylindrical nonlinear waveguides, *Sov. Phys. JETP*, **35**, No 1, 44–47. Original (in Russian): *Zh. Eksp. Teor. Fiz.* **62**, 81–88.

Eleonskii V. M., Oganes'yants L. G. and Silin V. P. (1972b) Structure of three–component vector fields in self–focusing waveguides, *Sov. Phys. JETP*, **36**, No 2, 282–285. Original (in Russian): *Zh. Eksp. Teor. Fiz.* **63**, 532–539.

Eleonskii V. M., Korolev V. G., Kulagin N. E. and Shil'nikov L. P. (1991) Branching bifurcations of vector envelope solitons and integrability, *Sov. Phys. JETP*, **72**, No 4, 619–623. Original (in Russian): *Zh. Eksp. Teor. Fiz.* **99**, 1113–1120.

Elgin J. N. (1992) Soliton propagation in an optical fiber with third- order dispersion, *Opt. Lett.*, **17**, No 20, 1409–1411,

Elgin J. N. (1993) Perturbations of optical solitons, *Phys. Rev. A*, **47**, No 5, 4331–4341.

Emplit P., Hamaide J. P., Reynaud F., Froehly C. and Barthelemy A. (1987) Picosecond steps and dark pulses through nonlinear single–mode fibers, *Opt. Commun.*, **62**, 374–379.

Evangelides S. G., Mollenauer L. F., Gordon J. P. and Bergano N. S. (1992) polarization multiplexing with solitons, *J. Lightwave Technol.*, **10**, No 1, 28–35.

Faddeev L. D. and Takhtadjan L. A. (1987) *Hamiltonian Methods in the Theory of Solitons*, Springer–Verlag, Berlin.

Fauve S. and Thual O. (1990) Solitary waves generated by subcritical instabilities in dissipative systems, *Phys. Rev. Lett.*, **64**, 282–285.

Feit M. D. and Fleck J. A. (1988) Beam nonparaxiality, filament formation, and beam breakup in the self–focusing of optical beams, *JOSA B*, **5**, No 3, 633–640.

Fibich G. (1996) Small beam nonparaxiality arrests self-focusing of optical beams, *Phys. Rev. Lett.*, **56**, 2276.

Finlayson N. and Stegeman G. I. (1990) Spatial switching, instabilities and chaos in a three-waveguide nonlinear directional coupler, *Appl. Phys. Lett.*, **56**, No 23, 2276.

Forest G. M. and Lee J. E. (1986) Geometry and modulation theory for the periodic nonlinear Schrödinger equation, in: *Oscillation Theory, Computation, and Methods of Compensated Compactness*, ed. C. Dafermos, J. L. Ericksen, D. Kinderlehrer, and M. Slemrod, Lecture Notes from the IMA, Springer–Verlag, New York.

Friberg S. R., Silberberg Y., Oliver M. K., Andrejco M. J., Saifi M. A. and Smith P. S. (1987) Ultrafast all–optical switching in a dual–core fiber nonlinear coupler, *Appl. Phys. Lett.*, **51**, 1135–1137.

Friberg S. R., Weiner A. M., Silberberg Y., Sfez B. G., and Smith P. S. (1988) Femtosecond switching in a dual–core–fiber nonlinear coupler, *Opt. Lett.*, **13**, 904–906.

Gagnon L. (1989) Exact travelling–wave solutions for optical models based on the nonlinear cubic–quintic Schrödinger equation, *JOSA B*, **6**, No 9, 1477–1483.

Gagnon L. (1990) Exact solutions for optical wave propagation including transverse effects, *JOSA B*, **7**, No 6, 1098–1102.

Gagnon L. (1993) Solitons on a continuous–wave background and collision between two dark pulses: some analytical results. *JOSA B*, **10**, No 3, 469–474.

Gagnon L. and Bélanger P. A. (1990) Soliton self–frequency shift versus Galilean–like symmetry, *Opt. Lett.*, **15**, No 9, 466–468.

Gagnon L. and Stiévenart N. (1994) N-soliton interaction in optical fibers: the multiple-pole case, *Opt. Lett.*, **19**, No 9, 619–621.

Gagnon L. and Winternitz P. (1989) Exact solutions of the cubic and quintic nonlinear Schrödinger equation for a cylindrical geometry, *Phys. Rev. A*, **39**, 297–306.

Gardner C. S., Greene J. M., Kruskal M. D., Miura K. M. (1967) Method for solving the Korteweg–de Vries equation, *Phys. Rev. Lett.*, **19**, 1095–1097.

Gatz S. and Herrmann J. (1992a) Soliton propagation and soliton collision in double–doped fibres with a non–Kerr like nonlinear refractive index change, *Opt. Lett.*, **17**, 484–486.

Gatz S. and Herrmann J. (1992b) Soliton collision and soliton fusion in dispersive materials with a linear and quadratic intensity dependent refraction index change, *IEEE J. Quant. Electron.*, **28**, No 7, 1732–1737.

Gidas B., Ni W.-M. and Nirenberg L. (1981) Symmetry of positive solutions of nonlinear elliptic equations in R^n, *Mathematical Analysis and Applications, A: Advances in Mathematics Supplementary Studies*, **7a**, 369–402.

Gizin B. V., Hardy A. A. and Malomed B. A. (1994) Stability of light beams in nonlinear antiwaveguides, *Phys. Rev. E*, **50**, No 4, 3274–3276.

Golovchenko E., and Pilipetskii A. N. (1994) Unified analysis of four–photon mixing, modulational instability, and stimulated Raman scattering under various polarization conditions in fibers, *JOSA B*, **11**, No 1, 92–101.

Gordon J. P. (1983) Interaction forces among solitons in optical fibers, *Opt. Lett.* **8**, No 11, 596–598.

Gordon J. P. (1992) Dispersive perturbations of solitons of the nonlinear Schrödinger equation, *JOSA* B, **9**, No 1, 91–97.

Gordon J. P. and Haus H. A. (1986) Random walk of coherently amplified solitons in optical fiber transmission, *Opt. Lett.*, **11**, No 10, 665–667.

Gorshkov K. A. and Ostrovsky L. A. (1981) Interactions of solitons in nonintegrable systems: direct perturbation method and applications, *Physica D*, **3**, 428–438.

Gradeskul S. A., Kivshar Yu. S. and Yanovskaya M. V. (1990) Dark–pulse solitons in nonlinear–optical fibers, *Phys. Rev. A*, **41**, 3994.

Gradshteyn I. S. and Ryzhik I. M. (1962) *Tables of Sums, Series and Integrals*, Nauka, Moscow.

Graham R. (1975) *Fluctuations, Instabilities and Phase Transitions*, ed. T. Riste, Springer-Verlag, Berlin.

Gray S., Grudinin A. B., Loh W. H. and Payne D. N. (1995) Femtosecond harmonically mode–locked fiber laser with time jitter below 1 ps, *Opt. Lett.*, **20**, No 2, 189–191.

Gregori G., Wabnitz S. (1986) New exact solutions and bifurcations in the spatial distribution of polarization in third–order nonlinear optical interactions, *Phys. Rev. Lett.*, **56**, No 6, 600–603.

Grillakis M., Shatah J., Strauss W. (1987) Stability theory of solitary waves in the presence of symmetry, *J. Functional Analysis*, **74**, 160-197.

Grimshaw R. H. J. (1994) Weakly non–local solitary waves in a singularly perturbed nonlinear Schrödinger equation, *Applied Mathematics Reports and Preprints* 94/1, Dept. of Mathematics, Monash University, Melbourne.

Grimshaw R. H. J., Malomed B. and Benilov E. (1994) Solitary waves with damped oscillatory tails: an analysis of the fifth–order Korteveg–de Vries equation, *Physica D*, **77**, 473–485.

Grudinin A. B., Richardson D. J. and Payne D. N. (1992) Energy quantization effect in figure eight fiber laser, *Electron. Lett.*, **28**, No 1, 67–68.

Grudinin A. B., Richardson D. J. and Payne D. N. (1993) Passive modelocking of fibre ring laser, *Electron. Lett.*, **29**, 1860–1861.

Haelterman M. and Sheppard A. P. (1994) Vector soliton associated with polarization modulation instability in the normal–dispersion regime, *Phys. Rev. E*, **49**, No 4, 3389–3399.

Haken H. (1983) *Synergetics*, Springer-Verlag, Berlin,.

Hakim V., Jakobsen P. and Pomeau Y. (1990) Fronts vs. solitary waves in nonequilibrium systems, *Europhys. Lett.*, **11**, No 1, 19–24.

Harkness G. K., Firth W. J., Geddes J. B., Moloney J. V. and Wright E. M. (1994) Boundary effects in large-aspect-ratio lasers, *Phys. Rev. A*, **50**, No 5, 4310.

Hart D. and Wright E. (1992) Stability of the TE_o guided wave of a nonlinear waveguide with a self–defocusing bounding medium, *Opt. Lett.*, **17**, No 2, 121–123.

Hasegawa A. and Brinkman W. F. (1980) Tunable coherent IR and FIR sources utilizing modulational instability, *IEEE J. Quant. Electron.*, **QE–16**, No 7, 694–697.

Hasegawa A. and Kodama Y. (1981) Signal transmission by optical solitons in monomode fiber, *Proc. IEEE*, **69**, 1145–1150.

Hasegawa A. and Kodama Y. (1991) Guiding–center soliton, *Phys. Rev. Lett.*, **66**, No 2, 161–164.

Hasegawa A. and Tappert F. (1973a) Transmission of stationary nonlinear optical pulses in dispersive dielectric fibers. I. Anomalous dispersion, *Appl. Phys. Lett.*, **23**, 142–144.

Hasegawa A. and Tappert F. (1973b) Transmission of stationary nonlinear optical pulses in dispersive dielectric fibers. II. Normal dispersion, *Appl. Phys. Lett.*, **23**, 171–172.

Hatami–Hanza H., Chu P. L. and Peng G. D. (1994) Optical switching in a coupler with saturable nonlinearity , *Opt. Quant. Electron.*, **26**, S365–S372.

Haus H. A. (1975) Theory of mode locking with a fast saturable absorber, *J. Appl. Phys.*, **46**, 3049–3058.

Haus H. A., Fujimoto J. G., and Ippen E. P. (1991) Structures for additive pulse mode locking, *JOSA B*, **8**, No 10, 2068–2076.

Haus J. W., Theimer J., and Fork R. L. (1995) Polarization distortion in bire-fringent fiber amplifiers, *IEEE Photonics Technol. Lett.*, **7**, No. 3, 296–298.

Hayata K. and Koshiba M. (1995) Algebraic solitary–wave solutions of a nonlinear Schrödinger equation, *Phys. Rev. E*, **51**, 1499–1502.

Herbst B. M. and Ablowitz M. J. (1989) Numerically induced chaos in the non-linear Schrödinger equation, *Phys. Rev. Lett.*, **62**, 2065–2068.

Herrmann J. (1992) Bistable bright solitons in dispersive media with linear and quadratic intensity–dependent refraction index change, *Opt. Commun.*, **87**, 161–165.

Hirota R. and Satsuma J. (1976) A variety of nonlinear network equations generated from the Bäcklund transformation for the Toda lattice, *J. Prog. Theor. Phys. Suppl.*, **59**, 64–100.

Hocking L. M. and Stewartson K. (1972) On the nonlinear response of a marginally unstable plane parallel flow to a two-dimensional disturbance, *Proc. Roy. Soc. London A*, **326**, 289–313.

Hofer M., Fernmann M. E., Haberl F., Ober M. H. and Schmidt A. J. (1991) Mode locking with cross-phase and self-phase modulation, *Opt. Lett.*, **16**, No 7, 502–504.

Höök A. and Karlsson M. (1993) Ultrashort solitons at the minimum-dispersion wavelength: effects of fourth-order dispersion, *Opt. Lett.*, **18**, 1388–1390.

Hutchings D. C., Aitchison J. S., Wherrett B. S., Kennedy G. T. and Sibbett (1995) Polarization dependence of ultrafast nonlinear refraction in an AlGaAs waveguide at the half-band gap, *Opt. Lett.*, **20**, 991–993.

Inoue Y. (1976) Nonlinear coupling of polarized plasma waves, *J. Plasma Phys.*, **16**, 439–459.

Ippen E. P., Haus H. A. and Liu L. Y. (1989) Additive pulse–mode–locking, *JOSA B*, **6**, 1736.

Islam M. N. (1992) *Ultrafast Fiber Switching devices and Systems*, Cambridge University Press, New York.

Islam M. N., Mollenauer L. F., Stolen R. H., Simpson J. R. and Shang H. T. (1987) *Opt. Lett.*, **12**, 625.

Islam M. N., Soccolich C. E., Gordon J. P. and Paek U. C. (1990) Soliton intensity–dependent polarization rotation, *Opt. Lett.*, **15**, No 1, 21 –23.

Its A. R. and Kotlyarov V. P. (1976) The explicit formulas for solutions of the Schrödinger nonlinear equation, *Doklady AN Ukr. SSR*, Ser. A, No 11, 965–968.

Its A. R., Rybin A. V., Sall' M. A. (1988) Exact integration of nonlinear Schrödinger equation, *Theor. Math. Phys. (USSR)*, **74**, No 1, 29–45.

Jain M., and Tzoar N. (1978a) Nonlinear pulse propagation in optical fibers, *Opt. Lett.*, **3**, 202–204.

Jain M., and Tzoar N. (1978b) Propagation of localized electromagnetic pulses in inhomogeneous media, *J. Appl. Phys.*, **49**, 4649–4654.

Jakobsen P. K., Moloney J. V., Newell A. C. and Indik R. (1992) Space–time dynamics of wide–gain–section lasers, *Phys. Rev. A*, **45**, No 11, 8129–8147.

Jensen S. M. (1982) The nonlinear coherent coupler, *IEEE J. Quant. Electron.*, **QE–18**, 1580–1583.

Jones C. K. R. T. and Moloney J. V. (1986) Instability of standing waves in nonlinear optical waveguides, *Phys. Lett. A*, **117**, No 4, 175–180.

Kang J., Stegeman G. I. and Aitchison J. S. (1995) Weak–beam trapping by bright spatial solitons in AlGaAs planar waveguides, *Opt. Lett.*, **20**, No 20, 2069 – 2071.

Kang J., Stegeman G. I., Aitchison J. S., Akhmediev N. N. (1996) Observation of Manakov spatial solitons in AlGaAs planar waveguides, *Phys. Rev. Lett.*, **76**, 3699–3702.

Kaplan A. E. (1985) Bistable solitons, *Phys. Rev. Lett.*, **55**, No 12, 1291–1294.

Karlsson M. and Höök A. (1994) Soliton-like pulses governed by fourth order dispersion in optical fibers, *Opt. Commun.*, **104**, 303–307.

Karpman V. I. (1993a) Radiation by solitons due to higher–order dispersion, *Phys. Rev. E*, **47**, No 3, 2073–2082.

Karpman V. I. (1993b) Stationary and radiation dark solitons of the third order nonlinear Schrödinger equation, *Phys. Lett. A*, **181**, 211–215.

Karpman V. I. (1994) Solitons of the fourth order nonlinear Schrödinger equation, *Phys. Lett. A*, **193**, 355–358.

Karpman V. I. (1995) Envelope solitons in gyrotropic media, *Phys. Rev. Lett.*, **74**, No 13, 2455–2458.

Karpman V. I. and Maslov E. M. (1977) Perturbation theory for solitons, *Sov. Phys. JETP*, **46**, No 2, 281–291. Original (in Russian): *Zh. Eksp. Teor. Fiz.*, **73**, 537–559.

Karpman V. I. and Solov'ev S. S. (1981) A perturbation approach to the two–soliton systems, *Physica D*, **3**, 487–502.

Kaup D. J. (1976) A perturbation expansion for the Zakharov-Shabat inverse scattering transform, *SIAM J. Appl. Math.* **31**, 121–133.

Kaup D. J. and Newell A. C. (1978) An exact solution for a derivative nonlinear Schrödinger equation, *J. Math. Phys.*, **19**, 798–801.

Kaw P. K., Nishikawa K., Yoshida Y. and Hasegawa A. (1975) Two–dimensional and three–dimensional envelope solitons, *Phys. Lett. Lett.*, **35**, No 2, 88–91.

Kawata T. and Inoue H. (1978) Inverse scattering method for the nonlinear evolution equations under nonvanishing conditions, *J. Phys. Soc. Japan*, **44**, No 5, 1722–1729.

Kay I. and Moses H. E. (1956) Reflectionless transmission through dielectrics and scattering potentials, *J. Appl. Phys.*, **27**, No 12, 1503–1508.

Khitrova G., Gibbs H. M., Kawamura Y., Iwamura I., Ikegami T., Sipe J. E., and Ming L. (1993) Spatial solitons in a self–focusing gain medium, *Phys. Rev. Lett.*, **70**, 920.

Kivshar Yu. S. (1990) Soliton stability in birefringent optical fibers: analytical approach, *JOSA B*, **7**, No 11 2204–2209.

Kivshar Yu. S. (1993) Dark solitons in nonlinear optics, *IEEE J. Quant. Electron.*, **29**, No 1, 250–264.

Kivshar Yu. S. and Królikowski W. (1995) Instabilities of dark solitons, *Opt. Lett.*, **20**, No 14, 1527–1529.

Klauder M., Laedke E. M., Spatschek K. H. and Turitsyn S. K. (1993) Pulse propagation in optical fibers near the zero dispersion point, *Phys. Rev. E*, **47**, No 6, R3844–3847.

Kodama Y. (1987a) On solitary-wave interaction, *Phys. Lett. A*, *123*, No 6, 276–282.

Kodama Y. and Hasegawa A. (1992a) Theoretical foundation of optical–soliton concept in fibers, in *Progress in Optics*, ed. E. Wolf, North–Holland, Amsterdam, vol.XXX, 205–259.

Kodama Y. and Hasegawa A. (1992b) Generation of asymptotically stable optical solitons and suppression of the Gordon–Haus effect, *Opt. Lett.*, **17**, No 1, 31–33.

Kodama Y. and Nozaki K. (1987) Soliton interaction in optical fibers, *Opt. Lett.*, **12**, No 12, 1038–1040.

Kodama Y. and Wabnitz S. (1991) Reduction of soliton interaction forces by bandwidth limited amplification, *Electron. Lett.*, **27**, No 21, 1931–1933.

Kodama Y. and Wabnitz S. (1993) Reduction and suppression of soliton interactions by bandpass filters, *Opt. Lett.* **18**, No 16, 1311–1313.

Kodama Y. and Wabnitz S. (1994) Analysis of soliton stability and interactions with sliding filters, *Opt. Lett.*, **19**, No 3, 162–164.

Kodama Y., Romagnoli M. and Wabnitz S. (1992) Soliton stability and interactions in fibre lasers, *Electron. Lett.*, **28**, No 21, 1981–1982.

Kodama Y., Romagnoli M., Wabnitz S. and Midrio M. (1994) Role of third–order dispersion on soliton instabilities and interactions in optical fibers, *Opt. Lett.* **19**, No 3, 165–167.

Kolodner P. (1991) Collisions between pulses of travelling–wave convection, *Phys. Rev. A*, **44**, No 10, 6466–6479.

Kolodner P., Bensimon D. and Surko C. M. (1988) Travelling–wave convection in an annulus, *Phys. Rev. Lett.*, **60**, No 17, 1723–1726.

Kolokolov A. A. (1973) Stability of the dominant mode of the nonlinear wave equation in a cubic medium, *Zh. Prikl. Mekh. Tekh. Fiz.* No **3**, 152–155. English translation pp.426–428.

Kolokolov A. A. and Sukov A. I. (1975) Instability in the higher modes of a nonlinear equation, *J. Appl. Mech. Tech. Phys.*, No 4, 519–522. Original (in Russian): *Zh. Prikl. Mekh. Tekh. Fiz.*, No 4, 56–60.

Korn G. A. and Korn T. M. (1961) *Mathematical Handbook for Scientists and Engineers*, McGraw Hill, NY .

Krausz F., Brabec T. and Spielmann Ch. (1991) Self–starting passive mode lock-ing, *Opt. Lett.*, **16**, No 4, 235–237.

Krökel D., Halas N. J., Giuliani G. and Grishkowsky D. (1988) Dark pulse prop-agation in optical fibers, *Phys. Rev. Lett.*, **60**, 2931.

Królikowski W. and Luther–Davies B. (1993) Dark optical solitons in saturable nonlinear media, *Opt. Lett.*, **18**, No 3, 188–190.

Królikowski W., Akhmediev N. and Luther–Davies B. (1993) Darker–than–black solitons: dark solitons with total phase shift greater than π, *Phys. Rev. E*, **48**, 3980–3987.

Kuehl H. H. and Zhang C. Y. (1990) Effects of higher–order dispersion on en-velope solitons, *Phys. Fluids B*, **2**, 889–900.

Kurokawa K. and Nakazawa M. (1992) Femtosecond soliton transmission charac-teristics in an ultralong erbium–doped fiber amplifier with different pumping configurations, *IEEE J. Quant. Electron.*, **28**, No 9, 1922–1929.

Kuznetsov E. A. and Rubenchik A. M. (1986) Soliton stability in plasmas and hydrodynamics, *Phys. Reports*, **142**, No 3, 103–165.

Kuznetsov E. A. and Turitsyn S. K. (1985) Talanov transformations in self–focusing problems and instability of stationary waveguides, *Phys. Lett. A*, **112**, Nos 6/7, 273–275.

Lakshmanan M. and Sahadevan R. (1985) Coupled quartic anharmonic oscil-lators, Painlevé analysis, and integrability, *Phys. Rev. A*, **31**, No 2, 861–876.

Li Q. Y., Pask C. and Sammut R. A. (1991) Simple model for spatial optical solitons in planar waveguides, *Opt. Lett.*, **16**, No 14, 1083–1085.

Litvak A. G. and Mironov V. A. (1968) *Izv. Vyssh. Uchebn. Zaved., Radiofiz.*, **11**, 1911.

Lugovoi V. N. and Prokhorov A. M. (1974) Theory of the propagation of high–power laser radiation in a nonlinear medium, *Sov. Phys. Usp.*, **16**, No 5, 658–679. Original (in Russian): *Usp. Fiz. Nauk*, **111**, 203–247.

Luther - Davies B. and Yang X. (1992a) Waveguides and Y–junctions formed in bulk media by using dark spatial solitons, *Opt. Lett.*, **17**, No 7, 496–498.

Luther - Davies B. and Yang X. (1992b) Steerable optical waveguides formed in self–defocusing media by using dark spatial solitons, *Opt. Lett.*, **17**, No 24, 1755–1757.

Ma Y. C. (1979) *Stud. Appl. Math.*, **60**, 43.

Ma Y. C. and Ablowitz M. J. (1981) The periodic cubic Schrödinger equation, *Stud. Appl. Math.*, **65**, 113 –158.

Maier A. A. (1984) Self–switching of light in a directional coupler, *Sov. J. Quant. Electron.*, **14**, 101–104.

Maker P. D., Terhune R. W. and Savage C. M. (1964) Intensity–dependent changes in the refractive index of liquids, *Phys. Rev. Lett.*, **12**, No 18, 507–509.

Malkin V. M. (1993) On the analytical theory for stationary self–focusing of radiation, *Physica D*, **64**, 251–266.

Malomed B. A. (1987) Evolution of nonsoliton and 'quasi–classical' wavetrains in nonlinear Schrödinger and KdV equations with dissipative perturbations, *Physica D*, **29**, 155–172.

Malomed B. A. (1991) Bound solitons in the nonlinear Schrödinger-Ginzburg-Landau equation, *Phys. Rev. A*, **44**, No 10, 6954–6957.

Malomed B. A., Skinner I. M., Chu P. L. and Peng G. D. (1996) Symmetric and asymmetric solitons in twin-core nonlinear optical fibers, *Phys. Rev. E*, **53**, No 4, 4084–4091.

Mamyshev P. V., Chernikov S. V. and Dianov E. M. (1991) Generation of fundamental soliton trains for high–bit–rate optical fiber communication lines, *IEEE J. Quant. Electron.*, **QE–27**, No 10, 2347–2355.

Mamyshev P. V., Wigley P. G. J., Stegeman G. I., Semenov V. A., Dianov E. M. and Miroshnichenko S. I. (1993) Adiabatic compression of Schrödinger solitons due to the combined perturbations of higher-order dispersion and delayed nonlinear response, *Phys. Rev. Lett.*, **71**, No 1, 73–76.

Manakov C. B. (1974) On the theory of two–dimensional stationary self–focussing of electromagnetic waves, *Sov. Phys. JETP*, **38**, 248. Original (in Russian): *Zh. Teor. Eksp. Fiz.*, **65**, No 2 (1973), 505–516.

Manassah J. T. and Gross B. (1992) Comparison of the paraxial–ray approximation and the variational method solutions to the numerical results for a beam propagating in a self–focussing Kerr medium, *Opt. Lett.*, **17**, No 14, 976–978.

Maradudin A. A., (1981a) in *'Surface Waves in Modern Problems of Surface Physics'*, ed. I. J. Labov. Bulgarian Academy of Sciences, Sofia, 11–399

Maradudin A. A. (1981b) s–polarized nonlinear surface polaritons, *Z. Phys. B*, **41**, 341–344.

Marburger J. H. and Dawes E. (1968) Dynamical formation of a small–scale filament, *Phys. Rev. Lett.*, **21**, No 8, 556–558.

Marcq P., Chaté H. and Conte R. (1994) Exact solutions of the one–dimensional quintic complex Ginzburg–Landau equation, *Physica D*, **73**, 305–317.

Marcuse D. (1973) *Light Transmission Optics*, Van Nostrand Reinhold, New York.

Marcuse D. (1980) Pulse distorsion in single–mode fibers, *Appl. Opt.*, **19**, No 10, 1653–1660.

Matsumoto M., Ikeda H., Uda T. and Hasegawa A. (1995) Stable soliton transmission in the system with nonlinear gain, *J. Lightwave Technol.*, **13**, 658–665.

Matveev V. B. and Salle M. A. (1991) *Darboux Transformations and Solitons*, Springer Series in Nonlinear Dynamics, Springer–Verlag, New York.

McLeod R., Wagner K. and Blair S. (1995) (3+1)–dimensional optical soliton dragging logic, *Phys. Rev. A*, **52**, No 4, 3254–3278.

Mecozzi A., Moores J. D., Haus H. A. and Lai Y. (1991) Soliton transmission control, *Opt. Lett.*, **16**, No 23, 1841–1843.

Mecozzi A., Moores J. D., Haus H.A. and Lai Y. (1992) Modulation and filtering control of soliton transmission, *J. Opt. Soc. Am. B*, **9**, No 8, 1350–1357.

Menyuk C. R. (1986) Origin of solitons in the 'real' world, *Phys. Rev. A*, **33**, No 6, 4367–4374.

Menyuk C. R. (1987) Stability of solitons in birefringent optical fibers. I. Equal propagation amplitudes, *Opt. Lett.*, **12**, No 8, 614–616.

Menyuk C. R. (1989) Pulse propagation in an elliptically birefringent Kerr medium, *IEEE J. Quant. Electron.*, **QE–25**, No 12, 2674–2682.

Menyuk C. R. (1993) Soliton robustness in optical fibers, *JOSA B*, **10**, No 9, 1585–1591.

Mihalache D. and Panoiu N. C., (1992a) Exact solutions of the nonlinear

Schrödinger equation for the normal–dispersion regime in optical fibers, *Phys. Rev.*, **45**, No 9, 6730–6733.

Mihalache D. and Panoiu N. C. (1992b) Exact solutions of nonlinear Schrödinger equation for positive group velocity dispersion *J. Math. Phys.*, **33**, No 6, 2323.

Mihalache D., Bertolotti M. and Sibilia C. (1989) Nonlinear wave propagation in planar structures, in *Progress in Optics*, Vol. XXVII, ed. E. Wolf, North–Holland, Amsterdam, 229–313.

Mihalache D., Lederer F. and Baboiu D. M. (1993) Two–parameter family of exact solutions of the nonlinear Schrödinger equation describing optical–soliton propagation, *Phys. Rev. A* **47**, No 4, 3285–3290.

Miller P. D. (1996) Zero-crosstalk junctions made from dark solitons. *Phys. Rev. E*, **76**, No 4, 4137–4142.

Miller P. D. and Akhmediev N. N. (1996a) Do solitons exchange conserved quantities during collisions?, *Phys. Rev. Lett.*, **76**, No 1, 38–41.

Miller P. D. and Akhmediev N. N. (1996b) Transfer matrices for multiport devices made from solitons, *Phys. Rev. E*, **76**, No 4, 4098–4106.

Mitchell D. J. and Snyder A. W. (1989) Modes of nonlinear couplers: building blocks for physical insight, *Opt. Lett.*, **14**, No 20, 1143–1145.

Mitchell D. J. and Snyder A. W. (1993) Stability of fundamental nonlinear guided waves, *JOSA B*, **10**, 1572.

Mitchell D. J., Snyder A. W. and Chen Y. (1990) Nonlinear triple core couplers, *Electron. Lett.*, **26**, No 15, 1164–1165.

Mitchell D. J., Snyder A. W., and Poladian L. (1996) Interacting self-guided beams viewed as particles: Lorentz force derivation, *Phys. Rev. Lett.*, **77**, No 2, 271–273.

Mollenauer L. F. and Stolen R. H. (1982) Solitons in optical fibers, *Fiberoptic Technol.* (April), 193–198.

Mollenauer L. F., Stolen R. H. and Gordon G. P. (1980) Experimental observation of picosecond pulse narrowing and solitons in optical fibers, *Phys. Rev. Lett.*, **45**, No 13, 1095–1098.

Mollenauer L. F., Gordon J. P., and Islam M. N., (1986) Soliton propagation in long fibers with periodically compensated loss, *IEEE J. Quant. Electron.*, **QE-22**, No. 1, 157–173.

Mollenauer L. F., Gordon J. P. and Evangelides S. G. (1992a) The sliding–frequency guiding filter: an improved form of soliton jitter control, *Opt. Lett.*, **17**, No 22, 1575–1577.

Mollenauer L. F., Lichtman E., Harvey G. T., Neubelt M. J. and Nyman B. M. (1992b) Demonstration of error–free soliton transmission over more than 15,000 km at 5 Gbit/s, single–channel, and over more than 11,000 km at 10 Gbit/s in two–channel WDM, *Electron. Lett.* **27**, No 8, 792–794.

Moloney J. V. (1987) Modulational instability of two–transverse–dimensional surface polariton waves in nonlinear dielectric waveguides, *Phys. Rev. A*, **36**, No 9, 4563–4566.

Moloney J. V., Ariyasu J., Seaton C. T. and Stegeman G. I. (1986) Stability of nonlinear stationary waves guided by a thin film bounded by nonlinear media, *Appl. Phys. Lett.*, **48**, 826–828.

Moon H. T. (1990) Homoclinic crossings and pattern selection, *Phys. Rev. Lett.*,

64, 412–414.

Moores J. D. (1993) On the Ginzburg–Landau laser mode–locking model with fifth–order saturable absorber term, *Optics Commun.*, **96**, 65–70.

Morse P. M. and Feshbach H. (1953) *Methods of Theoretical Physics*, McGraw–Hill, New York.

Nakamura A. and Hirota R. (1985) A new example of explode–decay solitary waves in one dimension, *J. Phys. Soc. Japan*, **54**, No 2, 491–499.

Nakazawa M. and Kubota H. (1992) Physical interpretation of reduction of soliton interaction forces by bandwidth limited amplification, *Electron. Lett.*, **28**, No 10, 958–960.

Nakazawa M. and Suzuki K. (1995) 10 Gbits/s dark soliton data transmission over 200 km, *Electron. Lett.*, **31**, 1076–1077.

Nakazawa M., Yamada E., Kubota H. and Suzuki E. (1991) 10 Gbit/s soliton data transmission over one million kilometers, *Electron.Lett.* **18**, No 14, 1270–1272.

Neugebauer G. and Meinel R. (1984) General N–soliton solution of the AKNS class on arbitrary background, *Phys. Lett.* A **100**, No 9, 467–470.

Newell A. C. (1985) *Solitons in Mathematics and Physics*, Society of Industrial and Applied Mathematics, Arizona.

Newell A. C. and Moloney J. V. (1992) *Nonlinear Optics*, Addison–Wesley Publishing Company, The Advanced Book Program, Redwood City.

Nicholson D. R. and Goldman M. V. (1976) Damped nonlinear Schrödinger equation, *Phys. Fluids*, **19**, No 10, 1621–1625.

Noether E. (1918) Invariante Variationsprobleme, *Nachr. König. Gesell. Wissen. Göttingen, Math.–Phys. Kl.*, 235–257. English translation: *Transport Theory and Stat.Phys.* (1971), 186–207.

Novikov S. P., Manakov S. V., Pitaevskii L. P. and Zakharov V. E. (1984) *Theory of Solitons–The Inverse Scattering Method*, Plenum, New York.

Nozaki K. and Bekki N., J. (1984) Exact solutions of the generalized Ginzburg–Landau equation, *Phys. Soc. Japan.*, **53**, No 5, 1581–1582.

Okamawari T., Hasegawa A. and Kodama Y. (1995) Analysis of soliton interactions by means of a perturbed inverse–scattering transform, *Phys. Rev. A.*, **51**, No 4, 3203–3220.

Olver P. J. (1986) *Applications of Lie Groups to Differential Equations*,(Springer-Verlag, New York).

Ostrovskaya E. A., Akhmediev N. N., Stegeman G. I., Kang J. U. and Aitchison J. S. (1996) Mixed-mode spatial solitons in semiconductor waveguides, *JOSA B*, in press.

Paré C. and Florjanczyk M. (1990) Approximate model of soliton dynamics in all-optical couplers, *Phys. Rev. A*, **41**, No 11, 6287–6295.

Peng G. D. and Ankiewicz A. (1992) Fundamental and second order soliton transmission in nonlinear directional couplers, *Int. J. Nonlinear Opt. Physics*, **1**, 135–150.

Peng G. D., Chu P. L. and Ankiewicz A. (1994) Soliton propagation in saturable nonlinear fibre couplers, *Int. J. Nonlinear Opt. Phys.*, **3**, 69–87.

Pereira N. R. (1977) Soliton in the damped nonlinear Schrödinger equation, *Phys. Fluids*, **20**, No 10, 1735–1743.

Pereira N. R. and Chu F. Y. F. (1979) Damped double solitons in the nonlinear

Schrödinger equation, *Phys. Fluids*, **22**, No 5, 874–881.

Pereira N. R. and Stenflo L. (1977) Nonlinear Schrödinger equation including growth and damping, *Phys. Fluids*, **20**, No 10, 1733–1734.

Pomeau Y., Ramani A. and Grammaticos B. (1988) Structural stability of the KdV solitons under a singular perturbation, *Physica D*, **31**, 127–134.

Porsezian K. and Nakkeeran K. (1996) Optical solitons in presence of Kerr dispersion and self-frequency shift, *Phys. Rev. Lett.* **76**, No 21, 3955 – 3958.

Powell J. A., Newell A. C., Jones C. K. R. T. (1991) Competition between generic and nongeneric fronts in envelope equations, *Phys. Rev. A* **44**, 3636–3652.

Previato E. (1985) Hyperelliptic quasi-periodic and soliton solutions of the nonlinear Schrödinger equation, *Duke Math. J.*, **52**, No 2, 329–377.

Pushkarov Kh. I. and Pushkarov D. I. (1980) Soliton solutions in some non-linear Schrödinger-like equations, *Rep. Math. Phys.*, **17**, No 1, 37–40.

Rasmussen J. J. and Ripdal K. (1986) Blow–up in nonlinear Schroedinger equations–I. A general review. *Physica Scripta*, **33**, 481–497.

Reynaud F. and Barthelemy A. (1990) Optically controlled interaction between two fundamental soliton beams, *Europhys. Lett.*, **12**, No 5, 401–405.

Richardson D. J., Laming R. I., Payne D. N., Matsas V. and Philips M. W. (1991) Self–starting, passively mode–locked erbium fibre ring laser based on the amplifying Sagnac switch, *Electron. Lett.*, **27**, No. 6, 542–544.

Ripdal K. and Rasmussen J. J. (1986) Blow–up in nonlinear Schroedinger equations–II. Similarity structure of the blow–up singularity. *Physica Scripta*, **33**, 498–504.

Romagnoli M., Trillo S. and Wabnitz S. (1992) Soliton switching in nonlinear couplers, *Opt. Quant. Electron.*, **24**, S1237–S1267.

Romagnoli M., Wabnitz S. and Midrio M.(1994) Bandwidth limits of soliton transmission with sliding filters, *Opt. Commun.*, **104**, 293–297.

Rowland D. R. (1991) All–optical devices using nonlinear fiber couplers, *IEEE J. Lightwave Technol.*, **9**, 1074.

Sall' M. A. (1982) Darboux transformation for nonabelian and nonlocal equations of Toda chain type, *Theor. Math. Phys. (USSR)*, **53**, No 2, 227–237.

Sammut R. A. and Pask C. (1991) Group velocity and dispersion in nonlinear–optical fibers, *Opt. Lett.*, **16**, No 2, 70–71.

Sammut R. A., Pask C. and Li Q. J. (1993) Theoretical study of spatial solitons in planar waveguides, *JOSA B*, **10**, No 3, 485–491.

Satsuma J. and Yajima N. (1974) Initial value problems of one–dimensional self–phase modulation of nonlinear waves in dispersive media, *Progr. Theor. Phys. Suppl.*, **55**, 284–306.

Schmidt–Hattenberger C., Trutschel U. and Lederer F. (1991) Nonlinear switching in multiple–core couplers, *Opt. Lett.*, **16**, No 5, 294–296.

Seaton C.T., Stegeman G.I. and Winful H. G. (1985) Nonlinear guided wave applications, *Opt. Eng.*, **24**, 593–599.

Segur H. (ed.) (1991) *Asymptotics Beyond All Orders*, Plenum, New York.

Shalaby M. and Barthelemy A. (1991) Experimental spatial soliton trapping and switching, *Opt. Lett.*, **16**, No 19, 1472–1474.

Silberberg Y. (1990) Collapse of optical pulses, *Opt. Lett.*, **15**, No 22, 1282–1284.

Snyder A. W. and Love J. (1983) *Optical Waveguide Theory*, Chapman & Hall,

London.

Snyder A. W., Mitchell D. J. (1993) Spatial solitons of the power–law nonlinearity, *Opt. Lett.*, **18**, No 2, 101–103.

Snyder A. W. and Sheppard A. P. (1993) Collisions, steering, and guidance with spatial solitons, *Opt. Lett.*, **18**, No 7, 482–484.

Snyder A. W., Chen Y., Rowland D. R. and Mitchell D. J. (1990) Unification of nonlinear–optical fiber devices, *Opt. Lett.*, **15**, No 3, 171–173.

Snyder A. W., Mitchell D. J., Poladian L. and Ladouceur F. (1991a) Self–induced optical fibers: spatial solitary waves, *Opt. Lett.*, **16**, No 1, 21–23.

Snyder A. W., Mitchell D. J., Poladian L., Rowland D. R. and Chen Y. (1991b) Physics of nonlinear fiber couplers, *JOSA B*, **8**, No 10, 2102–2118.

Snyder A. W., Poladian L., Mitchell D. J. (1992) Self–tapered beams, *Opt. Lett.*, **17**, 267–269.

Snyder A. W., Hewlett S. J. and Mitchell D. J. (1994) Dynamic spatial solitons, *Phys. Rev. Lett.*, **72**, No 7, 1012–1015.

Snyder A. W., Hewlett S. J. and Mitchell D. J. (1995) Periodic solitons in optics, *Phys. Rev. E*, **51**, No 6, 6297–6300.

Snyder A. W., Mitchell D. J. and Buryak A. (1996) Qualitative theory of bright solitons: the soliton sketch, *JOSA B*, **13**, No 6, 1146–1150.

Soto-Crespo J. M. and Akhmediev N. N. (1993) Stability of the soliton states in a nonlinear fiber coupler, *Phys. Rev. E*, **48**, 4710–4715.

Soto-Crespo J. M. and Wright E. M. (1991) All-optical switching of solitons in two-and three-core nonlinear fiber couplers, *J. Appl. Phys.*, **70**, No 12, 7240–7243.

Soto-Crespo J. M., Akhmediev N. N. and Ankiewicz A. (1995a) Soliton propagation in optical devices with two–component fields: a comparative study, *JOSA B*, **12**, No 6, 1100–1109.

Soto-Crespo J. M., Akhmediev N. N. and Ankiewicz A. (1995b) Stationary solitonlike pulses in birefringent optical fibers, *Phys. Rev. E*, **51**, No 4, 3547–3555.

Soto-Crespo J. M., Akhmediev N. N. and Afanasjev V. V. (1996) Stability of the pulselike solutions of the quintic complex Ginzburg-Landau equation, *JOSA B*, **13**, No 7, 1439.

Stegeman G. (1992) in: *'Contemporary Nonlinear Optics'*, ed. G. Agrawal and R. Boyd, Academic Press, Boston, 1992, p. 1.

Stegeman G.I. and Seaton C.T. (1985) Nonlinear integrated optics, *J. Appl. Phys.*, **58**, R57–R75.

Stegeman G. I., Wright E. M., Finlayson N., Zanoni R. and Seaton C. T. (1988) Third order nonlinear integrated optics, *J. Lightwave Technol.*, **6**, No 6, 953.

Stephani H. (1989) *Differential Equations–Their Solution Using Symmetries*, Cambridge University Press, Cambridge.

Taha T. P. and Ablowitz M. J. (1984) Analytical and numerical aspects of certain nonlinear evolution equations, *J. Comp. Phys.*, **55**, 192–230.

Tai K., Hasegawa A. and Tomita A. (1986a) Observation of modulational instability in optical fibers, *Phys. Rev. Lett.*, **56**, No 2, 135–138.

Tai K., Tomita A., Jewell J. L. and Hasegawa A. (1986b) Generation of sub-picosecond solitonlike optical pulses at 0.3 THz repetition rate by induced modulational instability, *Appl. Phys. Lett.*, **49**, No 5, 236–238.

Tajiri M. (1983) Similarity reductions of the one and two dimensional nonlinear Schrödinger equation, *J. Phys. Soc. Japan*, **52**, No 6, 1908–1917.

Talanov V. I. (1970) Focusing of light in cubic media, *JETP Lett.*, **11**, 199–201. Original (in Russian): *Pis'ma Zh. Eksp. Teor. Fiz.*, **11**, No 6, 303–305.

Thual O. and Fauve S. (1988) Localized structures generated by subcritical instabilities, *J. Phys. (France)*, **49**, 1829–1833.

Thurston R. N. and Wainer A. W. (1991) Collisions of dark solitons in optical fibers, *JOSA B*, **8**, No 2, 471–477.

Tien P. K., Ulrich R. and Martin R. J. (1970) Optical second harmonic generation in form of coherent Cherenkov radiation from a thin–film waveguide, *Appl. Phys. Lett.*, **17**, No 10, 447–450.

Tikhonenko V., Christou J. and Luther–Daves B. (1995) Spiraling bright spatial solitons formed by the breakup of an optical vortex in a saturable self–focusing medium, *JOSA B* **12**, No 11, 2046–2052.

Tomlinson W. J. (1980) Surface waves at a nonlinear interface, *Opt. Lett.*, **5**, 323–325.

Tomlinson W. J., Hawkins R. J., Weiner A. M., Heritage J. P. and Thurston R. N. (1989) Dark optical solitons with finite–width background pulses, *JOSA B*, **6**, No 3, 329–334.

Torres J. P. and Torner L. (1993) Universal diagrams for TE waves guided by thin films bounded by saturable nonlinear media, *IEEE J. Quant. Electron.*, **QE–29**, No 3, 917–925.

Torres-Cisneros G. E., Sánchez - Mondragón and Vysloukh V. A. (1993) Asymmetric optical Y-junctions and switching of weak beams by using bright spatial–soliton collisions, *Opt. Lett.*, **18**, No 16, 1299–1301.

Tracy E. R., Chen H. H. and Lee Y. C. (1984) Study of quasiperiodic solutions of the nonlinear Schrödinger equation and the nonlinear modulation instability, *Phys. Rev. Lett.*, **53**, No 3, 218–221.

Tran H. T. and Ankiewicz A. (1992) Instability regions of nonlinear planar guided waves, *IEEE J. Quant. Electron.*, **QE–28**, 488–492.

Tran H. T. and Sammut R. A. (1995) Families of multiwavelength spatial solitons in nonlinear Kerr media, *Phys. Rev. A*, **52**, No 4, 3170–3175.

Tran H. T., Mitchell J., Akhmediev N. N. and Ankiewicz A., (1992) Complex eigenvalues in stability analysis of nonlinear planar guided waves, *Opt. Commun.*, **93**, 227–233.

Tran H. T., Sammut R. A. and Samir W. (1994a) Interaction of self-guided beams of different frequencies, *Opt. Lett.*, **19**, No 13, 945–947.

Tran H. T., Sammut R. A. and Samir W. (1994b) Multi-frequency spatial solitons in Kerr media, *Opt. Commun.*, **113**, 292–304.

Tran H. T., Sammut R. A. and Samir W. (1994c) All-optical switching with multi-frequency spatial solitons, *Electron. Lett.*, **30**, No 13, 1080–1081.

Tratnik M. V. and Sipe J. E. (1987) Nonlinear polarization dynamics, *Phys. Rev. A*, **35**, 2965–2988.

Tratnik M. V. and Sipe J. E. (1988) Bound solitary waves in a birefringent optical fiber, *Phys. Rev. A*, **38**, No 4, 2011–2017.

Trillo S. and Wabnitz S. (1991) Dynamics of the nonlinear modulation instability in optical fibers, *Opt. Lett.*, **16**, No 13, 986–988.

Trillo S., Wabnitz S., Wright E. M. and Stegeman G. I. (1988) Soliton switching in fiber nonlinear directional couplers, *Opt. Lett.*, **13**, No 8, 672–674.

Trillo S., Wabnitz S., Wright E. M. and Stegeman G. I. (1989) Polarized soliton instability and branching in birefringent fibers, *Opt. Commun.*, **70**, No 2, 166–172.

Ueda T. and Kath W. L. (1990) Dynamics of coupled solitons in nonlinear optical fibers, *Phys. Rev. A*, **42**, 563–571.

Ulrich R. and Simon A. (1979) Polarization optics of twisted single–mode fibers, *Appl. Opt.*, **18**, No 13, 2241–2251.

Uzunov I. M., Muschall R., Gölles M., Kivshar Yu. S., Malomed B. A. and Lederer F. (1995) Pulse switching in nonlinear fiber directional couplers, *Phys. Rev. E*, **51**, 2527–2537.

Vach H., Seaton C. T., Stegeman G. I. and Khoo I. C. (1984) Observation of intensity–dependent guided waves, *Opt. Lett.*, **9**, 238–240.

Vakhitov N. G. and Kolokolov A. A. (1974) *Izv. Vuzov, Radiofizika*, **16**, 1020.

Vallée R. and Essambre R. - J. (1994) Long distance soliton transmission with a nonlinear twin–core fiber, *Opt. Lett.*, **19**, 2095.

Van Saarloos W. and Hohenberg P. C. (1990) Pulses and fronts in the complex Ginzburg–Landau equation near a subcritical bifurcation, *Phys. Rev. Lett.*, **64,** No 7, 749–752.

Van Saarloos W. and Hohenberg P. C. (1992) Fronts, pulses, sources and sinks in generalized complex Ginzburg–Landau equations, *Physica D*, **56**, 303–367.

Villeneuve A., Kang J. U., Aitchison J. S. and Stegeman G. I. (1995) Unity ratio of cross- to self-phase modulation in bulk AlGaAs and AlGaAs/GaAs multiple quantum well waveguides at half the band gap, *Appl. Phys. Lett.*, **67**, No 6, 760–762.

Vinogradova M. B., Rudenko O. V. and Sukhorukov A. P. (1979) *Teoriya Voln (Theory of Waves)*, Moscow, Nauka (in Russian).

Vlasov V. N., Petrishev I. A. and Talanov V. I. (1971) *Izv. Vuzov, Radiofizika*, **14**, 1353.

Vysloukh V. A. and Cherednik I. V. (1986) Modelling the self-interaction of supershort pulses in optical fibers using the inverse scattering method, *Sov. Phys. Dokl.*, **31**, No 7, 532–534.

Vysloukh V. A. and Cherednik I. V. (1988) Modelling of the reconstruction of the envelope of supershort optical pulses from the characteristics of their nonlinear interaction with probing single-soliton pulses, *Sov. Phys. Dokl.*, **33**, No 3, 182–184.

Vysotina N. V., Rozanov N. N. and Smirnov V. A. (1987) Small–scale self–focusing of nonlinear surface waves, *Sov. Phys. Tech. Phys.*, **32**, 104–105.

Wabnitz S. (1988) Modulation polarization instability of light in a nonlinear birefringent dispersive medium, *Phys. Rev.*, **38**, No 4, 2018–2021.

Wabnitz S. (1995) Effect of frequency sliding and filtering on the interaction of polarization–division–multiplexed solitons, *Opt. Lett.*, **20**, No. 3, 261–263.

Wabnitz S., Wright E. M. and Stegeman G. I. (1990) Polarization instabilities of dark and bright coupled solitary waves in birefringent optical fibers, *Phys. Rev. A*, **41**, 6415–6424.

Wai P. K. A. and Menyuk C. R. (1994) Polarization decorrelation in optical fibers with randomly varying birefringence, *Opt. Lett.*, **19**, No 19, 1517–1519.

Wai P. K. A., Menyuk C. R., Lee Y. C. and Chen H. H. (1986) Nonlinear pulse propagation in the neighbourhood of the zero-dispersion wavelength of monomode optical fibers, *Opt. Lett.*, **11**, 464–466.

Wai P. K. A., Menyuk C. R., Chen H. H., and Lee Y. C. (1987) Soliton at the zero-group-dispersion wavelength of a single-mode fiber, *Opt. Lett.*, **12**, No 8, 628–630.

Wai P. K. A., Chen H. H., and Lee Y. C. (1990) Radiations by 'solitons' at zero group–dispersion wavelength of single–mode optical fibers, *Phys. Rev. A*, **41**, No 1, 426–439.

Wai P. K. A., Menyuk C. R. and Chen H. H. (1991) Stability of solitons in randomly varying birefringent fibers, *Opt. Lett.*, **16**, No 16, 1231–1233.

Weiland J., Ichikawa Y. H. and Wilhelmsson H. (1978) A perturbation expansion for the NLS with application to the influence of nonlinear Landau damping, *Physica Scripta*, **17**, 517–522.

Weiner A. M., Heritage J. P., Hawkins R. J., Thurston R. N., Kirschner E. M., Leaird D. E. and Tomlinson W. J. (1988) Experimental observation of the fundamental dark soliton in optical fibers, *Phys. Rev. Lett.*, **61**, 2445–2448.

Winful H. G. (1985) Self–induced polarization changes in birefringent optical fibers, *Appl. Phys. Lett.*, **47**, No 3, 213–215.

Wright E. M., Stegeman G. I. and Wabnitz S. (1989) Solitary-wave decay and symmetry-breaking instabilities in two-mode fibers, *Phys. Rev. A*, **40**, No 8, 4455–4466.

Yankauskas Z. K. (1966) *Sov. Radiophys.*, **9**, 261. Original (in Russian): *Izv. Vuzov, Radiofiz.*, **9**, 412.

Zabusky N. J. and Kruskal M. D. (1965) Interaction of 'solitons' in a collisionless plasma and the recurrence of initial states, *Phys. Rev. Lett.*, **15**, 240–243.

Zakharov V. E. and Rubenchik A. M. (1973) Instability of waveguides and solitons in nonlinear media, *Sov. Phys. JETP*, **38**, 494–500. Original (in Russian): *Zh. Eksp. Teor. Fiz.*, **65**, 997–1011.

Zakharov V. E. and Shabat A. B. (1971) Exact theory of two dimensional self focusing and one dimensional self modulation of nonlinear waves in nonlinear media, *Sov. Phys. JETP*, **34** 62–69]. Original (in Russian): *Zh. Eksp. Teor. Fiz.*, **61**, 118.

Zakharov V. E. and Shabat A. B. (1973) Interaction between solitons in a stable medium [*Sov. Phys. JETP*, **37**, 823–828 (1973)]. Original (in Russian): *Zh. Eksp. Teor. Fiz.*, **64**, 1627–1639.

Zakharov V. E. and Synakh V. S. (1975) The nature of the self–focusing singularity, *Sov. Phys. JETP*, **41**, 465–468. Original (in Russian): *Zh. Eksp. Teor. Fiz.*, **68**, 940–947.

Zakharov V. E., Kuznetsov E. A. and Rubenchik A. M. (1986) Soliton stability, in *Modern Problems in Condensed Matter Sciences*, **17**, (North-Holland, Amsterdam, 503–554.

Zhao W. and Bourkoff E. (1989) Propagation properties of dark solitons, *Opt. Lett.*, **14**, 703–705.

Zharova N. A. and Sergeev A. M. (1989) Steady state nonlinear effects in whistlers, *Sov. J. Plasma Phys.*, **15**, No 10, 681–683.

Index